# THE FIRST HELICOPTER WAR

## Logistics and Mobility in Algeria, 1954–1962

CHARLES R. SHRADER

PRAEGER

Westport, Connecticut
London

**Library of Congress Cataloging-in-Publication Data**

Shrader, Charles R.
The first helicopter war : logistics and mobility in Algeria, 1954–1962 / Charles R. Shrader.
p. cm.
Includes bibliographical references and index.
ISBN 0-275-96388-8 (alk. paper)
1. Algeria—History—Revolution, 1954–1962. 2. Military helicopters—Algeria—History. I. Title.
DT295.S487 1999
965'.046—dc21 98-40913

British Library Cataloguing in Publication Data is available.



Library of Congress Catalog Card Number: 98-40913
ISBN: 0-275-96388-8

First published in 1999

Praeger Publishers, 88 Post Road West, Westport, CT 06881
An imprint of Greenwood Publishing Group, Inc.
www.praeger.com

Printed in the United States of America

The paper used in this book complies with the Permanent Paper Standard issued by the National Information Standards Organization (Z39.48–1984).

10 9 8 7 6 5 4 3 2 1

*In memory of my parents*

# Contents

# Illustrations

## TABLES

## MAPS

## FIGURES

# Note on Translation, Acronyms, and Measurements

An extended discussion of any military organization necessarily involves frequent repetition of terms, abbreviations, and acronyms, particularly those relating to unit types and designations. The problem is compounded when the discussion involves more than one military establishment, each of which is organized on a different pattern. Additional complexity is added by the necessity to translate from two languages into yet a third. The welter of lengthy designations and repeated acronyms can be confusing as well as distracting. The following procedures have been adopted in this volume in an attempt to simplify the identification of political (territorial) entities and military organizations.

At first mention in each chapter, French political entities and military organizations are identified in full by an English translation of the original French designation, which follows immediately in parentheses, along with the corresponding French abbreviation. Thus, one finds the 10th Military Region (*10e Région Militaire*; 10e RM); the Constantine Corps Area (*Région Territoriale et Corps d'Armée Constantine*; RT/CAC, or simply CAC); or the 5th Armored Division (*5e Division Blindée*; 5e DB). For subsequent references within the same chapter, any of the three forms may be used, usually the French acronym. The most frequently used French acronyms and their English equivalents include:

| | | | |
|---|---|---|---|
| 10e RM | 10th Military Region | DI | Infantry Division |
| RT/CAC | Constantine Corps Area | DIA | Alpine (Mountain) Division |
| RT/CAA | Algiers Corps Area | DIM | Motorized Infantry Division |
| RT/CAO | Oran Corps Area | DB | Armored Division |
| CSI | Joint Sahara Command | DP | Parachute Division |

Algerian nationalist territorial divisions and military units are identified in English when first mentioned and subsequently by the transliterated form of the appropriate Arabic or Berber term. Thus, one finds the 1st Political-Military Region (*Wilaya* 1), the 2nd Political-Military Zone (*Mintaka* 2), a battalion (*faïlek*), or a company (*katiba*). North African place names and Arabic or Berber proper names follow the transliterations and forms used by French authorities in the period under examination. Exceptions to the rule are countries (Algeria, Morocco, Tunisia, Egypt, etc.) and the city of Algiers (Alger), for which the familiar Anglicized names are used.

It is assumed that most readers are familiar with the metric measurements for weight, distance, and area but that they may be less certain about others. Thus, most measurements are given in the metric system except for temperature, for which the Fahrenheit system is used, and volume, in which case liters and cubic meters have been converted into gallons or barrels. Unless otherwise noted the term "ton" refers to the metric ton of 2,204.62 pounds. It may be useful to recall the following equivalencies:

1 meter = 39.37 inches

1 kilometer (km) = 0.6214 miles

1 square meter ($m^2$) = 1.196 square yards

1 hectare = 2.471 acres

1 kilogram (kg) = 2.2046 lbs

1 metric ton (tonne) = 2,204.62 lbs

1 short ton (ST) = 2,000 lbs (907.2 kg)

1 liter = .2641 U.S. gallons

1 cubic meter ($m^3$) = 201.987 U.S. gallons

1 barrel (bbl) = 42 U.S. gallons

# Preface

A key element in counterinsurgent warfare, as indeed in any type of military conflict, is mobility—the movement of men and supplies. Mobility at all levels—strategic, operational, and tactical—is regulated by several factors. Basic geography and the nature of the terrain determine the distances that must be traversed and the physical difficulties that must be overcome. Climate and weather affect visibility and trafficability and thus influence the conduct of air and naval as well as ground operations. The availability of transportation facilities, such as roads, railroads, ports, and airfields, and the nature and location of storage facilities, and their distance from the units using them, also have a profound impact on mobility. The means of transport available—the numbers and types of ships and boats, aircraft and helicopters, locomotives and rolling stock, trucks, pack and draft animals, even porters—are of surpassing importance, as are the availability of trained and experienced personnel and the effectiveness of petroleum distribution, parts supply, and the maintenance of vehicles and supporting equipment. Moreover, the level of technological sophistication of the means of transport plays an important role in its adaptability and suitability for a given situation. Operational factors also enter into the equation. The number of men and the types and amounts of matériel that must be moved are key elements, movement requirements versus movement capabilities being always a major consideration. The nature of operations—general or limited, nuclear or nonnuclear, offensive or defensive, static or mobile—also determines mobility requirements and limitations as do both the security of plans, facilities, and routes and the efforts of the enemy to impede an army's mobility through active ground, naval, and air interdiction. Finally, the effectiveness of command and control systems, particularly communications and both movement and traffic control, impact directly on the effectiveness of mobility and countermobility operations.

This study addresses the logistical requirements, organization, methods, and operations of the French Army and the armed forces of the Algerian National Liberation Front (FLN) during active combat operations in Algeria from November 1954 to May 1962. Although due attention is given to such logistical functions as the determination of supply requirements and the acquisition, storage, issue, maintenance, and disposal of equipment and supplies, this study focuses primarily on the strategic and operational mobility (transportation) systems of the opponents and on their respective efforts to deny mobility to their enemy. The key question that this comparative analysis seeks to answer is: To what degree was the military outcome of the war in Algeria from 1954 to 1962 determined by the relative logistical strengths and weaknesses of the two sides, particularly with respect to operational mobility and the effectiveness of countermobility operations? Thus, movement requirements, transportation facilities and capabilities at the strategic and operational level, and the interdiction efforts of the French forces and the Algerian Army of National Liberation (ALN) receive detailed consideration, while other aspects of operational logistics, such as supply and maintenance, receive less. Moreover, important political, diplomatic, social, and even military aspects of the Algerian rebellion, such as revolutionary warfare per se, subversion, terrorism, psychological warfare, population control, and the internal political struggles of the two opponents, are discussed only insofar as they had a direct impact on the distribution of men and supplies to forces in the field.

This study is based primarily on the records of French and FLN/ALN logistical activities assembled by the French during and immediately after the Algerian war of 1954–1962. The declassified French records have been supplemented by the reports of contemporary Western military and civilian observers and intelligence agencies. Secondary works by French and other Western scholars have also been used extensively. For the most part, access to such FLN/ALN documentation as may have survived the war remains difficult, if not impossible. The one-sided nature of the sources used in this study thus requires that the results be used with some caution.

Given its restricted scope, this study is important mainly as a demonstration of the importance in any military operation of the movement of men and supplies and of efforts to restrict such movements. More specifically, this study provides insight into the impact of transportation and of interdiction capabilities and operations on the course and outcome of an insurgency/counterinsurgency conflict in which the opposing sides possess unequal levels of transportation technology and disparate degrees of sophistication in the organization of logistical resources.

I am particularly indebted to the late Sterling Hart for his assistance in the preparation of this study. His keen editorial eye and sound advice will be sorely missed. I am grateful also to the staff of the *Service Historique de l'Armée de Terre* at the Château de Vincennes for their patience and aid to a novice *client*, the more so in that the records of the French Army in Algeria during the war of 1954–1962 had just been opened to outside researchers at the time of my visit. As always, the staff of the United States Army Military History Institute at Carlisle Barracks, Pennsylvania, particularly Mr. John Slonaker, Mr. Dennis Vetock, and Mrs. Louise Arnold-Friend, were most helpful in pointing out and locating interesting and pertinent materials. I also gratefully acknowledge the patience and love of my wife, Carole, who cheerfully supported an often reclusive and grouchy writer over an extended period of time. Of course, such sins of commission or omission as the reader may find are mine alone.

# Introduction

The shocking victory of the Viet Minh over the French Union forces at Dien Bien Phu on 8 May 1954 led directly to the French withdrawal from Indochina after a decade of hard-fought war. Moreover, the defeat at Dien Bien Phu seriously damaged the prestige and morale of the French Army and created the myth that nationalist insurgencies were all but unstoppable.[1] Yet by May 1958, after only three and a half years of conflict, the same French Army had defeated decisively another nationalist rebellion, on the battlefields of Algeria, and had proven that militarily at least, nationalist insurgency could be contained. Why were the French armed forces able to achieve in Algeria what they had been unable to achieve in Indochina only a few years earlier? An important part of the answer can be found by comparing the logistical strengths and weaknesses of the opposing forces, particularly in the key area of mobility and counter-mobility.

During the Algerian war of 1954–1962, the French forces faced many of the same problems they had experienced in Indochina: an elusive and determined enemy supported by outside patrons and a large part of the indigenous population; an indecisive and unstable government at home; a shortage of trained men and suitable matériel; and above all a lack of a clear, integrated political-military strategy for winning the war. The crucial difference was that in Algeria the French had a clear advantage in mobility and counter-mobility at all levels over their insurgent opponent. At the strategic level the French forces in Algeria were much closer to metropolitan France, their ultimate source of men and supplies, than they had been in Indochina. The lines of communication were accordingly shorter and less complicated and therefore more effective. Despite internal political dissension that amounted to open revolt, the French forces in Algeria also had the advantage of effective central command and control systems and established logistical services. At the operational level, the French also had significant advantages over their Algerian opponent. From the beginning the

French in Algeria controlled such transportation routes and facilities as existed to a degree they had not in Indochina. They were thus able to maximize the advantages of their superior rail and motor transport as well as coastal shipping. Given the superior French technology, terrain and climate favored them rather than the Algerian rebels. Air transport mastered the great distances and rugged terrain in a way men on foot or pack animals could not. Moreover, the French in Algeria possessed an effective means of operational mobility not available to them to any significant degree in Indochina—the helicopter. Furthermore, in Algeria the French efforts at air, sea, and ground interdiction of their enemy's movements were far more effective than they had been in Indochina, principally because the terrain and climate facilitated aerial observation and also because improved and more numerous means of air and sea interdiction were available. In short, the French military forces in Algeria possessed most of the advantages in mobility that had been denied to them in Indochina, and, having recognized the importance of mobility and counter-mobility to their loss of Indochina, the French were able to exploit these advantages fully.

For their part, the Algerian rebels of the National Liberation Front (FLN) enjoyed many of the advantages that had fallen to the Viet Minh in Indochina: powerful patrons abroad to provide equipment and supplies as well as safe bases outside the battle area; the support of a large segment of the population; and skilled and determined leadership. However, the Algerian Army of National Liberation (ALN) also had to cope with the many serious difficulties of creating a logistical system from scratch and of operating that system in the face of superior French military power without the advantages in mobility the Viet Minh had enjoyed. The chief disadvantage faced by the ALN was the lack of a porous border over which to move men and supplies at will. Although both Tunisia and Morocco provided safe bases for massing supplies and training men for the ALN, the French were able to construct effective barriers against the movement of those supplies and men to where they were most needed, something they had been unable to achieve in Indochina. Given the rebels' relative lack of air and motor transport, the difficult mountainous terrain and desert wastelands that cover most of Algeria were a serious obstacle to their movement within the country, and effective French air and sea interdiction generally denied to the ALN the effective use of Algeria's long coastline. Moreover, very poor communications systems hampered ALN efforts to resupply and reinforce its forces in the interior of Algeria. A seriously fragmented command system and internal dissension within the top political and military leadership of the FLN only added to the problem.

Although at first confused and hesitant in their efforts to counter the growing power of the rebellion in Algeria, the French soon managed to exploit their overwhelming advantages in mobility and interdiction at all levels—strategic, operational, and tactical—to achieve a clear-cut military victory over the ALN by mid-1958. The effectiveness of the eastern and western barrages and of the French interdiction program, coupled with a vigorous effort to search out and destroy the remaining rebel bands within Algeria itself, all but destroyed the ALN by the end of 1959. With its major combat units and supply facilities isolated outside Algeria and unable to pass significant numbers of troops, supplies, or equipment through the French barriers, the ALN was thenceforth unable to maintain formations of larger than platoon size inside Algeria. Consequently the rebels were able to mount only the most feeble, localized operations, and they largely reverted to terrorism as their only offensive method.

As A. A. Heggoy has noted, "By 1958 the rebels had lost the military struggle. But they had managed to sow doubt in the minds of many Frenchmen and political leaders in France about the desirability of imposing a French solution to the Algerian question."[2] The Challe offensive in 1959 in fact finished the ALN as a significant military force, but the war would not end until the signing of the Evian agreements in March 1962 and the declaration of Algerian independence on 3 July 1962. Meanwhile, despite their defeat on the battlefield, the leaders of the FLN/ALN, inspired by the dream of national independence, vowed to continue the struggle at all costs and by any means. The political disarray and war weariness of their French opponent made it possible for the rebels to achieve on the political and diplomatic front what they had been denied by force of arms—an end to the war on their terms and the total independence of Algeria from French rule.

## NOTES

1. The psychological and political impact on the French Army of the defeat in Indochina is a central theme of the novels of Jean Lartéguy and has been outlined succinctly by George A. Kelly in *Lost Soldiers: The French Army and Empire in Crisis, 1947–1962* (Cambridge, MA: MIT Press, 1965). Kelly has also noted the influence of the Indochina experience on French military organization and tactics in Algeria as have Jean-Marc Marill in "L'Héritage Indochinois: Adaptation de l'Armée Française en Algérie (1954–1956)," *Revue Historique des Armées* 187 (June 1992), 26–32; and Jean-Charles Jauffret in "L'Armée et l'Algérie en 1954," *Revue Historique des Armées* 187 (June 1992), 17–19, among others. The myth of insurgent invincibility is mentioned by Edgar O'Ballance, *The Algerian Insurrection, 1954–62* (Hamden, CT: Archon Books, 1967), 9; and by Douglas Pike, *Viet Cong: The Organization and Techniques of the National Liberation Front of South Vietnam* (Cambridge, MA: MIT Press, 1966), 49, who states: "The success of the Chinese revolution and the Viet Minh Resistance resulted in ascribing to the writings of Mao Tse-tung and Vo Nguyen Giap a pretension they do not have. Their tactics, and that is all they offer, can be defeated. The French proved this in Algeria, where militarily they won, largely by applying the lessons learned in Indochina."

2. Alf Andrew Heggoy, *Insurgency and Counterinsurgency in Algeria* (Bloomington: Indiana University Press, 1972), xii.

# 1

# The Physical Environment

The physical environment plays a major role in military operations, but seldom does it affect both sides equally, in part due to inevitable differences in the resources and technology available to the opposing forces. In general, the belligerent better able to overcome the limitations imposed by the physical environment possesses a distinct advantage over an opponent. Such was the case in the Algerian war of 1954–1962, in which the French, controlling the principal transportation facilities and possessing superior transportation resources, including air transport, held a major advantage over the less well equipped Algerian rebels, who often found their military plans thwarted by a hostile environment.

## THE GEOGRAPHY OF ALGERIA

Algeria, a land of diversity and extremes, is a large and difficult arena for military operations.[1] Bounded on the north by the Mediterranean Sea and on the east, south, and west by long, ill defined, and isolated borders, Algeria has an extensive surface area, about four times that of France or three and one-half times that of the state of Texas, and 90 percent is desert. Algeria's Mediterranean coastline extends for some 1,100 kilometers; from Algiers on the Mediterranean coast south to Tamanrasset in the Hoggar massif is a distance of some 2,200 kilometers. In 1954 the political boundaries of Algeria incorporated some 2,204,864 square kilometers, divided into two major regions: the four administrative departments (Algiers, Oran, Constantine, and Bône) of northern Algeria (209,630 square kilometers), and the southern territories of the Sahara (1,995,234 square kilometers). In 1954 the extensive land borders of Algeria, which today total some 6,343 kilometers, were indeterminate and disputed in many areas, with few recognizable landmarks in the southern regions. The two most important international boundaries were those with Tunisia (today 965 kilometers) and with Morocco (today 1,559 kilometers). Other bordering states

**Map 1**
**Algeria, 1954–1962**

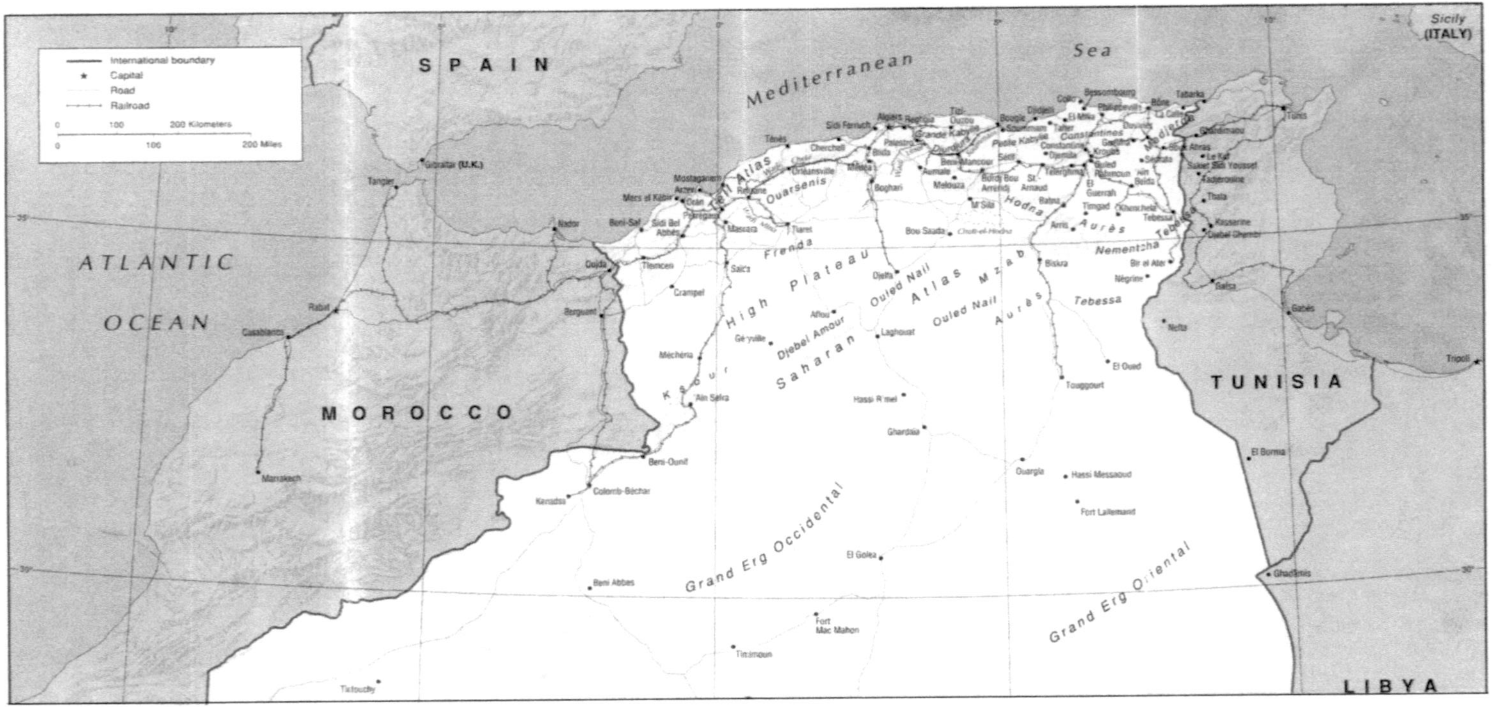

Spanish Sahara
MAURITANIA
Erg Iguidi
Erg Chech
Tanezrouft
S A H A R A
Hoggar
MALI
NIGER
Tropic of Cancer
Tamanrasset
In Salah
Reggane
Adrar
Tindouf
Djanet
Fort Polignac
Fort Flatters
Edjeleh
Bidon
SPAIN
PORTUGAL
FRANCE
ITALY
YUGOSLAVIA
ROMANIA
BULGARIA
GREECE
TURKEY
ALB.
TUNISIA
LIBYA
U.A.R.
SUDAN
CHAD
NIGER
NIGERIA
UPPER VOLTA
MALI
MAURITANIA
SENEGAL
MOROCCO
ALGERIA
ATLANTIC OCEAN
Mediterranean Sea
Sardinia
Sicily
Malta (U.K.)
Algiers
Oran
Tangier
Rabat
Marrakech
Constantine
Bône
Tunis
Gabès
Marseille
Spanish Sahara

included the present nations of Libya, Niger, Mali, the Western Sahara, and Mauritania.

The terrain of Algeria, mostly high plateau and desert, ranges from flat coastal plains to steep mountains. Except for the narrow coastal strip and the highland plateau between the two principal mountain chains, the Algerian terrain is mostly very rugged and difficult to traverse, even on foot. The great distances and jumbled topography pose significant obstacles to overland movement. The average elevation throughout the country is nine hundred meters, but some 68 percent of the land lies between eight hundred and 1,600 meters above sea level. The highest point is Mount Tahat in the Sahara, which rises to three thousand meters. Only about 21 percent of Algeria consists of productive land (versus 61 percent of metropolitan France), and only 3 percent of the total surface is truly arable.

Perhaps the most striking geographical feature of Algeria is its compartmentalization, which only adds to the difficulty of movement and communication. Algeria is divided into three major east-west zones by two chains of the Atlas mountains, the Tellian and the Saharan. Both mountain chains are fairly narrow and comprise several distinct secondary ranges, separated by valleys and plains. The narrow coastal plain is fertile and heavily populated but broken into compartments by spurs of the Tellian Atlas that run down to the sea. The Tellian Atlas chain is characterized by steep ridges, narrow valleys, and gorges, the plains being few and quite small. There are few rivers or constantly flowing large streams, and the area is characterized by draws (*wadis*) and rubble-strewn gorges. Between the two mountain chains lies the High Plateau, which gradually declines from west to east. The higher and more rugged Saharan Atlas chain forms a discontinuous chain in the west, becoming more massive and even higher in the Aurès and Nementcha mountains of eastern Algeria. South of the Saharan Atlas lies the true Saharan desert of windblown sand and rock, practically devoid of vegetation and sedentary population except in the few oases.

Water is a precious commodity throughout Algeria. Most water is obtained from wells, which are often contaminated. In the northern section of Algeria the water supply is normally sufficient to meet the needs of the population, but in the south water is generally scarcer. In the Sahara itself water is obtained from subsurface sources, which include shallow wells (*shadoofs*) and deep wells (*birs*). Around the oases underground water channels known as *foggaras* are quite common. The *foggaras* generally cover a distance of from ten to twelve miles from the underground source and have openings every ten meters for maintenance and removal of sediment. Well heads are constructed to prevent dust and sand from entering the channel.

The vegetation of Algeria conforms to the topography, the available water supply, and general climatic conditions. Forests cover about 11 percent of the total land area of Algeria, with another 13 percent covered by scrub forest (*maquis*). The narrow coastal strip, once marshy, is now well watered and fertile, supporting extensive agriculture and forests. The mountains of the Tellian Atlas are generally well forested (about 20 percent), but the semi-arid expanse of the High Plateau is generally treeless (only about 3 percent forested), and agriculture in the latter is possible only in artificially irrigated areas or in narrow valleys inundated by seasonal torrential streams (*oueds*). The Saharan Atlas is even more barren, with only isolated pockets of forest. The Sahara desert itself is practically devoid of vegetation, and only in the oases is limited agriculture, such as the cultivation of the date palm, possible.

Despite its length the 1,100-kilometer Mediterranean coastline of Algeria is largely inaccessible. Good natural harbors and anchorages are few, the three main ones being Bône, Algiers, and Oran. The eroded, indented coastline, once home to the Barbary pirates, has some beautiful, isolated sandy beaches and coves, but it is generally hilly with rather high cliffs, especially in the east from Bougie to Djidjelli, where the Kabylie mountains descend directly into the sea. The narrow coastal plain was largely marshland when the French arrived but it has since been drained and now provides agricultural land excellent for citrus fruit and grapes in particular. The coastal strip is the most Europeanized part of Algeria and contains the major cities, including the important ports of Algiers, Oran, Bougie, Philippeville, and Bône, as well as the numerous towns of the Mitidja region. The principal cities are studies in contrast, with ancient, squalid, labyrinthine native quarters amidst the broad boulevards and modern office buildings of the European sections.

The Tellian Atlas chain limits the fertile coastal plain, and its spurs, which extend into the Mediterranean, divide the coastal plain into isolated pockets. Including the coastal strip, the Tellian Atlas mountains constitute a 150-kilometer-wide east-west zone consisting in the interior of narrow, isolated plains connected only by passes or narrow valleys. In general the soil is poor and eroded, and the region is crossed by numerous streams, which are torrential in the rainy season.

The coastal mountains on either side of Algiers are different. To the east they attain heights of 2,300 meters in the limestone ridges of the Djurdjura and continue to Bône in the form of the winding ridges of the Kabylie range, well watered and covered by forests of cork trees. Separated from the eastern portion of the chain by the Wadi Leser east of Algiers, the western portion of the Tellian Atlas chain stretches from Algiers to the Moroccan border, broken only by the Wadi Mina. Here the mountains are less rugged, although the coast itself is quite wild. Scored by *wadi*s, the mountains of the western Tell are lower than and not so broken as the eastern Kabylies. Nor are they as fertile. Sparsely settled, they are suitable only for grazing and low-yield farming in the wetter ground around the *wadi*s. Except in a few places the dryness of the climate permits only a meager vegetation: dwarf pines and jujube trees interspersed with green oak and Aleppo pines at higher elevations, and scrub forests and grass at lower elevations. The slopes are often barren, and there is little stable population, although the interior plains are richly endowed with fruit, vineyards, wheat, and tobacco.

The Kabylies Mountains of northeastern Algeria are the most massive of the Tellian Atlas chain. Dominated by the great jagged spine of the Djurdjura, the Kabylies reach 2,100 meters and are mantled with snow until early summer. Dropping rapidly to the coastal plain in places, the Kabylies are broken by jumbled, rocky gorges with overhanging cliffs pierced by many caves. In some respects Kabylia is a bucolic paradise, with picturesque villages (*douars*) perched on narrow ridges overlooking clear streams and forests of cedar, cork, and Spanish chestnut. The region is heavily populated, and the hills and valleys are fairly fertile despite the thin arable soil, the cold winters, and the scanty and unpredictable rainfall. Heavily forested in some places, the Kabylies can also be bare or covered with grass and six-foot-high brush called *lentisque*.

South of the Tellian Atlas chain is the High Plateau, consisting of plains and grassy tablelands with scattered and isolated fertile valleys that constitute the granary of Algeria. More elevated in the west (over a thousand meters above sea level near the Moroccan frontier), the High Plateau slopes gradually downward toward the east, being less than four hundred meters above sea level in the Chottel-Hodna depression. The region is about three-quarters steppe, with eroded summits and humid lowland areas. Without irrigated agriculture, dry, cold in winter, and hot in summer, the High Plateau is also subject to strong winds that inhibit the growth of trees, which exist only in protected areas. In general the eastern and western sections of the High Plateau region are quite different. The western steppes are larger, reaching a depth of 350 kilometers north to south at the Moroccan frontier. Although drier than those to the east, the western plains are better cultivated and more open to the influences of the south due to the dispersion of the mountains of the Saharan Atlas chain. The high plains of the east are narrower (about two hundred kilometers at the Tunisian frontier) but less compartmented and more open to Mediterranean influences. Bounded by the Tunisian border to the east and the line of 4° west longitude to the west, the eastern plain extends about 260 kilometers west to east. It is bounded on the north by the coastal hills and mountains of the Grande Kabylies to the northwest, the Petite Kabylies to the north, the Constantines to the northeast, and the Medjerda along the northeast Tunisian border. On the south the eastern plain is bordered by three rugged mountain ranges of the Saharan Atlas chain: the Hodna in the southwest, the Aurès in the south, and the Nementcha in the southeast. The Tebessa mountains extend across the Tunisian border in the southeast corner. These ranges are rather barren, quite rocky, reaching up to 2,300 meters in elevation and dropping rapidly to the south to the Sahara desert. Rain is rare on the High Plateau, and the region is poorly supplied with water sources. Perennial rivers are few; only the Cheliff reaches the sea—with difficulty. Most of the streams are torrential and flow only during periods of heavy rainfall. Scattered depressions become extensive salt lakes (*chotts*) and marshes (*sebkhas*) during the rainy season.

Parallel to the Tellian Atlas and separating the High Plateau from the Sahara desert is the southern branch of the Atlas chain, the Saharan Atlas. The mountains of the Saharan Atlas chain are more rugged and somewhat drier than the coastal mountains but are more regular. From west to east the Saharan Atlas chain consists of a number of distinct mountain ranges: the Ksour, the Djebel Amour, the Ouled Naïl, the Mzab, and the Aurès. Characterized by steep and barren slopes, long and lean ridges, and *wadis* subject to torrential flooding, the region has few roads. It is very hot in summer and very cold in winter. The central portion of the Saharan Atlas chain is one of pastureland and nomadic flock tenders, while in the Ksour and the Aurès mountains at either end of the chain there exist sedentary populations, with villages sited on dominating terrain. The Aurès mountains of the east form a land of savage, inhospitable grandeur, a treeless wilderness relieved by a few narrow, fertile valleys and occasional scrub oak and ivy forests. Rocky, hot in summer, and bitterly cold in winter, the Aurès have been historically a refuge for bandits and guerrillas.

South of the Saharan Atlas chain lie the wastelands of the Sahara desert itself. The northern portion of the Sahara, close to the mountains, provides good forage and supports a nomadic population on the mountain slopes. Further south is the true Sahara, a great, desolate expanse of wind-eroded bare rock and large sand areas, or *ergs*, with constantly shifting dunes. Broken by scattered oases, the Sahara boasts only sparse and stunted vegetation, although the infrequent rains can produce a temporary blooming of the desert. In fact the Sahara consists of a number of regions of very different characters, all of which, however, are characteristically desert: dry, eroded by violent winds, almost completely without rain, and subject to vast temperature variations. Centered just north of Tamanrasset (Fort Lapperine), the Hoggar massif and the high plateaus associated with it dominate the entire southeastern quarter of the Algerian Sahara. Generally desolate and rugged, the Hoggar averages over 1,525 meters above sea level and includes numerous peaks over two thousand meters. West of the Hoggar and at a lower elevation is the sterile pebble desert of the Tanezrouft, where nothing grows and no one lives. Farther west are the great sand deserts of the *Erg Chech* and *Erg Iguidi*.

## THE CLIMATE OF ALGERIA

Algeria can be divided into four climatic zones. The coastal strip and Kabylia are under Mediterranean influence. The inland hills, plains, and valleys of the Tellian Atlas chain experience a mixed Mediterranean and continental climate. The High Plateau and the Saharan Atlas regions are under continental climate influences. The Sahara, of course, has a desert climate. In addition, the compartmented nature of the Algerian terrain creates what are called "mesoclimates," regional variations that make for dramatic differences in the weather between points only a short distance apart. For example, in one area the sky can be clear while only fifty miles away clouds and fog completely obscure visibility.

In general, the climate of northern Algeria, including the coastal strip and the

Tellian Atlas chain, is quite temperate, being modulated by the Mediterranean Sea. Winters in the coastal region are mild and rainy, and summers are hot and rainless but with very high humidity. As one proceeds south, daily and seasonal climatic variations are more pronounced. The High Plateau region experiences drier, colder winters and hotter summers, and climatic conditions in the mountains of the Saharan Atlas and the Sahara desert can be severe. The variation in seasonal and regional temperatures throughout Algeria can be significant. In general, the annual variation in temperature in the coastal region is quite small. In the coastal plain the Mediterranean influence is greatest during the winter months, when the mean temperature in January is between 50° and 52° Fahrenheit. In summer the mean temperature ranges from 75° to 77°, but in August, the hottest month, maximum temperatures of over 100° are common. At the other extreme is the Sahara desert, where temperatures can vary from 27° to 122° Fahrenheit. In the desert the average daytime temperature of 95° drops precipitously at night, especially in the winter. The seasonal temperature fluctuations in the mountainous and High Plateau regions of Algeria are substantial. Available weather statistics note record lows of 23° and record highs of 115° in the Tellian Atlas, record lows of 21° and highs of 118° in the High Plateau region, and record lows of 17° and highs of 109° in the Saharan Atlas.

Rainfall in Algeria is seasonal and varies widely from region to region. In general the rainy season occurs from November through January; Algiers receives half its annual rainfall during that period. During the rainy season torrential storms are common throughout the country and can cause serious flooding and soil erosion. However, only northern Algeria, the strip under Mediterranean climatic influence, consistently receives sufficient rainfall to support continuous agriculture and a sedentary population. Rainfall in the coastal region varies widely from west to east. Around Oran, for example, the average annual rainfall is less than two inches, but around Algiers it is about twenty-six inches, and in parts of Kabylia it reaches sixty inches. In mountainous areas the northern slopes facing the Mediterranean receive the rain, while the southern slopes are much drier. The Tellian Atlas, also under Mediterranean climatic influence, experiences generally abundant rainfall, averaging thirty-five inches in 102 days each year. Some areas receive considerable rain. For example, the area around Bessombourg receives an average of seventy inches per year, about the same as Seattle. Snow is common in winter in Kabylia, and the Djurdjura range averages twenty snow days per year between November and May. Heavy snowfalls in the mountains frequently block the roads, although generally the snow does not remain long below 1,500 meters elevation. In the High Plateau annual precipitation is substantially less, averaging only 8.8 inches in thirty-nine days per year. The Saharan Atlas chain receives even less precipitation, only about 7.6 inches in thirty-seven days per year. Of course, the Sahara desert is characterized by the rarity of any rainfall at all, some areas going for years without measurable precipitation.

The weather throughout Algeria is generally excellent most of the year, with the characteristic clear skies typical of a Mediterranean climate. Daily rainstorms of a few hours may occur, but seldom are more than two days per month overcast. However, the "mesoclimates" of Algeria can create locally poor weather conditions for extended periods. In the Kabylie mountains and the area around Bône the weather is unpredictable, and the season of intermittent poor visibility lasts about five months—in contrast to the area around Algiers, which has about three or four months of intermittent poor visibility per year. In western Algeria, early morning fog and clouds in some areas last until about noon. Moreover, the entire country is subject to the *sirocco*, a hot dust and sand-laden wind, especially common between February and September. In the desert area of the Sahara sandstorms are common.

## THE POPULATION OF ALGERIA

In October 1954, just before the Algerian war began, the total population of Algeria was 8,945,800: 974,200 non-Moslems (Europeans), of whom about 80 percent were born in Algeria; and 7,971,600 indigenous Moslems, of whom about 800,000 were ethnically distinct Berbers living in the Kabylia region. About 55 percent of the Moslem population was of military age (15–59), with women outnumbering men by a ratio of 1,000:958. Both the Moslem and non-Moslem populations grew during the course of the war, and the 1960 census profiled a total population of 9,925,400: 8,850,000 Moslems and 1,075,000 non-Moslem Europeans.

Although the population density of the Sahara region was less than one person per square mile in 1954, northern Algeria was highly urbanized, some 27.7 percent of the total population (and over 85 percent of the European population) living in urban areas. By the time of the 1960 census the percentage of Algerians living in urban areas had increased to 30.3 percent. Some 22 percent of the total population resided in the thirteen cities and towns of over fifty thousand population. In rural areas the European and Moslem residents lived together with little friction; in urban areas the two communities were more separate, with clear Moslem majorities in most Algerian cities. The urbanization of Algeria continued during the war, and by 1960 the population of the thirteen major cities of Algeria had reached the totals shown in Table 1.1.

Ethnically the Moslem population of Algeria is divided into two main groups, Arabs and Berbers. The Arab population stems from the Moslem invaders of the seventh century A.D., augmented by later arrivals from Arab lands to the east. The Berbers, the dominant ethnic group in Kabylia, the Aurès and Nementcha mountains, and the Sahara, speak their own language and were Christians before being conquered and forced to convert to Islam by Arab invaders in the seventh through eleventh centuries. Always fiercely independent, the Berbers have traditionally resisted Arab political dominance, a fact of some importance during and after the 1954–1962 war for independence.

In 1954 the European population, mostly of French extraction but including

**Table 1.1**
**Population of Major Algerian Cities, 1960**

| City | Non-Moslem | Moslem | Foreign | Other | Total |
|---|---|---|---|---|---|
| Algiers | 307,287 | 542,918 | 9,670 | 24,004 | 883,879 |
| Oran | 183,686 | 175,770 | 23,084 | 10,097 | 392,637 |
| Constantine | 36,486 | 184,025 | 338 | 2,410 | 223,259 |
| Bône | 47,906 | 113,896 | 476 | 2,566 | 164,844 |
| Sidi-bel-Abbès | 23,237 | 64,387 | 2,389 | 5,344 | 105,357 |
| Sétif | 6,845 | 84,104 | 85 | 2,527 | 93,561 |
| Blida | 19,070 | 68,358 | 213 | 5,605 | 93,246 |
| Philippeville | 25,363 | 58,934 | 631 | 2,994 | 87,922 |
| Tlemcen | 10,843 | 67,853 | 470 | 3,361 | 82,527 |
| Mostaganem | 18,541 | 48,487 | 733 | 1,160 | 68,921 |
| Bougie | 6,434 | 55,285 | 128 | 1,074 | 62,921 |
| Biskra | 2,218 | 52,722 | 133 | 299 | 55,372 |
| Mascara | 8,297 | 43,003 | 227 | 1,370 | 52,897 |

*Source*: Norman C. Walpole and others, *U.S. Army Area Handbook for Algeria*, DA Pamphlet No. 550–44 (Washington, DC: Headquarters, Department of the Army, January 1965), 74 (Table 1). ''Other'' includes ''population counted separately,'' such as members of the military and persons in institutions.

significant numbers of Spanish, Italian, and Maltese immigrants, controlled the richest farmland and dominated both agriculture and trade. Europeans owned about 90 percent of the industry and about 40 percent of the best land in Algeria. Of the 974,200 European inhabitants of Algeria in 1954 only about 19,400 lived on the land. Of these, about 7,500 had holdings of less than ten hectares (about twenty-five acres), and perhaps only three hundred could be classed as major landed proprietors (*gros colons*). The ranks of the super-rich comprised only about a dozen families, who dominated the political and economic life of Algeria. At the opposite end of the scale were some ten thousand unskilled urban Europeans, who were little better off than their Moslem neighbors. In between lay the great mass of the urban and small-town French Algerians, the *pieds noirs*—schoolteachers, shopkeepers, clerks, and mechanics—a group imbued with an innate sense of superiority over all Moslems but fearful of being submerged in a Moslem sea.

The Moslem population occupied the less fertile regions and the more remote and rugged mountainous areas of northern Algeria, worked on the farms of the Europeans, or occupied shantytowns (*Bidonvilles*) in the metropolitan areas. Of the 7,971,600 Moslems in 1954, the majority lived in rural conditions, mainly on small farms. About three million of them were landless, and another million were unable to subsist on their small private plots and were obliged to earn their keep by performing casual labor for their European neighbors.

In 1954 the total Algerian labor force totaled some 3.5 million persons, of whom about 3.2 million, or 90 percent, were Moslems, the great majority of whom—about 91.5 percent—were engaged in agricultural occupations. Although Moslem unemployment was relatively low—about 4.2 percent—a significant portion of the Moslem workforce was seriously underemployed. A substantial number of Algerian Moslem males found employment in metropolitan France. The practice began in 1904, and by 1954 some four hundred thousand Algerians (one-fifth the employable male Moslem population) worked in metropolitan France to support some two million dependents in Algeria. Women constituted about 23 percent of the European labor force and about 32 percent of the Moslem labor force. Almost all Moslem women employed outside their homes were unpaid family agricultural workers. In 1954 Algerians of European descent dominated the professions and white collar, skilled labor, and supervisory positions. Unemployment among European Algerians was about 4 percent.

## PUBLIC HEALTH

In the 1950s and early 1960s health and sanitation conditions varied widely throughout Algeria; the contrast between areas of Moslem and European habitation and influence was particularly dramatic. In general, the areas occupied by non-Moslem Europeans were relatively clean and free of disease; those occupied by the Moslem population, particularly the squalid, overcrowded urban districts and the rural villages lacking adequate water supplies, were pestilential. The most important factors affecting health and sanitation in Algeria in the 1950s were malnutrition, a generally insufficient water supply, overcrowded living conditions, inadequate education in personal hygiene and public health, indigenous custom and superstition, and a lack of adequate health care services in many regions. In particular, the absence of an adequate water supply in much of Algeria greatly magnified such health problems as trachoma and diseases resulting from poor personal hygiene and unsanitary practices. The fluctuating and often stagnant watercourses and marshes of northern Algeria provided ample breeding areas for mosquitoes.

Disease vectors were plentiful and widespread in Algeria. Flies, lice, ticks, sandflies, fleas, and other insect vectors of disease were particularly abundant, and the varieties of mosquito known to carry yellow fever, malaria, and filariasis were all present. Among the diseases of most concern to public health officials were eye disorders, tuberculosis, malaria, and venereal diseases. Enteric, respiratory, and helminthic infections as well as the usual childhood diseases were prevalent. Louse-born typhus, schistosomiasis, and relapsing fever occurred sporadically, but the more serious epidemic diseases, such as smallpox, plague, cholera, and yellow fever, were generally under control.

Nutritional diseases resulting from inadequate intake and vitamin deficiencies were not uncommon among the Moslem population of Algeria. Algerian government officials estimated the daily caloric intake of most Algerians at 2,100 calories, although some experts put the figure as low as 1,250 calories. Famine and periods of near starvation, particularly between harvests (in January and

February), were not unknown. The usual diet of most Algerian Moslems consisted primarily of grain—in the form of bread or *couscous*—and olive oil supplemented by some fish, fresh vegetables, and fruit (dates in the desert areas). Meat, milk, and eggs were seldom consumed. The European population, being generally better off economically, was able to maintain an adequate, varied diet similar to that enjoyed in metropolitan France. French military forces, of course, were generally well fed.

Civilian medical care in Algeria in the 1950s and early 1960s was generally inadequate, although northern Algeria was somewhat better provided with medical facilities and personnel than were the southern territories. In 1956 there were approximately 28,500 hospital beds in some 150 hospitals and clinics in northern Algeria—about one bed for every three hundred inhabitants—but only about 1,500 beds in two military hospitals and twenty-seven dispensaries in southern Algeria. Mobile medical teams combating malaria, eye diseases, and tuberculosis, as well as some thirty-eight ambulances, were scattered throughout the country. At the beginning of 1959 there were in northern Algeria 1,954 physicians, 606 midwives, 523 dentists, and 710 pharmacists, compared to thirty-five physicians, thirteen midwives, five dentists, and eleven pharmacists in the southern territories in 1957. Although French military forces were adequately supplied with their own hospitals and medical personnel, the small number of civilian doctors in Algeria, most of whom were Europeans in any event, posed a problem of some magnitude for the rebels.

## THE ALGERIAN ECONOMY

From 1830 onward, French government policy emphasized the role of Algeria as an agricultural producer and as a market for goods manufactured in metropolitan France. Consequently, the Algerian economy focused on agriculture and exploitation of its limited natural resources, and Algerian industry was little developed until after World War II. In 1959 agriculture accounted for one-fourth of the Algerian national income, while industry and public works accounted for another one-fourth, and services for the remainder.[2]

During the period 1954–1962, over three-quarters of the Algerian population was directly employed in working the land. The best agricultural land was in the hands of the Algerians of European descent (*colons*), but even on marginal land the hard work of the *colons* created flourishing farms, in contrast to the generally primitive and unproductive Moslem holdings. Of a total surface area of 221 million hectares only about 150,000 hectares in the southern territories and 12.9 million in northern Algeria were used for agriculture. In 1955 some 5.6 million hectares were pasturage and grasslands, which supported 6.3 million sheep, 3.35 million goats, 80,000 pigs, 207,000 horses, 230,000 mules, 365,000 donkeys, 913,000 cows, and 221,000 camels. About 200,000 hectares were artificially irrigated, some 2.5 million hectares were left fallow every other year, approximately 20,500 tractors and 3,700 combines were in use, and some 45,000

**Table 1.2**
**Crop Area and Production of Selected Crops in Algeria, 1959–1960**

| Product | Area Cultivated | Production |
|---|---|---|
| Wine | 372,330 hectares | 418 million gallons |
| Hard Wheat | 1,531,790 hectares | 1,116,000 tons |
| Soft Wheat | 410,770 hectares | 345,000 tons |
| Barley | 1,180,510 hectares | 847,000 tons |
| Oats | 62,730 hectares | 49,000 tons |
| Citrus Fruit | 230,000 hectares | 398,000 tons |
| Olives | not available | 131,000 tons |
| Figs | not available | 65,000 tons |
| Dates | not available | 94,000 tons |
| Dry Legumes | not available | 43,000 tons |
| Tobacco | 22,260 hectares | 16,000 tons |

*Source*: Norman C. Walpole and others, *U.S. Army Area Handbook for Algeria*, DA Pamphlet No. 550–44 (Washington, DC: Headquarters, Department of the Army, January 1965), 381 (Table 9).

tons of chemical fertilizer were used. Table 1.2 shows the area under cultivation and the production of selected crops in Algeria in 1959–1960.

With the exception of oil and natural gas, Algeria is not rich in mineral resources. Such deposits of iron ore, phosphates, coal, and other minerals (zinc, lead, manganese, and copper) as exist are located in remote parts of the country and present enormous difficulties in mining and transportation. Production of minerals in 1962 amounted to 2,062,000 tons of iron ore, 389,900 tons of phosphates, and 53,100 tons of coal, down from 3,388,000 tons, 618,700 tons, and 295,000 tons, respectively, in 1953.[3] Proven mineral reserves in 1963 included a rich deposit of some three billion tons of high-grade iron ore at Gara Djebilet, near Tindouf on the Moroccan border, and scattered deposits elsewhere; some forty million tons of low-grade coal near Colomb-Béchar, on the Moroccan border; and a large deposit of manganese ore in the Sahara.[4] Rock salt, clay, building stone, gravel, and sand were also available. The greater part of these natural resources remain unexploited today.

The first major discoveries of Algeria's principal natural resources, oil and natural gas, came in 1956, and these resources were only beginning to be exploited in 1962 when Algeria gained its independence. Algerian crude oil production totaled only 3.4 million barrels in 1958, 9.5 million barrels in 1959, 66.3 million barrels in 1960, 121.4 million barrels in 1961, and 158.8 million barrels in 1962.[5] By the end of 1963 proven crude oil reserves totaled some 6.5 billion barrels, and production had reached 183.3 million barrels.[6] Located mostly in remote parts of the Sahara, Algerian oil and natural gas resources are difficult to exploit and far from markets.

Energy resources were a constraint on industrial production and development throughout the period of the Algerian war. In 1955 the collieries at Kénadza supplied about one-third of Algeria's coal requirements, most of which was used for the production of electrical power and manufactured gas.[7] Following World War II, efforts were made to develop hydroelectric power resources to replace thermal generating plants using coal and gas produced from coal. By 1956 Algeria had eight thermal generating plants and thirty-two hydroelectric plants.[8] The production of electricity for public use was divided almost equally between thermal and hydroelectric generating plants. Development of hydroelectric power was halted when petroleum and natural gas became available to fuel power plants.

French government policy and the lack of adequate power retarded the development of Algerian manufacturing industries until after 1945. The isolation of Algeria from metropolitan France during World War II highlighted the need to make it able to provide a substantial part of its own requirements in manufactured goods; between 1945 and 1956 some 120 new factories were established to produce finished metals, chemicals, plastics, pharmaceuticals, paper, textiles, construction materials, processed foods, fat products, and a variety of mechanical and electrical items.[9] A large proportion of Algerian industry was focused on food processing, both for local consumption and export. The production of building materials was also quite important. The industrial sector of Algeria's economy, dominated (as might be expected) by Europeans, grew steadily throughout the period of the Algerian war. By 1961 industry accounted for about one-third of Algeria's gross domestic product and employed some 300,000 persons.[10] The production of industrial goods in Algeria in 1961 included 1,071,700 tons of cement, 15,000 tons of fabricated metal goods, 1,050,100 tons of milled grain, 73,000 tons of refined vegetable oils, 36,958,500 gallons of beer and soft drinks, 378,000 pairs of shoes, and 1,230 motor vehicles.[11]

Algeria's export-import trade in 1955 amounted to some 11,133,300 metric tons of goods worth 405 billion French francs: 244 billion FF in imports and 161 billion FF in exports.[12] Except for food products, Algeria was a net importer from France. In 1960, Algerian imports from France had a value of some 5,395.2 million FF, and exports to France 2,534.8 million FF.[13]

## TRANSPORTATION FACILITIES

When French troops first arrived in Algeria in 1830, their landing and subsequent movements were made extremely difficult by the lack of adequate ports and roads. By 1954, however, the ports of Algeria were well developed and were linked to the numerous cities and towns by an excellent system of well maintained modern roads and railroads, supplemented by a network of usable airfields. Even the northern Sahara desert was penetrated by rail and highway routes, although the vast southern Sahara remained a mostly trackless waste, and the mountainous areas had few roads. Originally constructed in part for

**Table 1.3**
**Capacity of Major Algerian Ports, 1957**

| Port | Length of Quais (meters) | Maximum Draft (meters) | Number of Berths | Ships Discharged at Once | Daily Capacity (metric tons) |
|---|---|---|---|---|---|
| Algiers | 8,400 | 10 | 54 | 44 | 20,000 |
| Oran | 4,900 | 10 | 49 | 31 | 10,000 |
| Bône | 3,140 | 10 | 19 | 10 | 4,000 |
| Mostaganem | 685 | 8 | 7 | 5 | 1,500 |
| Némours | 967 | 10 | 10 | 3 | 1,500 |
| Bougie | 1,850 | 10 | 14 | 4 | 1,200 |
| Philippeville | 1,100 | 10 | 12 | 10 | 1,000 |
| Djidjelli | 440 | 9 | 4 | 2 | 900 |
| Arzew | 958 | 10 | 7 | 6 | 700 |

*Source*: Map 10, "Infrastructure Portuaire d'A.F.N." (Alger, 1 janvier 1957), in folder "Aide-Major Général en AFN—Organisation du commandement et coordination des services et moyens logistiques, 1946–57," Dossier 1H2665 d. 1, Fonds Algérien, SHAT. The naval base at Mers-el-Kébir also had 3,500 meters of quais and thirty-two berths capable of handling vessels of ten meters draft.

military reasons, the Algerian transportation infrastructure remained almost entirely under French control throughout the Algerian war of 1954–1962 and gave the French a decided advantage over the Algerian rebels, who had to rely on more primitive facilities.

### Water Transport

Algeria has few good natural harbors, but when the Algerian war for independence began in 1954 there were twenty-one modern ports of varying importance. In 1955 they handled 29,474 ships of forty-five million deadweight tons, carrying some 13,382,883 tons of import-export cargo and 908,278 passengers.[14] Algiers was the major cargo and passenger port, followed by Oran, Bône, Mostaganem, Némours, Bougie, Philippeville, Djidjelli, and Arzew. In 1955 Algiers handled 9,820 vessels (over twenty million deadweight tons), 4,377,549 tons of cargo, and 483,944 passengers; the next-largest port, Oran, handled 6,247 vessels (over ten million deadweight tons), 2,214,796 tons of cargo, and 271,837 passengers.[15] Table 1.3 shows the physical characteristics of the nine largest Algerian ports in 1957.

Despite Algeria's extensive coastline, coastal shipping was of little importance in the years of French domination, being used mainly for bulk commodities such as coal, fertilizer, and construction materials. Economically, it was cheaper, and thus preferable, to import and export most goods directly between metropolitan France and an Algerian port rather than to transfer cargo by means of coastal shipping. However, in 1960 the Algerian merchant marine had 135 ships and some 9,350 seamen, one-third of them operating between Algeria and France.[16] The rugged terrain and lack of any usable rivers precluded the development of inland waterways in Algeria.

## Railroads

In 1954 Algeria's modern rail system totaled some 4,375 kilometers, almost all of which was single tracked, standard gauge, and passed over difficult terrain requiring numerous bridges and tunnels.[17] The principal trunk route paralleled the Mediterranean coast from the Moroccan border at Tralimet via Tlemcem, Sidi-bel-Abbès, Ste. Barbe-du-Tlelat, Perrégaux, Orléansville, Blida, Maison Carrée, Ménerville, Beni-Mancour, Sétif, La Kroubs, Duvivier, and Souk-Ahras to the Tunisian border at Sidi-el-Nemessi. Algiers was connected to the main line by a short spur. West of Algiers other spurs led northward to the major port cities of Arzew and Mostaganem. There were two major north-south lines west of Algiers: from Perrégaux via Saida, Mécheria, Ain-Sefra, Tiout, and Colomb-Béchar to Kénadza and Abadla; and from Blida via Médéa and Boghari to Djelfa. East of Algiers the trunk line connecting Algiers, Sétif, Constantine, and Souk-Ahras was supplemented by spurs to Tizi-Ouzou and the ports of Bougie, Philippeville, and Bône. There were two north-south lines east of Algiers: from El-Guerrah near Constantine via Batna to Biskra and on to Tougourt on meter gauge, and from Souk-Ahras to Tebessa. East of Constantine two additional east-west lines connected St. Charles with La Calle via Bône, and Ouled Rahmoun with Le Kouif via Ain-Beida, Youks-les-Bains, and Tebessa. There was a modern remote-control signaling system, and in 1960 a radiotelephone communications system, identical to that in use in metropolitan France, was installed. In 1962, there were over two hundred locomotives (all diesel except for a few antique steam locomotives and about forty electric ones), 450 passenger cars, ten thousand freight and baggage cars, and over a thousand special cars. Private companies owned another thousand special cars, mostly petroleum tankers and ore cars. In 1955 the Algerian rail system handled 850,000,000 passenger/kilometers and 1,506,000,000 ton/kilometers of freight.[18]

## Highways

In 1954 there were some eighty thousand kilometers of roadways in Algeria: in the north, 54,000 kilometers, including 11,000 kilometers of *Routes Nationals* and 15,000 kilometers of other modern routes, in the departments of the Sahara, 26,000 kilometers, including 1,500 kilometers of modern paved highway.[19] The highway network in northern Algeria consisted of three main east-west routes (*rocades*): the Coastal (from Oran to Taborka), the Northern (from Oujda to Ghardimaou), and the Southern (from Berguen to Tebessa). The *rocades* were complemented by three main north-south routes (*penetrants*): from Oran/Mostaganem via Mascara, Mécheria, and Ain Sefra to Ben Ounit; from Algiers via Blida and Médéa to Djelfa; and from Philippeville/Bône via Constantine, Batna, and Biskra to Touggourt. There were also three east-west *rocades* across the northern Sahara: from Colomb-Béchar via Tabelbala and Tinfouchy to Tindouf; from Adrar via Timimoun, El Goléa, and Ghardaia to Ouargla; and from Reggan via Aoulef, In Salah, and Fort Mirabel to Fort Flatters. Three north-south highway *penetrants* linked the terminals of the Algerian railroads with the railroads of sub-Saharan Africa: from Beni-Ounif via Bech, Adrar, and Tessalit to Gao in Mali; from Djelfa via Laghouat, El Goléa, and Tamanrasset to Aeades in Niger; and from Touggourt via Ouargla and Fort Flatters to In Ezzane.

The Algerian highway system was generally adequate for commercial and military purposes, although there were few adequate routes in the mountains of the Tellian and Saharan Atlas or in the Sahara. However, difficult terrain, seasonal torrential rains and heavy snows, and desert conditions in the Sahara made road construction and maintenance difficult.[20] Although the Algerian rebels frequently sabotaged the roadway, bridges, and culverts and mounted ambushes on military convoys and civilian highway traffic, the Algerian highways remained under the effective control of the French throughout the 1954–1962 war. French military and civilian traffic were consistently able to use the highway system, and French military guardposts and patrols were generally able to deny its use to the rebels. Such had not been the case in Indochina, where at least in Tonkin, the Viet Minh were able to seize control of major portions of the highway net and use them for their own purposes.

### Air Transport

Commercial air service to Algeria from metropolitan France began in the 1920s, but it was not until the early 1950s that regular scheduled commercial air service was available within Algeria itself.[21] The discovery of oil in the Sahara and the insecurity of land transportation routes during the 1954–1962 war did much to spur the development of commercial air transport in Algeria. In 1955 Algeria had four international airports—at Algiers (Maison Blanche, now known as Dar-el-Beida), Oran, Bône, and Aoulef—plus fourteen regional and twenty lesser airfields; Algerian airports handled 21,380 aircraft, 600,000 passengers, and 12,500 tons of cargo.[22] By 1960 there were some ninety-three airfields of one sort or another in northern Algeria and another 117 regional airports and private airstrips in the Sahara.[23] At that time Air France, the French national carrier, based about 20 percent of its flying equipment in Algeria, and Air Algérie, a privately owned company, serviced thirty-two airports, with 80 percent of its business consisting of trans-Mediterranean traffic. In addition, there were a number of nonscheduled airlines and some thirty-five private flying clubs, with one hundred aircraft.

In addition to the commercial and private airfields in Algeria there were a number of military airfields occupied by the French Air Force, Army, and Navy. When the war began in 1954 there were four military air installations in Algeria capable of supporting modern military aircraft: two Air Force bases near Algiers and one Air Force and one Navy air base near Oran. As demands for air support increased during the war, the number of bases was expanded substantially. A major Air Force base was established at Telergma near Constantine, and the Air

Force bases at Slida (near Algiers) and Oran were built up to provide support for units and aircraft in their area. Over thirty airfields capable of handling fighters and light bombers were constructed in northern Algeria during the war, and in the eastern Sahara alone there were some two hundred strips capable of handling light aircraft. Airfields able to handle C-47s were constructed throughout the Sahara in about three weeks' time as required to support operations.[24]

## MILITARY IMPLICATIONS OF THE PHYSICAL ENVIRONMENT

The French military forces and the Algerian nationalist rebels shared the physical environment of Algeria, but the special benefits and special problems of that environment fell unequally on the two sides. In most respects the major advantages accrued to the French, who had the resources to overcome the military problems created by a generally hostile environment. The strategic location of Algeria on the southern rim of the Mediterranean Sea opposite metropolitan France, the ultimate source of men and supplies, conferred on the French forces an advantage they had not enjoyed in the Indochina war against the Viet Minh. The lines of communication were shorter and less complicated and therefore more effective.

From the point of view of military operations, the size of Algeria, and particularly the length and remoteness of her borders, posed special problems for both the French and the rebels. The Mediterranean coast and the southern desert borders constituted natural limits that, while extensive, were fairly easy to protect. Such was not the case with the long borders with Tunisia and Morocco. However, through the skillful application of modern technology, the French were ultimately successful in sealing those borders and thus in denying to the rebels the ability to move men and supplies into and out of Algeria at will, as the Viet Minh had been able to do in Tonkin. Even so, the vastness of Algeria stretched French logistical resources to their limits. Support units were often required to cover a much larger area than their Tables of Organization and Equipment (TOE) prescribed, and the geography of Algeria exerted a marked influence on the positioning of stocks and direct-support units; it also required the continuous support of isolated, far-flung outposts, careful prestocking before operations, and massive resupply efforts during combat.

The diversity of the Algerian topography also posed military challenges that only the French were prepared and equipped to overcome, principally by the use of aviation, in particular by the introduction of the helicopter for transport and combat assault operations. The rugged terrain provided ample cover and concealment for rebel bands but also presented obstacles to the movement of men afoot and pack animals. Given the rebels' relative lack of air and motor transport, the difficult mountainous terrain and desert wastelands that cover most of Algeria were thus serious obstacles to rebel movements within the country. Only in the crowded cities did the topography of Algeria favor the rebels, who could take refuge in the old Turkish quarters (the Casbah in Algiers and La Moune in Oran, for example); these zones were very difficult to isolate, screen, and search due to the complex, narrow streets and numerous underground passageways. Even on the plains the rebels were frequently able to conceal themselves in the orange groves and other vegetation in the cultivated areas, but on the whole effective aerial observation, possessed only by the French, canceled out such advantages.

The effects of extreme climatic conditions in the mountains and deserts of Algeria struck Frenchmen and Algerian nationalists with equal force, causing injuries and disease and adversely affecting mobility. Desert sandstorms and the seasonal *sirocco* wore out men, animals, and machines without regard to their political alignment, and in desert areas unacclimatized personnel suffered heat injuries and were limited in their endurance for work outside during daylight hours. Both Frenchmen and Algerian nationalists experienced frostbite and other cold injuries in the mountainous regions in winter. Temperature fluctuations, the *sirocco* and desert sandstorms, contaminated water, and the general lack of sanitation caused intestinal, respiratory, and eye diseases. The singular advantage possessed by the French was that they had the medical personnel and equipment to minimize such effects.

The highly mechanized French forces were more vulnerable to the impact of climatic extremes on equipment, but they had the means to overcome all but the most pronounced challenges of the Algerian climate. While torrential seasonal rainstorms frequently washed out transportation routes and limited cross-country mobility in plains areas, the French had the skilled labor and special equipment needed to repair roads, bridges, and culverts. The rebels did not. The French forces also had the capability of transporting water to areas where there was none, and they could provide the clothing and specialized equipment necessary to operate in the Sahara in high summer and the high mountains in the grip of winter. Moreover, the generally good weather over most of Algeria, mesoclimates notwithstanding, favored the extensive air operations and aerial observation that only the French had the ability to conduct. Only rarely were the rebels cloaked by fog, and even the rainy season did not substantially reduce French air operations.

Algerian demographic and economic factors also favored the French. Though overwhelmingly outnumbered by the Moslem population of Algeria, the French were able to retain the support of a large part of the indigenous population. More important was the fact that most of the skilled manpower of Algeria was of European descent and thus not inclined to the rebel cause. The rebels were thus limited in their ability to produce their own military equipment and supplies, as had the Viet Minh. French control of Algerian industry and commerce also conveyed a special advantage by denying, or at least restricting, rebel access to those goods and services available on the commercial market.

Effective French control of transportation facilities of all types throughout the 1954–1962 war also constituted a significant advantage over the rebels. Air transport in particular, extremely important in a country as large as Algeria, was a benefit from which the rebels were almost totally excluded. Effective French air, land, and sea interdiction was able to the deny to the rebels the use of all but the most remote and primitive tracks through the wilderness, and even then good observation in the desert areas frequently permitted French forces to locate and destroy rebel convoys and caravans.

In sum, the physical environment of Algeria conferred several clear advantages on the French forces, particularly with respect to the crucial factor of mobility. The proximity of the resources of metropolitan France was of strategic importance. At the operational level, given French superiority in technology, the geography, topography, and climate favored the French more than they did the Algerian rebels. Air transport and the helicopter dealt with the great distances and rugged terrain in a way men on foot or pack animals could not. Aerial observation and air interdiction were much easier over Algeria than over the jungles of southeast Asia. Moreover, from the beginning the French controlled the means of production and the transportation facilities in Algeria to a degree they had not in Indochina. Although endowed with the very significant advantage of fighting on and for their native soil, the Algerian nationalist rebels lacked many of the technological and organizational advantages held by their French opponent.

## NOTES

1. See Map 1. The details of Algeria's physical geography and climate are described in a number of sources, the more important of which include: Commandement en Chef des Forces en Algérie, État-Major Interarmées, 3e Bureau, Antenna de Documentation Géographique, *Aperçu géographique de l'Algérie*, No. 182/EMI/3/Doc. Géo., Alger, mai–novembre 1960, in folder "10e Région Militaire, État-Major Interarmées, 3e Bureau—Documentation sur l'Algérie, 1954–62," Dossier 1H2100 d. 1, Fonds Algérien, Service Historique de l'Armée de Terre, Château de Vincennes, France [cited hereafter as SHAT]; Gouvernement-Général de l'Algérie, *Notions essentielles sur l'Algérie* (Paris: Georges Lang, [ca. 1956]); Lieutenant Colonel Nougues, *Caracteristiques générales des opérations en Algérie* (Algiers: SDRR/EMI, [1960]), in folder "10e Région Militaire, État-Major Interarmées, 3e Bureau—Enseignements de la lutte contre les rebelles d'AFN-ESG, 10 mars 1958," Dossier 1H1942 d. 3, Fonds Algérien, SHAT; and Norman C. Walpole and others, *U.S. Army Area Handbook for Algeria*, DA Pamphlet No. 550–44 (Washington, DC: Headquarters, Department of the Army, January 1965). The same works also provide details regarding the human resources of Algeria, as do: Commandement en Chef des Forces en Algérie, État-Major Interarmées, 3e Bureau, Antenna de Documentation Géographique, *Note sur la population Algérienne d'après l'estimation du 1er janvier 1960*, No. 26/EMI/3/Doc. Géo. (Alger, 2 février 1961), Tableaux I and II, in folder "10e Région Militaire, État-Major Interarmées, 3e Bureau—Documentation sur l'Algérie, 1954–62," Dossier 1H2100 d. 1, Fonds Algérien, SHAT; Gouvernement-Général de l'Algérie, Direction Générale des Finances, Service de Statistique Générale, *Résultats Statistiques du Dénombrement de la Population effectué le 31 octobre 1954,*

*Volume I: Population Légale ou de résidence habituelle*, "Répertoire Statistique des Communes d'Algérie" (Algiers: Service de Statistique Générale, [1954]); and United States Army Medical Service, Medical Information and Intelligence Agency, *Health and Sanitary Data for Algeria (A Special Study)*, Intelligence Research Project No. PO 2306 (Washington, DC: Office of the Assistant Chief of Staff for Intelligence, Headquarters, Department of the Army, 11 January 1960).

2. Arslan Humbaraci, *Algeria: A Revolution That Failed* (New York: Frederick A. Praeger, 1966), 282 (Table 8). Additional details on Algerian agriculture, industry, and trade can be found in: *U.S. Army Handbook for Algeria*, 381–411; *Notions essentielles sur l'Algérie*, 7–9; *Les grands secteurs de l'agriculture algérienne* (Paris: Crété, 1958).

3. *U.S. Army Area Handbook for Algeria*, 400 (Table 11). Production of most major minerals declined during the course of the war in Algeria. In 1954 Algeria accounted for about three percent of world phosphate production (see table entitled "Resources Minères de l'Algérie [Production en tonnes en 1954]" [Alger, 1954], in folder "10e Région Militaire, État-Major Interarmées, 3e Bureau—Documentation sur l'Algérie, 1954–62," Dossier 1H2100 d. 1, Fonds Algérien, SHAT).

4. *U.S. Army Area Handbook for Algeria*, 400–401.

5. Ibid., 397 (Table 10). There were no petroleum refineries in Algeria during the 1954–1962 period.

6. Ibid., 396.

7. *Notions essentielles sur l'Algérie*, 7.

8. Ibid., 8.

9. Ibid.

10. *U.S. Army Area Handbook for Algeria*, 391. The capital, management, and skilled labor in Algerian industry was almost entirely European, as were the markets for manufactured goods. As a result of the departure of European capital and skilled workers and the new Algerian government's flirtation with socialism, industrial employment and production fell dramatically after independence in 1962, being offset only by development of the oil and natural gas industry.

11. Ibid., 411 (Table 12). Algerian industrial production declined somewhat with independence in 1962. For example, cement production in 1963 was only 872,300 tons versus 1,071,000 tons in 1961, and beer and soft drink production declined from 36,958,500 gallons in 1961 to only 11,940,800 gallons in 1963.

12. *Notions essentielles sur l'Algérie*, 8–9.

13. Humbaraci, 280–281 (tables 5 and 6).

14. *Notions essentielles sur l'Algérie*, 5. The major French naval base at Mers-el-Kébir also handled substantial amounts of military (naval and naval aviation) cargo and personnel.

15. Ibid., 5 and 7.

16. *U.S. Army Area Handbook for Algeria*, 403.

17. The Algerian railroad system is discussed in ibid., 405–406.

18. *Notions essentielles sur l'Algérie*, 7.

19. Commandement en Chef des Forces en Algérie, Commandement du Train et Direction des Transports, *L'Arme du Train Algérie-Sahara*, No. 94/CTDT.AL/1/ORG ([Alger], January 1960), 22, in folder "10e Région Militaire, État-Major Interarmées, 3e Bureau, Commandement du Train et Direction des Transports—Le Train en Algérie et au Sahara, 1960–61," Dossier 1H1917 d. 6, Fonds Algérien, SHAT. An additional 7,400

kilometers of roadway were under construction in 1955 (see *Notions essentielles sur l'Algérie*, 5).

20. *L'Arme du Train Algérie-Sahara*, 23. The tracks (*pistes*) across the Sahara required constant, difficult maintenance. French authorities found it advantageous to transform a *piste* into an improved road if annual traffic reached thirty thousand tons (or about fifteen trucks of ten tons each during two hundred days per year). The cost of such improvement was high: about twenty million FF per kilometer.

21. *U.S. Army Area Handbook for Algeria*, 408.

22. *Notions essentielles sur l'Algérie*, 7.

23. *U.S. Army Area Handbook for Algeria*, 408.

24. Concepts Division, Aerospace Studies Institute, Air University, *Guerrilla Warfare and Airpower in Algeria, 1954–1960*, Project No. AU-411-62-ASI (Maxwell Air Force Base, AL: Concepts Division, Aerospace Studies Institute, Air University, March 1965), 111–112.

# 2

# The French Armed Forces in Algeria

Perhaps the most obvious advantage possessed by the French forces in Algeria at the start of the war in November 1954 was that there were already in place in Algeria an established political and military administration as well as modern combat forces with a full range of supporting weaponry, including air and naval power, and fully functioning command and control and logistical support systems. Although the forces initially available proved inadequate in number and configuration for the task at hand, they were soon reinforced and reorganized so as to meet every requirement imposed upon them. The problem was one of fine-tuning an existing and already very good machine rather than creating one from scratch, as the Algerian nationalist rebels were obliged to do. Furthermore, the French forces, and in particular the elite units on which the main offensive combat burden fell, were composed in large part of experienced, combat-seasoned troops led by capable, experienced officers and staffs. A special advantage accrued to the French forces through the existence of a flexible and reasonably efficient logistical system. Although subject to the usual constraints of insufficient funding and lack of critical trained personnel, the resources of the French in Algeria far surpassed anything available to their opponent, a condition that had not existed in Indochina, where the disruption of civil and military administration caused by the Second World War, the disarray and obsolescence of the French armed forces, and the postwar penury of the French government had combined to level the field with the Viet Minh.

## POLITICAL ORGANIZATION OF FRENCH ALGERIA

Algeria was legally a part of metropolitan France. Before the rebellion began in November 1954 the civil administration of Algeria was organized on the same pattern as metropolitan France, with the addition of a senior government official at the head of the civil hierarchy. Known successively as the *Gouverneur-*

*Général* (Governor-General), *Ministre-Résidant* (Resident Minister), and *Délégué-Général du Gouvernement* (Delegate-General of the Government), this official was responsible for all governmental functions, including the maintenance of order and security. He was advised and assisted by an Algerian Assembly of two "colleges," each consisting of sixty members, and by a government council. For the purposes of civil administration Algeria was divided into three *départements* (Oran, Algiers, and Constantine), which were further subdivided into *arrondissements, préfectures, sous-préfectures*, and *communes mixtes*. The vast Sahara region was administered directly by the French Minister of War, through the French military commander in Algeria.

Once the rebellion was underway, various changes were made in the civil structure. The principal thrust of these changes was to align better the civil with the military administration and to accommodate the progressive transfer of power and responsibility for the maintenance of law and order to the military authorities. In April 1955 a state of emergency for six months was declared, and on 7 August a fourth department, that of Bône, and twelve new *arrondissements* were added to the existing civil administration.[1] On 16 March 1956 the emergency decree was replaced by a governmental decree regarding the reestablishment of order, a decree that in effect extended the state of emergency to the end of the war.[2] The same year, the four existing departments were reconfigured into twelve, and three "super-departments"—known as IGAMEs, from the title of the officials who headed them, the *Inspecteurs-Général d'Administration en Mission Extraordinaire* (Inspectors-General of Administration on Extraordinary Mission)—were created, with headquarters at Oran, Algiers, and Constantine. On 17 March 1958, the civil administration was reorganized into fifteen departments (as shown on Map 2, with the addition of the departments of Aumale and Bougie) plus the Saharan territories under direct military control.[3] The fifteen-department structure endured until Algerian independence in 1962.

## HIGHER MILITARY COMMAND AND STAFF ORGANIZATION

On 1 November 1954 the senior French military commander in Algeria was General Paul ("Babar") Cherrière, the commander of the 10th Military Region (*10e Région Militaire*; 10e RM), which comprised all of Algeria. Only four days before, Cherrière had been designated Joint Commander (*Commandant Interarmées*) of all French ground forces in Algeria. On 1 November he was tasked to restore order in the Aurès (the initial seat of the rebellion), and the following day he assumed operational control of French air and naval forces in Algeria. As the rebellion ran its course, the authority and responsibilities of the *Commandant Interarmées* expanded, and the title of the position changed as well: in March 1956 to Senior Joint Commander (*Commandant Supérieur Interarmées*; CSI) and in June 1958 to Commander in Chief of Forces in Algeria

**Map 2**
**Civil Departments of Algeria, ca. 1957**

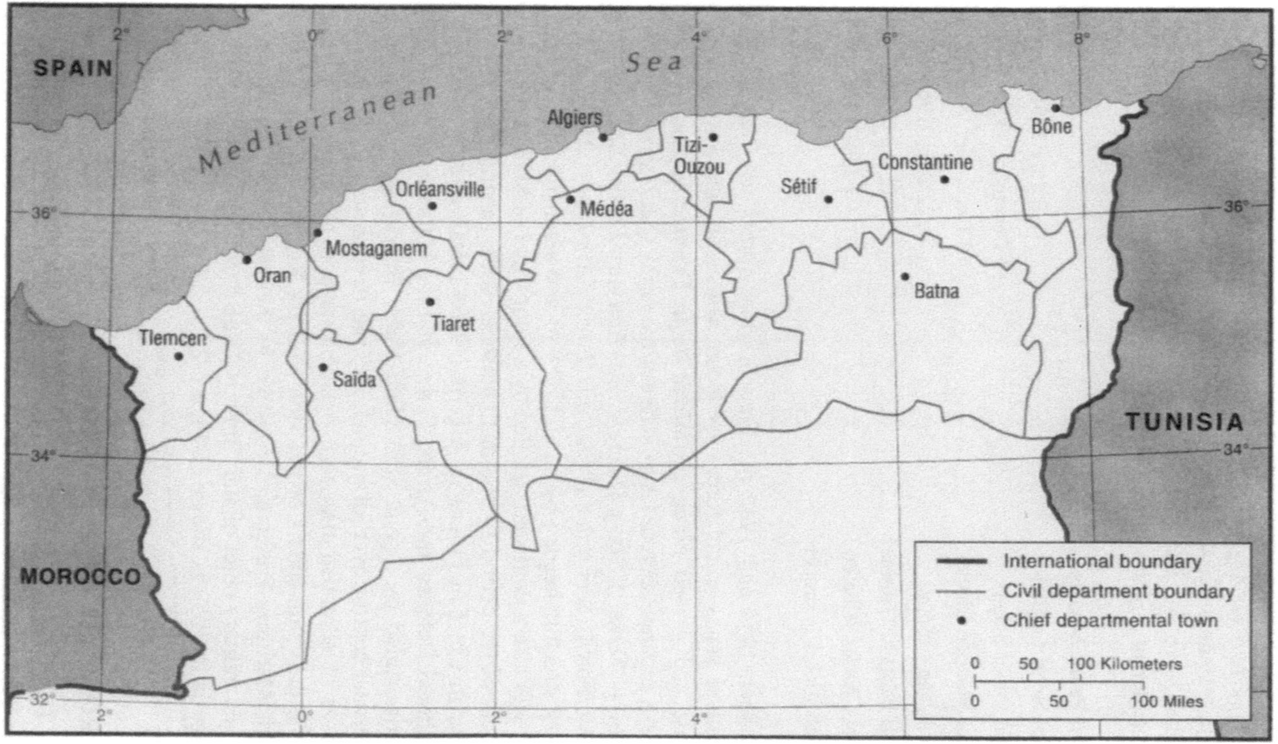

(*Commandant en Chef des Forces en Algérie*; CCFA).[4] On 15 December 1961 the title was changed once again, to Senior Commander in Algeria (*Commandant Supérieur en Algérie*; CSA). Throughout the war the senior command in Algeria was held by Army generals, with two exceptions: generals Challe and Fourquet were Air Force officers. The officers who served as senior military commander in Algeria during the war included:

| | |
|---|---|
| General Paul Cherrière | To May 1955 |
| General Henri Lorillot | 3 July 1955–December 1956 |
| General Raoul Salan | 14 December 1956–November 1958 |
| General Maurice Challe | 19 December 1958–22 April 1960 |
| General Jean Crépin | 22 April 1960–12 April 1961 |
| General Fernand Gambiez | 12 April 1961–7 June 1961 |
| General Charles Ailleret | 7 June 1961–18 April 1962 |
| General Michel Fourquet | From 18 April 1962 |

Between 1954 and 1962 the French armed forces were organized much as the United States armed forces are today, with a civilian Secretary of National Defense and Armed Forces (*Ministre de la Défense Nationale et des Forces Armées*), separate secretaries of state for the Army (*Secrétaire d'État aux Forces Armées—"Terre"*), Navy (*Secrétaire d'État aux Forces Armées—"Marine"*), and Air Force (*Secrétaire d'État aux Forces Armées—"Air"*), and a parallel military structure headed by a chief of staff and a general staff for each service. There was also a joint staff, headed by a Joint Chief of Staff, a position occupied for most of the period by the distinguished Army general Paul Ely. The Commander in Chief of Forces in Algeria (CCFA) dealt directly with the Minister of National Defense and Armed Forces and the higher echelons of government in Paris. Aided by a joint staff (*État-Major Interarmées*; EMI), he exercised operational control over all ground, air, and naval forces in Algeria, but for the purposes of administration and logistical support the component commanders (that is, the commanders of the 10th Military Region, the 5th Air Region, and the 4th Maritime Prefecture) reported to their respective civilian secretary of state and military chief of staff in Paris.[5] The overall military command structure in Algeria in 1959 is shown in Figure 2.1.

## TERRITORIAL ORGANIZATION

On 1 November 1954 the military command in Algeria was organized territorially under laws going back to the Third Republic.[6] Before World War II the entire Algerian territory had been organized as the 19th Army Corps (*19e Corps d'Armée*); in 1946 it became the 10th Military Region (*10e Région Militaire*; 10e RM), controlling the three subordinate territorial divisions (*Division Terri-*

**Figure 2.1**
**Military Command Structure in Algeria, 1959**

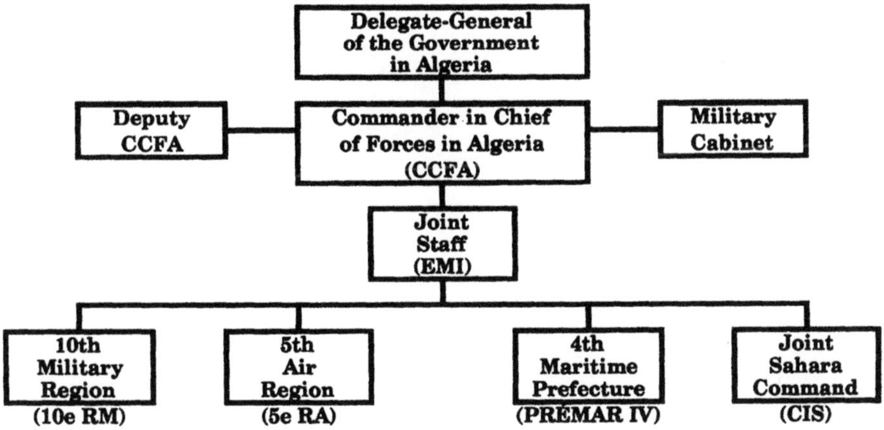


*toriale*; DT) of Oran, Algiers, and Constantine plus the autonomous Inspectorate of Troops and Services of the Southern Territories (*Inspection des Troupes et des Services des Territoires du Sud*; ITSTS) in the Sahara. In 1952 the three territorial divisions had ten subdivisions: Tlemcen and Mascara in the Territorial Division of Oran; Algiers, Dellys, Blida, and Orléansville in the Territorial Division of Algiers; and Constantine, Sétif, Bône, and Batna in the Territorial Division of Constantine. To these ten subdivisions were added on the eve of the rebellion two additional subdivisions: Tiaret in the Territorial Division of Oran and Aumale in the Territorial Division of Algiers.

During the course of the Algerian war the territorial structure of the 10th Military Region underwent constant modification as further subdivisions were created and geographical boundaries were modified in an effort to meet the demands of a developing enemy, the transition from a peacetime to a war footing, an increasing force structure, and the imposition of the defensive *quadrillage* program which involved the division of an area into sectors ("squares" or "quads") and subsectors for the purpose of control and administration.[7] The pace of such changes was dizzying and often threatened to disrupt totally the effort to subdue the rebels. In time the territorial organization of three major divisions grew from twelve principal subdivisions (zones) to fifteen operational zones and some seventy-five sectors.[8] At the height of the shooting war in May 1959, the 10th Military Region comprised the Saharan territories and three army corps areas as shown in Figure 2.2: the Oran Corps Area, with five zones (twenty-one sectors and three independent subsectors); the Algiers Corps Area, with four zones (twenty-two sectors); and the Constantine Corps Area, with four zones (twenty-nine) sectors.[9]

In theory, the military command structure was aligned with the civil administrative structure in Algeria, each civil administrator having his military coun-

**Map 3**
**Military Organization of Algeria, ca. 1957**

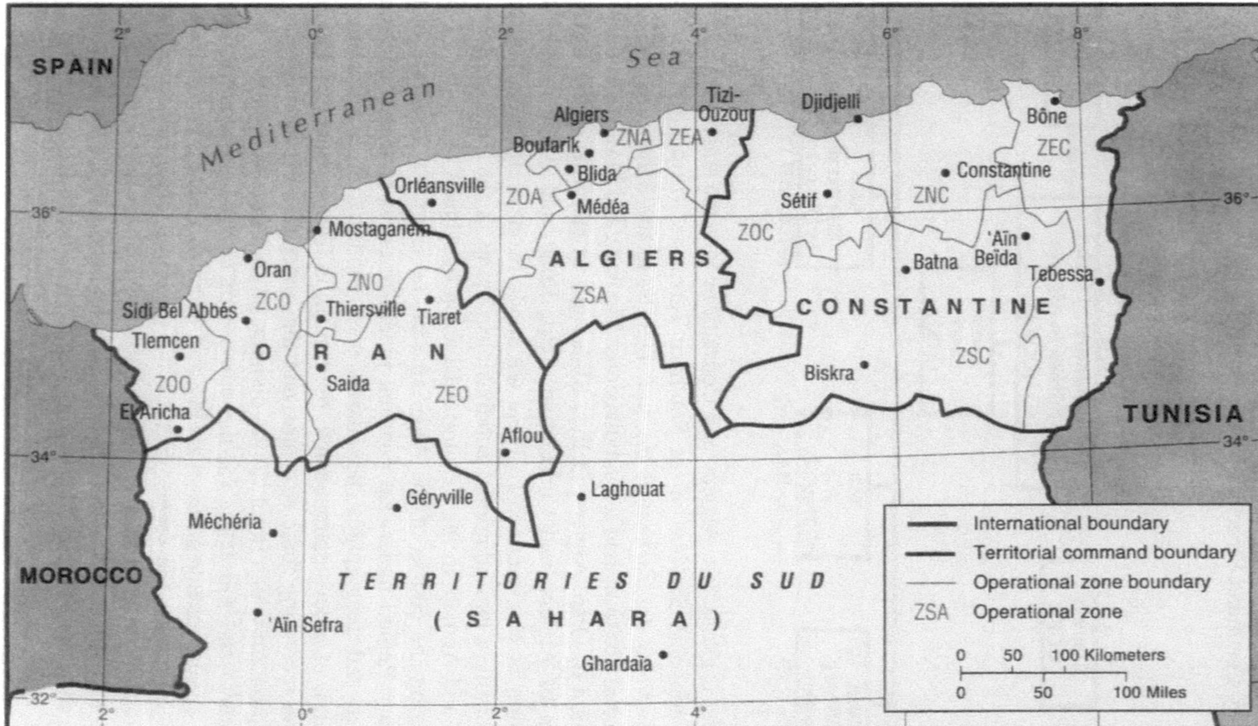

**Figure 2.2**
**Organization of the 10th Military Region, May 1959**

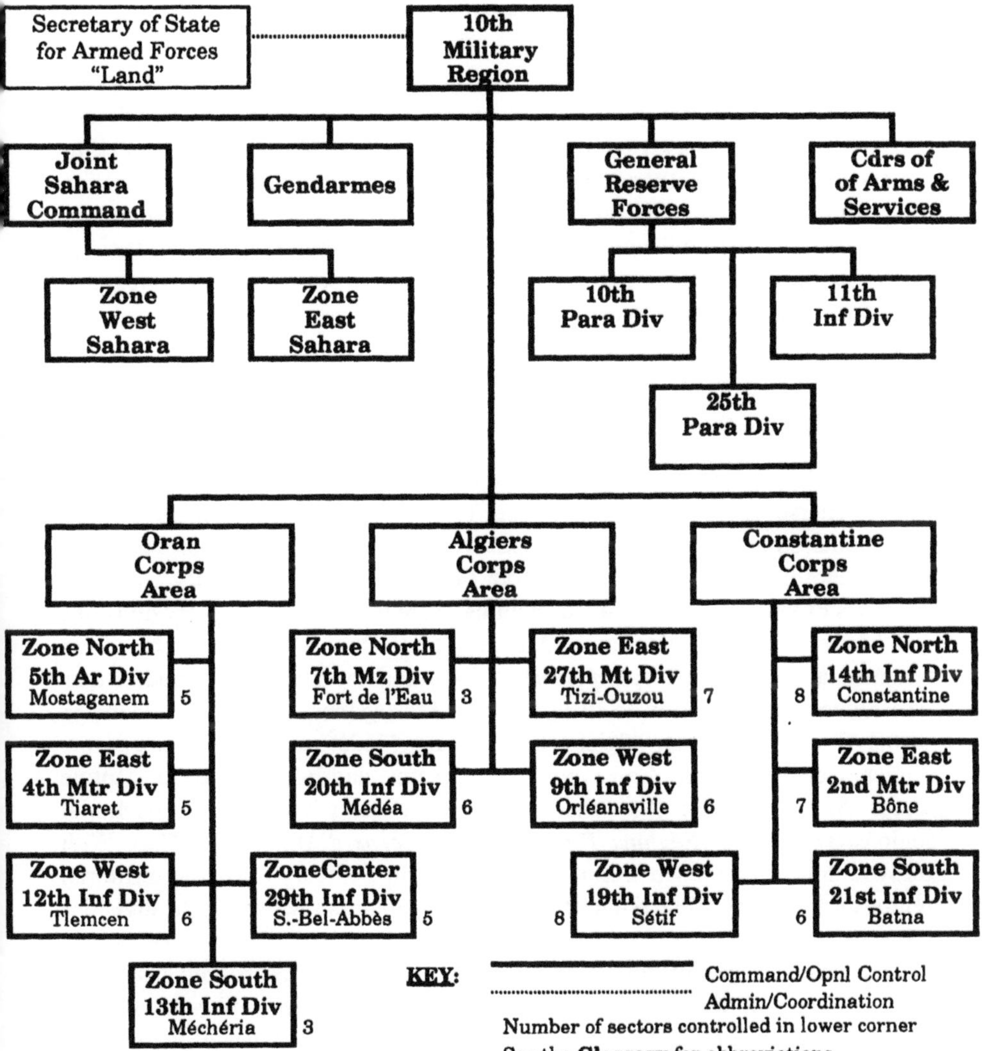


*Source*: 10e Région Militaire, État-Major Interarmées, 3e Bureau, *Organigrammes—Organisation du C.C.F.A. (Mai 59)* (Alger, mai 1959), in folder "10e Région Militaire, État-Major Interarmées, 3e Bureau—Organigrammes du Commandement Supérieur Interarmées/Commandant en Chef des Forces en Algérie, février 1959–février 1960," Dossier 1H1881 d. 3, Fonds Algérien, SHAT.

terpart and each civil subdivision its corresponding military structure, as shown in Table 2.1. Although every effort was made to retain this correspondence of the civil and military structure, the operational demands of the war against the rebels made it very difficult to maintain. The influx of additional military units, fluctuating personnel levels, and the need to designate certain areas as operational zones (*zones opérationnels*) meant that frequently military subdivisions encompassed parts of two or more civil subdivisions. The problem was compounded by the fact that the relationship of the civil and military administration in wartime was predicated on the assumption that a state of siege would be proclaimed and the civil power would become subordinate to the military. In fact no such declaration was forthcoming, and the necessary relationships thus tended to vary from area to area.[10]

## Territorial Divisions

The military command structure in Algeria underwent numerous changes during the course of the war, most of them to accommodate the imposition of an operational structure on top of the existing territorial organization. The three major subordinate commands of the 10th Military Region in 1954, the territorial divisions of Oran, Algiers, and Constantine, were redesignated military divisions (*Divisions Militaires*; DM) on 17 March 1956.[11] On 8 March 1957 they were renamed army corps (*Corps d'Armées*; CA), and the number of subdivisions was reduced from sixteen to twelve (Oran, Mostaganem, Tiaret, and Tlemcen in the Oran Corps Area; Algiers, Orléansville, Médéa, and Tizi-Ouzou in the Algiers Corps Area; and Constantine, Sétif, Batna, and Bône in the Constantine Corps Area). In April 1959 the army corps were redesignated as territorial regions/army corps (*Régions Territoriales et Corps d'Armées*; RT/CA). With the withdrawal of French troops from Algeria in February 1962 the territorial regions/army corps were reduced to army corps and subsequently to divisions (*Divisions*).

The territorial commands performed important functions in both peacetime and wartime.[12] In time of peace the three territorial commands (plus the Saharan command) were responsible for security and for the maintenance of the military infrastructure (caserns, ranges, transport, signals, etc.). In time of war these territorial organizations became operational support units, responsible for surface defense and the resupply and evacuation of personnel and supplies. Their efforts were focused in both peace and war on such matters as garrison service, military police and military justice, administration of reserve personnel, preparation for mobilization, manning of military organizations, and the organization of the territory against air and ground attack. The logistical functions of the territorial command were carried out by the various services (ordnance, quartermaster, petroleum, engineers, medical, etc.).

**Table 2.1**
**Civil Government Hierarchy in Algeria with Corresponding Military Structure**

| Civil Sub-Division | Chief Civil Authority | Chief Military Authority | Military Sub-Division |
|---|---|---|---|
| Algeria | Resident Minister | Commander in Chief | Algeria |
| | | | |
| IGAME | Inspector-General of Administration on Extraordinary Mission | Army Corps Commander | Territorial Region and Army Corps |
| | | | |
| Department | Prefect | Division Commander | Zone |
| | | | |
| Arrondissement | Sub-Prefect | Sector Commander | Sector |
| | | | |
| Prefecture | Sub-Prefect | Sub-Sector Commander | Sub-Sector |
| | | | |
| Mixed Commune | Sub-Prefect | Quarter Commander | Quarter |

*Source*: 10e Région Militaire, État-Major Interarmées, 3e Bureau, *Le problem militaire en Algérie* (Alger, avril 1957), 1 and 5, in folder ''10e Région Militaire, État-Major Interarmées, 3e Bureau—Le problem militaire en Algérie, avril 1957,'' Dossier 1H1933 d. 2, Fonds Algérien, SHAT. The civil-military structure is shown in its developed form ca. 1959–1960.

## The Sahara

The French territories in the Sahara under military administration evolved somewhat separately during the course of the Algerian war.[13] While shifts in the three northern corps areas were clearly dictated by the changing operational environment, it appears that changes in the Sahara territorial commands may have been made more for bureaucratic reasons, inasmuch as the operational situation there altered very little during the course of the war. In 1954 the French military command in the Sahara (ITSTS) was divided into four military territories: Aïn-Sefra, Ghardaia, Touggourt, and des Oasis. In 1956 the ITSTS became the Command of Troops and Services of the Territories of South Algeria (*Commandement des Troupes et des Services des Territoires de Sud Algérien*; CTSTSA). At the end of 1957 the CTSTSA was reorganized into two zones: the Western Sahara (*Zone Ouest Saharien*; ZOS), with three sectors (Méchéria, Colomb-Béchar, and Palmeraies), and the Eastern Sahara (*Zone Est Saharien*; ZES), with four sectors (Touggourt, El Oued, Ouargla, and Fort Flatters), plus the autonomous sector of Laghouat. In 1958 the CTSTSA became the Military Command of the Departments of the Sahara and the Oases (*Commandement Militaire des Départements de la Saoura et des Oasis*; CMDSO), and the sectors were reorganized as two departments (Saoura and des Oasis). In April 1959 the name was again changed to the Joint Sahara Command (*Commandement Interarmées au Sahara*; CIS), and the departments of Saoura and des Oasis became territorial regions. The title was changed once more on 15 December 1961, to the Senior Command in the Sahara (*Commandement Supérieur au Sahara*).

## Zones

Over the course of the war in Algeria it became necessary to modify the peacetime subdivisions of the four major territorial commands (Oran, Algiers, Constantine, and the Sahara) with the addition of "operational zones" (*zones opérationnels*; ZO) in areas subject to rebel action.[14] This operational structure overlaying, and in some cases replacing, the normal peacetime administrative subdivisions was progressively extended and regularized from 1954 to 1962. Early on, five ad hoc operational zones were created in areas where the rebellion flared: one in the Territorial Division of Algiers; that of the 2nd Motorized Infantry Division (*2e Division d'Infanterie Motorisée*; 2e DIM) in the Grande Kabylie; and three in the Territorial Division of Constantine (ZO North Constantine, ZO South Constantine, and ZO West Constantine). The creation of operational zones was institutionalized by a decree of 4 April 1956 that increased the number of normal military zones to sixteen. The decree provided that each of these sixteen normal zones could, if necessary, be converted to an operational zone. The operational zones created in April 1956 included: Tlemcen (the zone of Tlemcen in the Territorial Division of Oran); Kabylie (the zone of Tizi-Ouzou in the Territorial Division of Algiers); North Constantine (the zone of Constantine and the *commune mixte* of El Milia in the zone of Sétif), West Constantine (the zone of Sétif less the *commune mixte* of El Milia), East Constantine (the zone of Bône and the *commune mixte* of Morsott in the zone of Tebessa), and Aurès-Nementcha (the zones of Batna and Tebessa less the *commune mixte* of Morsott). On 8 March 1957 the operational zones became actual territorial entities, with staffs furnished by the tactical divisions assigned to them.

The demands of the operational situation in some cases precluded the establishment of operational zones corresponding exactly to the existing administrative areas, but every effort was made to ensure unity of command within the operational zone. The operational zone commander had both operational and territorial command authority, and in territorial (normal) zones not converted to an operational zone the territorial (normal) zone commander exercised both territorial and operational command. When a normal (territorial) zone was converted to an operational zone the territorial commander usually became the deputy for administration of the operational zone commander.

The commander of an operational zone was assisted in carrying out his missions by the staff of the major tactical unit assigned. In most cases the staff structure was inadequate, since the staff of a tactical unit was not designed to handle the administration of territory, and tactical unit headquarters were not staffed to handle the size and variety of forces found in the usual operational zone. For example, the staff of an infantry division was meant to control a force of eight or nine thousand men, not the sixteen to thirty thousand men (twenty–thirty battalions) found in some operational zones in Algeria.[15] The tactical unit staffs assigned zonal responsibilities were especially deficient when it came to such matters as liaison with civil authorities and civil affairs in general, participation in joint matters, operational intelligence centers (*Centres de Renseignements Opérationnels*; CRO), regional interrogation centers, and psychological warfare.

The combat forces assigned to a given operational zone varied over time with the availability of forces and the waxing and waning of rebel activity in a given operational zone. In most cases an operational zone constituted the area of operations for one or two tactical divisions plus a varying number of nondivisional combat and service units, but it was not unusual for a tactical unit (division or regiment) to be parceled out over two or even more operational zones. Given the usual pattern of the assignment of one or two combat divisions to an operational zone, the common practice was for the subordinate sectors to be manned by about two regiments; the subsectors by one regiment or a mobile group with various reinforcements; and the quarter by one or two battalions.

## Sectors

In 1957 and 1958 the operational zones in Algeria underwent a thorough process of sectorization to support better the defensive *quadrillage* program and to absorb the many new reinforcements being received from metropolitan

France, Germany, and Indochina. The sectors varied in size and population, but a typical example incorporated some two hundred square miles and over a hundred thousand people.[16] In early 1959 General Jacques Allard, the commander of the 10th Military Region, initiated a program to create seventy-five sectors, organized into three standard types: Type A (large sectors manned by two regiments); Type B (medium sectors manned by one regiment); and Type C (small sectors, or those of less dense population, manned by fewer than three battalions).[17] Some twenty-five sectors were reorganized by 1 September 1959, and in the end seventy-five sectors were created as envisaged by General Allard: twenty-two in the Algiers Corps Area, twenty-four in the Oran Corps Area, and twenty-nine in the Constantine Corps Area.[18]

The sector commander was responsible for the maintenance of order and the suppression of rebel activity in his sector and for the administration of personnel, units, and military establishments in his sector as well as their integration into the defensive plans. In an emergency he exercised operational control over all military and quasimilitary formations stationed in his sector. To assist him in the performance of his duties the sector commander had a deputy, a sector staff, and troops organized into a sector battalion headquarters company (*Compagnie de Commandement de Bataillon du Secteur*; CCBS), a sector commando (*commando de secteur*), and one or more pacification quarter support companies (*Compagnie Support de Quartier de Pacification*; CSQP).[19] The sector battalion headquarters company was organized with a command section, a services section, a signal section, an escort and protection section, an automobile section, and a pacification section. The sector commando had a command section (*Section du Commandement et d'Appui*; SCA) and four combat sections of thirty-one men each. The pacification quarter support company consisted of a command section (command and services group, pacification group, signal group, intelligence group, and medical group); a group of *harkis* (Algerian troops), consisting of a command section and three sections of *harkis*; and a varying number of reinforcement teams from the special administrative sections (*Sections Administratives Specialisées*; SAS).

## STRENGTH OF THE FRENCH ARMED FORCES IN ALGERIA

The primary mission for French forces in Algeria once the rebellion began in November 1954 was the restoration of order. This required the accomplishment of five main tasks: protection of persons and property; prevention of the spread of the rebellion to unaffected areas; control of the lines of communications; security of urban centers; and destruction of the rebel forces in the field.[20] Given the enormous size of Algeria, the simultaneous accomplishment of all five tasks required a large number of ground combat troops and an equally large number of service troops and supporting air and naval forces. Although the French government committed a large portion of its available military forces to

Algeria, to the detriment of its commitments to NATO and elsewhere throughout the French overseas territories, the number of effectives available was never sufficient to meet the perceived needs of operations in Algeria.[21] The insufficiency of combat forces was compounded by the operational doctrines employed. In particular, the *quadrillage* of Algeria into sectors with permanent contingents of defensive forces tied down the majority of available troops, required the dispersion of tactical units, and until modifications were made in 1959, all but eliminated the possibility of offensive action to destroy the rebel bands. As a consequence, French military commanders were forced to "make do" and frequently found that they had to accept less than 100 percent achievement of essential tasks.[22]

The two principal factors influencing the availability personnel were the budget and higher-level political considerations.[23] After an initial period of reaction to the threat of the Algerian nationalist rebellion, which generated substantial increases in the Algerian troop lists, budgetary constraints and competing political priorities began to exert pressure for reductions. By the beginning of 1959 the achievement of some military success against the rebels had already precipitated cuts in the number of French troops, particularly draftees, committed to the Algerian war.[24] The process began in February 1959, when the Minister of National Defense and Armed Forces announced a reduction of 6,500 positions in each of the Algerian corps areas. The reductions had to be absorbed while maintaining the overall *quadrillage* and thus led to reduced effectiveness in certain zones and sectors. The process was accelerated after 1960 by the desire of General Charles de Gaulle to liquidate the Algerian war and to modernize the French armed forces and orient them toward other tasks. However, even "*le grand Charles*" found it impossible to reduce substantially the number of regular ground troops in Algeria until after the cease-fire in March 1962.

## Ground Forces

At the beginning of the Algerian war on 1 November 1954 overall French military strength in Algeria was about seventy thousand, including the *Gendarmerie*, Algerian auxiliaries, and about eight thousand Air Force and 1,500 Navy personnel.[25] French ground forces in Algeria totaled some fifty-eight thousand men, of which only fourteen thousand—two parachute battalions of the *41e Demi-Brigade*, five infantry battalions, and ten squadrons of the *Garde Mobile*—were available for immediate deployment to the seat of the rebellion in the Aurès.[26] Of the remainder, twenty-seven thousand were on guard or training duties, and seventeen thousand were headquarters and service troops.

The French authorities in Algeria and in Paris at first failed to recognize that the rebellion in Algeria required more than the existing complement of forces. However, once the seriousness of the situation was recognized, steps were taken to reinforce the meager forces in Algeria with troops drawn from metropolitan France, forces committed to NATO in Germany, and units returning from Indochina. On 5 July 1955 General Henri Lorillot, the overall French commander in Algeria, urgently requested the immediate deployment to Algeria of an additional thirty infantry battalions, eighteen armored squadrons, and ten transport companies.[27] The government in Paris reacted positively, and between June 1955 and June 1956 six infantry, two motorized infantry, one alpine infantry, one armored, and one mechanized division arrived in Algeria.[28] Among the first reinforcements to arrive were the 3rd Foreign Legion Infantry Regiment (*3e Régiment d'Infanterie Étrangère*; 3e REI), the 13th Demi-Brigade of the Foreign Legion (*13e Demi-Brigade de la Légion Étrangère*; 13e DBLE), and the 1st Foreign Legion Parachute Battalion (*1e Bataillon Étrangère Parachutiste*; 1e BEP). The divisions deployed included the 2nd Motorized Infantry Division (*2e Division d'Infanterie Motorisée*; 2e DIM) (June 1955); the 4th Motorized Infantry Division (*4e Division d'Infanterie Motorisée*; 4e DIM) and the 27th Alpine Infantry Division (*27e Division d'Infanterie Alpine*; 27e DIA) (September 1955); the 19th Infantry Division (*19e Division d'Infanterie*; 19e DI) (December 1955); the 12th Infantry Division (*12e Division d'Infanterie*; 12e DI), the 5th Armored Division (*5e Division Blindée*; 5e DB), and the 7th Light Mechanized Division (*7e Division Mechanisée Rapide*; 7e DMR) (April 1956); the 29th Infantry Division (*29e Division d'Infanterie*; 29e DI) (May 1956); and the 9th, 13th, and 20th infantry divisions (*9e, 13e, and 20e Divisions d'Infanterie*; 9e, 13e, and 20e DI) (June 1956). The 10th Parachute Division (*10e Division Parachutiste*; 10e DP) was created in Algeria in 1956. Of course, a large number of independent combat and service units were also deployed to Algeria during the same period. Table 2.2 shows the evolution of French ground combat strength in Algeria during the course of the war.

Within eighteen months the strength of French armed forces in Algeria had reached 260,000 men, and by April of 1957 the total had risen to 370,000 men, comprising 223 battalions of infantry, twenty-six regiments of armor, twenty-nine artillery groups, and fifty squadrons of *Gendarmes*.[29] At the peak in 1958–1959 there were over four hundred thousand French troops in Algeria plus another 176,000 men in the *Gendarmerie*, auxiliaries, and police.[30] Of the total regular forces around 350,000 were Army, forty thousand were Air Force, and eight thousand were Navy. Auxiliary forces numbered about 150,000 and the *Gendarmerie* and police another twenty-six thousand. The distribution of French ground forces in Algeria in February 1959 is shown in Table 2.3.

All told, between 1 January 1952 and 1 July 1962 some 1,447,200 French officers, noncommissioned officers (NCOs), regular soldiers, and draftees saw service in Algeria. The normal tour for French regular officers, NCOs, and enlisted personnel was three years, and on average some 9,960 officers, 31,800 NCOs, and 54,300 soldiers arrived in French North Africa (Algeria, Tunisia, and Morocco) each year to begin their tour.[31]

Although the forces deployed to Algeria provided welcome reinforcement for the establishment and maintenance of the *quadrillage* and even afforded some capability for offensive action to destroy the rebel forces, they also created a

**Table 2.2**
**Growth of French Regular Ground Forces in Algeria, 1954–1962**

| Date | Effectives |
|---|---|
| 1 November 1954 | 58,000 |
| 1 January 1955 | 73,500 |
| 1 January 1956 | 180,000 |
| 1 January 1957 | 355,300 |
| 1 January 1958 | 395,000 |
| 1 January 1959 | 400,000 |
| 1 January 1960 | 395,000 |
| 1 January 1961 | 400,000 |
| 1 January 1962 | 372,000 |

*Source*: Lettre (No. 1135/DEF/EMAT/SH/DEP), Chef du Service Historique de l'Armée de Terre [Général Delmas] au Chef d'État-Major des Armées, Division Organisation-Logistique, Vincennes, 11 février 1983, sujet: Statistiques des effectifs et pertes en A.F.N., Annexe 1E ("Effectifs de l'Armée de Terre [sauf Gendarmerie] engagés pendant les périodes de combat au 1er janvier de chaque annee"). Note that the number of ground troops cited does not include *Gendarmerie*.

number of problems. The effort to absorb the rapid and enormous influx of troops of a variety of types, coupled with the continuing efforts to make the transition from peacetime to wartime structures and methods, and the levy of units for the Suez intervention in late 1956, caused enormous confusion that never really abated.[32]

## Air and Naval Forces

French air and naval forces committed to the war in Algeria also experienced significant expansion during the course of the war. In 1954 the French air forces were heavily committed in Indochina, and the 5th Air Region (*5e Région Aérienne*; 5e RA) in Algeria was limited in personnel and equipment. The 5th Air Region operated only ten bases (three major ones), all in northern Algeria, with one wing of fighters, one transport squadron, a liaison squadron, a number of support aircraft, and a total of about eight thousand men.[33] By 1958–1959 Air Force resources in Algeria had grown to nearly fifty thousand personnel and over seven hundred aircraft including a hundred helicopters, organized as shown in Figure 2.3.[34] The fifty thousand men represented 40 percent of the total strength of the French air forces. Another seventeen thousand men (12 percent of the total) in metropolitan France and six thousand men (6.5 percent of the total) in training were supporting the French air forces in Algeria or enroute there. Thus nearly 60 percent of all French Air Force personnel were directly involved in the Algerian war. The six hundred fixed-wing aircraft and one hundred helicopters represented about 20 percent of total French Air Force assets.

**Table 2.3**
**Distribution of French Ground Forces in Algeria, 1 February 1959**

| Function | Number | % |
|---|---|---|
| Sector Defense (*quadrillage*) Units | 148,000 | 39.0 |
| Mobile Units (General, Corps Area, and Operational Zone Reserves) | 56,000 | 15.0 |
| Operation of the Barrages | 26,000 | 7.0 |
| Close Protection of Infrastructure | 27,000 | 7.0 |
| Services and Training Center Cadre | 41,500 | 11.5 |
| Headquarters and General Staff | 13,500 | 3.0 |
| Detached to Non-Military Organizations (SAS, etc.) | 5,000 | 1.5 |
| Normal Absences (Leave, Hospital, etc.) | 52,500 | 14.0 |
| Individuals Involved in Central Administration | 7,000 | 2.0 |
| **Total** | **376,000** | **100** |

*Source*: 10e Région Militaire, État-Major Interarmées, 3e Bureau, *Exposé du Général Allard sur la situation militaire en Algérie (Forces Terrestres) à la date du 9 février 1959* (Alger, 9 février 1959), Charte 15 (''Répartition des Effectifs Instruits au 1 février 1959''), in folder ''10e Région Militaire, État-Major Interarmées, 3e Bureau—Le problem militaire en Algérie, avril 1957,'' Dossier 1H1933 d. 2, Fonds Algérien, SHAT.

French naval forces in Algeria in November 1954 were also limited. French naval forces assigned to the Algerian Maritime Command (*Commandement Maritime d'Algérie; COMAR Algérie*) included only three coastal escort vessels, one patrol boat, and about 1,500 men.[35] These forces were concentrated at the naval base at Mers-el-Kébir and the naval air station at Lartigue, both near Oran. However, French naval forces in Algeria were quickly reinforced by the dispatch of additional vessels, aircraft, and personnel from metropolitan France, and by 1958–1959 some eight thousand men were assigned directly to the 4th Maritime Prefecture, which was organized in May 1959, as shown in Figure 2.4.[36]

## Auxiliary Forces

The regular ground, sea, and air forces available to the French command in Algeria were supplemented by indigenous ''regular'' and self-defense forces, Algerian reservists of European descent, and other personnel recruited on a temporary basis. The pool of available trained manpower was quite large. There were in Algeria perhaps as many as 640,000 French Army veterans, both Europeans and Moslems, of whom around two hundred thousand served in the regular and auxiliary forces of the French Army during the Algerian war.[37] Of these, at one time or another about 180,000 Algerian Moslems served as auxiliary combat troops, *harkis*, many of whom were mounted.[38] In April 1956 the Resident Minister in Algeria established regulations for the creation, organization, and armament of such units ''to participate actively in the operations for the maintenance of order,'' and by 1 January 1959 some twenty-eight thousand Algerian Moslems had been enrolled.[39] Other auxiliary formations included: the *maghzen* (lightly armed forces controlled by the officers assigned to the Bureau of Algerian Affairs [*Bureau des Affaires Algériennes*]); reservists, mostly Euro-

**Figure 2.3**
**Organization of the 5th Air Region, May 1959**

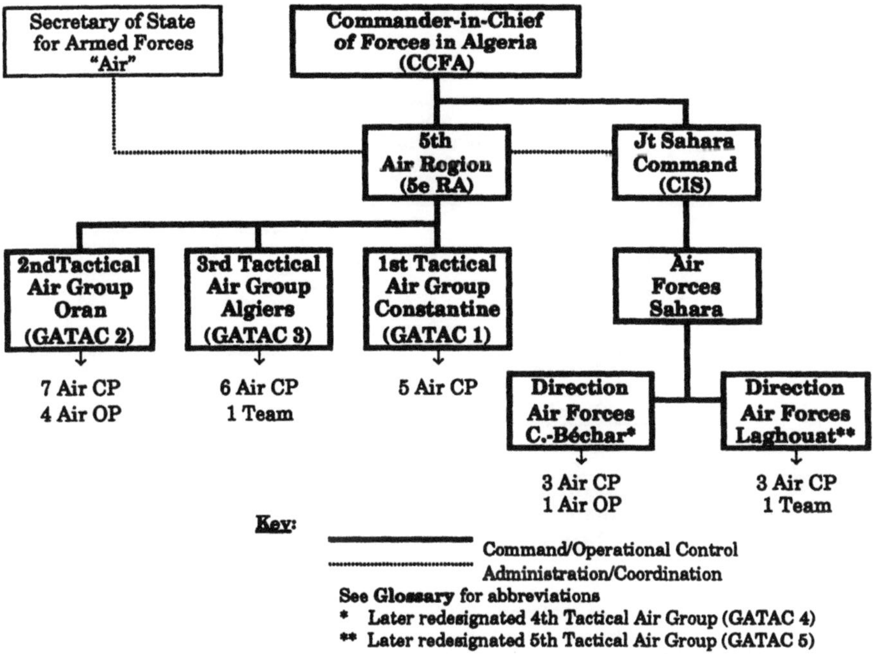


*Sources*: 10e Région Militaire, État-Major Interarmées, 3e Bureau, *Organigramme—E.M.I.* (Alger, no date), in folder "10e Région Militaire, État-Major Interarmées, 3e Bureau—Organigrammes du Commandement Supérieur Interarmées (CSIA) et Commandant en Chef des Forces en Algérie (CCFA), février 1959–février 1960," Dossier 1H1881 d. 3, Fonds Algérien, SHAT; Philippe Vial and Pascal Tanchoux, "Les Archives 'Algérie' de l'Armée de l'Air," *Revue Historique des Armées*, 187, no. 2 (June 1992), 66–76 passim.

peans, of the Territorial Units (*Unités Terrtoriales*; UT); Mobile Security Groups of the *Garde Mobile*; and the rural defense companies (*Compagnies Auto-Défense Rurale*) and Protection Battalions (*Bataillons de Protection*) of the self-defense forces.[40] Table 2.4 shows the growth of such auxiliary forces during the critical period from 1 January 1957 to 1 January 1959.

## TACTICAL ORGANIZATION

In general, the tactical organization of the conventional French ground combat forces in Algeria at the beginning of the conflict was unsuited to the task at hand.[41] Most of the forces already in Algeria, as well as those transferred from Germany and Indochina, were organized under antiquated Tables of Organization and Equipment, ill suited to a mobile, counter-guerrilla war. As a result there had to be a thorough internal reorganization of French ground combat

**Figure 2.4**
**Organization of the 4th Maritime Prefecture, May 1959**

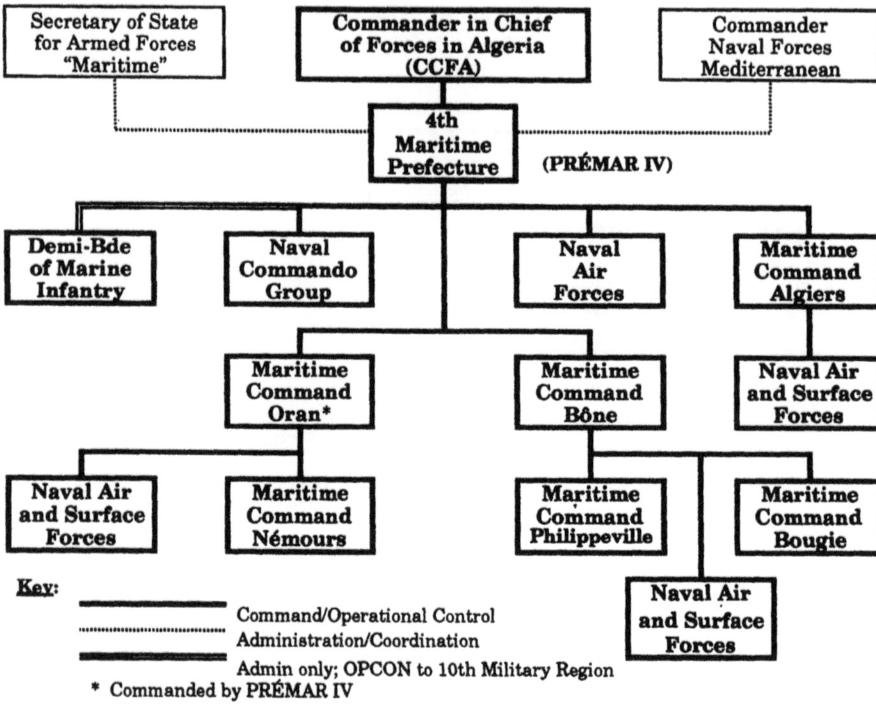


*Sources*: 10e Région Militaire, État-Major Interarmées, 3e Bureau, *Organigrammes—Organisation du C.C.F.A. (Mai 59)* (Alger, mai 1959), in folder "10e Région Militaire, État-Major Interarmées, 3e Bureau—Organigrammes du Commandement Supérieur Interarmées/Commandant en Chef des Forces en Algérie, février 1959–février 1960," Dossier 1H1881 d. 3, Fonds Algérien, SHAT; Jean Kessler, "Témoignage: La surveillance des frontières maritimes de l'Algérie, 1954–1962," *Revue Historique des Armées*, 187, no. 2 (June 1992), 94–101 passim; Jean-Pierre Dubois, "L'Aéronautique navale et les opérations d'Algérie, 1954–1962," *Revue Historique des Armées*, 187, no. 2 (June 1992), 109.

units, which further contributed to the inefficiency of many. The number and types of French combat battalions in Algeria at the peak of the war in March 1959 are shown in Table 2.5.

In practice the TOE of units in Algeria were constantly being adjusted throughout the war to better suit the terrain and prevailing modes of combat. However, since Algeria was not a "classic" war, little consideration was given to the formal amendment of the tables of allowances for personnel and equipment.[42]

The basic combat unit was the infantry battalion. By the end of 1956 most infantry battalions in Algeria had been reorganized and reequipped as light infantry battalions of four rifle companies and about 650 men, somewhat less than

**Table 2.4**
**French Auxiliary Forces, 1957–1959**

| Type Force | 1 January 1957 | 1 January 1958 | 1 January 1959 |
|---|---|---|---|
| *Harkis* | 4,700 | 16,902 | 28,021 |
| *Maghzens* | 3,500 | 12,218 | 17,191 |
| Self-Defense Groups | 3,400 | 6,277 | 16,855 |
| Mobile Security Groups | 2,220 | 6,200 | 8,614 |
| Territorial Reservists | 250 | 300 | 4,100 |
| **Total** | **14,070** | **41,897** | **74,781** |

*Source*: 10e Région Militaire, État-Major Interarmées, 3e Bureau, *Exposé du Général Allard sur la situation militaire en Algérie (Forces Terrestres), à la date du 9 février 1959* (Alger, 9 février 1959), Charte 7 (''Forces Suppletives''), in folder ''10e Région Militaire, État-Major Interarmées, 3e Bureau—Le problem militaire en Algérie, avril 1957,'' Dossier 1H1933 d. 2, Fonds Algérien, SHAT. Reliable figures for French auxiliary forces in Algeria are not available for all periods of the war.

the number prescribed for the normal Type 107 French infantry battalion, the organization of which is shown in Figure 2.5.[43] The new Type 130 infantry battalions had fewer heavy weapons and were thus more mobile, and they sought to enhance their advantage in mobility over the rebels through the use of air power. The additional companies in the Type 130 battalions had a negative impact on efficiency, in that the number of trained French commanders, staff officers, and NCOs was limited, and the creation of additional companies spread the available trained French personnel too thin. Some of the infantry units composed of draftees sent to Algeria had as few as nineteen officers to seven hundred men, one factor in their generally poor performance.[44]

French ground forces in Algeria also included parachute and motorized infantry battalions. The standard parachute infantry battalion was only slightly larger than a standard infantry battalion and had thirty-one officers, 104 NCOs, 765 men, and fifty-five vehicles (sixteen motorcycles, twenty-five jeeps, fourteen trucks of less than 1.5 tons, and fifteen trucks of between 1.5 and three tons).[45] The standard motorized infantry battalion was considerably larger than either the regular infantry or parachute infantry battalion and had twenty-eight officers, 143 NCOs, 876 enlisted men, and 155 wheeled and tracked vehicles (seven motorcycles, twenty-eight jeeps, five trucks of less than 1.5 tons, seventeen trucks of between 1.5 and three tons, one wrecker, and 104 half-tracks).[46]

Given the chronic shortage of personnel and the requirement to maintain the defensive *quadrillage* throughout Algeria, the creation and maintenance of an operational reserve to be used offensively to pursue and destroy the rebel military forces was extremely difficult. Until significant numbers of reinforcements began to arrive in late 1955 the only reserves consisted of units withdrawn for rest and refitting. However, such units were generally worn out, and commanders took the opportunity to permit leave for up to 25 percent of their forces at a time; thus, such units were effectively unavailable.[47] The commanders of zones and sectors attempted to set aside forces as a regional reserve but found that

Table 2.5
French Combat Battalions in Algeria as of 13 March 1959

| Area | Infantry and Other Dismounted Combat Troops | | | | | Armor | *Garde* Sqdns | Artillery | Engr |
|---|---|---|---|---|---|---|---|---|---|
| | Type 107 | Type 130 | Mtrz | Para | Naval Cmdos | | | | |
| **Oran Corps Area** | | | | | | | | | |
| Oran Zone West | 16 | | | | | 4 | 3 | 3 | 2 |
| Oran Zone Center | 15 | 2 | | | | 1 | 5 | 3 | 2 |
| Oran Zone North | 11 | | | | 1 | 1 | 2 | 2 | 1 |
| Oran Zone South | 3 | | 10 | | | 2 | 1 | 3 | |
| MéchériaAutoZone | 4 | 2 | 2 | | | 1 | | | 3 |
| Total | 44 | 4 | 12 | 0 | 1 | 9 | 11 | 11 | 8 |
| **Algiers Corps Area** | | | | | | | | | |
| Algiers Zone West | 15 | | | | 1 | 5 | 5 | 2 | 1 |
| Algiers Zone North | 11 | | 2 | 6 | | 2 | 15 | 6 | 2 |
| Algiers Zone South | 17 | | | | 1 | 4 | 2 | 2 | 1 |
| Algiers Zone East | 23 | | | | | 2 | 2 | 1 | 1 |
| Total | 66 | 0 | 2 | 6 | 2 | 13 | 24 | 11 | 5 |
| **Constantine Corps Area** | | | | | | | | | |
| Const. Zone West | 21 | 2 | | | | 3 | 6 | 2 | 2 |
| Const. Zone North | 18 | | | 1 | | 3 | 6 | 4 | 1 |
| Const. Zone South | 15 | | 1 | 1 | | 2 | 3 | 1 | 1 |
| Const. Zone East | 11 | 2 | 9 | 3 | 1 | 8 | 6 | 6 | 6 |
| Total | 65 | 4 | 10 | 5 | 1 | 16 | 21 | 13 | 10 |
| **Joint Sahara Command** | | | | | | | | | |
| Sahara Zone West | 7 | | | | | | 1 | 1 | 2 |
| Sahara Zone East | 2 | | | | | | 1 | | 1 |
| Total | 9 | 0 | 0 | 0 | 0 | 0 | 2 | 1 | 3 |
| **TOTAL** | **191** | **8** | **24** | **11** | **4** | **38** | **58** | **36** | **26** |

NB: Troops in the Joint Sahara Command included and additional twenty-eight independent companies and three groups of Saharan *Méharistes*.

*Source:* 10e Région Militaire, État-Major Interarmées, 3e Bureau, *Ordre de Bataille (13-3–59)* (Alger, 13 mars 1959), in folder ''10e Région Militaire, État-Major Interarmées, 3e Bureau—Ordre de Bataille, 1955–63,'' Dossier 1H1881 d. 1, Fonds Algérien, SHAT.

**Figure 2.5**
**Standard French Infantry Battalion Organization**

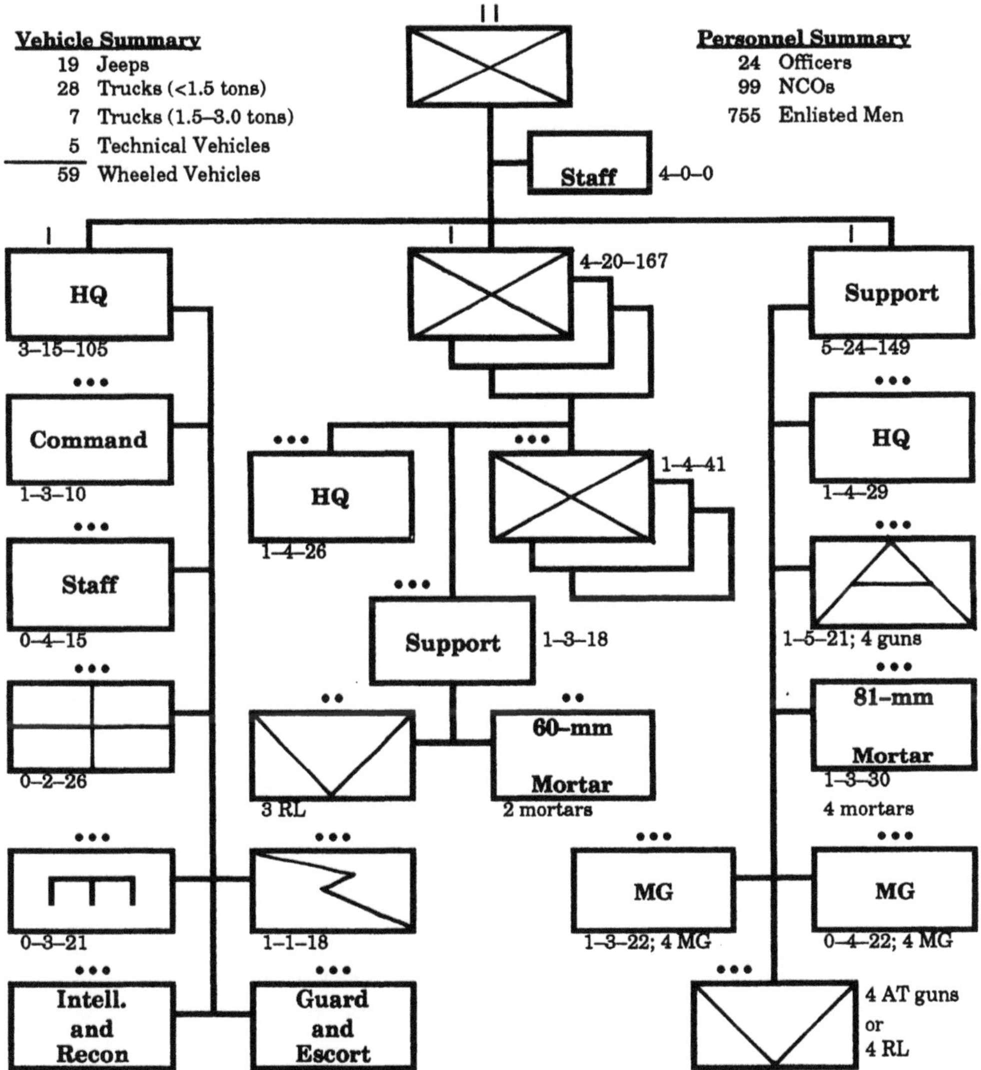


*Source*: France, Ministére de l'Armée de Terre, École d'État-Major, École Supérieure de Guerre et École d'État-Major, *Aide-Mémoire pour les Travaux d'État-Major*, provisional ed. (Paris: École Supérieure de Guerre et École d'État-Major, February 1952), I, ¶212a ("Bataillon d'Infanterie-Type Normal-Schéma d'Organisation") and I, ¶212b ("Tableau Numerique [Personnels et Véhicules]"). See Common French Map Symbols and Glossary for abbreviations.

because such reserves could not be relieved entirely of their defensive responsibilities and could be committed only with the approval of the commander of the 10th Military Region, they could not be employed effectively in operations requiring immediate response.[48] For that reason most offensive missions were left to the "theater reserve," consisting of the 10th Parachute Division, the 25th Parachute Division, and the 11th Infantry Division, which divisions were officially designated as the General Reserve *(Réserve Générale)* in September 1959.[49]

## CONCLUSION

While the coordination of political and diplomatic policies and initiatives with the conduct of military operations in Algeria left something to be desired, the high command structure and territorial organization of the French military forces in Algeria were adequate to the task of conducting a counter-guerrilla campaign in which the objective was to seal Algeria's international borders, clear populated areas of rebel forces, and protect the population against renewed rebel attack and exploitation. The French were also reasonably successful in adapting the somewhat heavy, Europe-oriented tactical organization of their combat forces to meet the needs of the unique operational situation in Algeria. On the other hand, the French experienced significant difficulty in absorbing the rapid and somewhat haphazard influx of combat troops between 1955 and 1957. The necessity for wide dispersion of combat forces throughout Algeria also placed considerable strain on both the command and control system and the supporting logistical system of the French Army in Algeria. However, the principal problem faced by the French command in Algeria arose from the fact that despite the commitment of over four hundred thousand troops to the war, the number of French combat troops in Algeria was never sufficient to maintain the defensive *quadrillage* and carry out effective offensive operations against the rebels simultaneously until after 1958, by which time the full complement of reinforcements had arrived, indigenous forces had been mobilized, the barrages were in place, and operational procedures had been refined. However, after 1958 the decisive actions of the war were taking place not on the battlefield but at the conference table.

## NOTES

1. Bruno Waraschitz, "Le Cadre territorial et Opérationnel," in *Introduction à l'étude des archives de l'Algérie*, ed. Service Historique de l'Armée de Terre (Château de Vincennes: Service Historique de l'Armée de Terre, 1992), 76; Gouvernement-Général de l'Algérie, *Notions essentielles sur l'Algérie* (Paris: Georges Lang, [ca. 1956]), 19.

2. Waraschitz, 76.

3. Commandement Supérieure Interarmées/10e Région Militaire, État-Major Interarmées, 3e Bureau, Antenna de Documentation Géographique, *Analyse: Tableau et*

*Calque relatifs à la nouvelle organisation départmentale de l'Algérie fixée par la Décret du 17.3.1958* (No. 453/RM10/3.Plan, Alger, 21 mars 1958), in folder "10e Région Militaire, État-Major Interarmées, 3e Bureau—Documentation sur l'Algérie, 1954–62," Dossier 1H2100 d. 1, Fonds Algérien, SHAT.

4. For the sake of clarity, the senior military commander in Algeria is referred to hereafter as the Commander in Chief (CCFA), regardless of the title actually in effect at the time.

5. Until the appointment of Air Force general Maurice Challe as CCFA in December 1958 the post of CCFA and that of commander of the 10th Military Region were combined. General Challe initiated an overall program of devolution of power to his subordinate territorial commanders and at the same time delegated his responsibilities as the ground component commander in Algeria to *Général de Corps d'Armée* Jacques Allard, who assumed command of the 10th Military Region (see Waraschitz, 85).

6. The pre-1954 territorial organization of the French forces in Algeria is discussed by Waraschitz, 72–74. The organization of the military forces in Algeria was prescribed by two laws in particular: the law of 16 March 1882 concerning the administration of the Army and the law of 13 July 1927 concerning the general organization of the Army.

7. The *quadrillage* process had been tried in Indochina with mixed results and was imposed in Algeria soon after the start of the rebellion. The program was essentially static and defensive, and it required substantial manpower. On the whole, the results obtained in Algeria were much better than in Indochina, in part because the French had by then greater experience with such programs and the terrain of Algeria was more favorable.

8. Waraschitz, 108. Map 3 shows the organization as it existed in 1957, with three corps areas each with four operational zones. The joint command in the Sahara constituted a fourth major division.

9. Ibid., 85.

10. Ibid., 105.

11. The evolution of the territorial commands is described ibid., 73–86 and passim.

12. The functions of the territorial commands are described ibid., 73–74.

13. The evolution is described ibid., 74–87 and passim. The joint command in the Sahara is generally referred to as the Joint Sahara Command (CIS) throughout this study.

14. The organization and functions of the *zones opérationnels* are discussed by Waraschitz, 75–79 and 92–93.

15. Ibid., 97.

16. Air University, Aerospace Studies Institute, Concepts Division, *Guerrilla Warfare and Airpower in Algeria, 1954–1960*, Project No. AU-411-62-ASI (Maxwell Air Force Base, AL: Concepts Division, Aerospace Studies Institute, Air University, March 1965), 25.

17. Waraschitz, 103.

18. Ibid., 103 n. 81.

19. Ibid., 80–81.

20. Ibid., 94. In 1957 a sixth major task was added: the operation of the defensive barrages on Algeria's Moroccan and Tunisian borders.

21. Ibid., 98. In January 1956 General Callies, the Inspector of French Ground, Naval, and Air Forces in North Africa, estimated that the simultaneous execution of the four principal missions of the French armed forces (participation in NATO; maintenance of order in North Africa; maintenance of order in overseas territories; and maintenance of

order in metropolitan France) would require 650,000–700,000 men, but that the budgetary realities permitted only around 450,000, of whom only 150,000 were available for all of French North Africa. See also Major Etcheverry, "Refléxions sur la guerre subversive d'Algérie," *Revue des Forces Terrestres* 17 (July 1959), 46, and Alistair Horne, *A Savage War of Peace* (New York: Viking Press, 1977), 112–113.

22. Waraschitz, 89–90.

23. Ibid., 98.

24. The reduction program is discussed ibid., 96–99 and passim.

25. 10e Région Militaire, État-Major, 4e Bureau, *Dossier 4e Bureau, 24-5-56* (Alger, 24 mai 1956), "Dossier VIII—Organisation et Fonctionnement des Services: Progression Comparée des Effectifs et des Services en 10RM," in folder "Aide-Major Général en AFN—Organisation du commandement et coordination des services et moyens logistiques, 1946–57," Dossier 1H2665 d. 1, Fonds Algérien, SHAT. About 17 percent of the total were service troops. The *Gendarmerie*, a normal part of the French Army, was spread rather thin. For example, in the Aurès, where the rebellion began, there were only six gendarmes for a population of sixty thousand (see Constantin Melnik, *The French Campaign against the FLN*, RAND Memorandum RM-5449-ISA [Santa Monica, CA: Rand Corp. for the Office of the Assistant Secretary of Defense-International Security Affairs, September 1967], 11). Melnik was a former member of the staff of Prime Minister Michel Debré.

26. Waraschitz, 95; Olivier Hamon, "Chronique du Conflit Algérien, 1954–1962," in *Introduction à l'étude des archives de l'Algérie*, ed. Service Historique de l'Armée de Terre (Château de Vincennes: Service Historique de l'Armée de Terre, 1992), 34 and n. 53.

27. Waraschitz, 95.

28. Ibid.

29. 10e Région Militaire, État-Major Interarmées, 3e Bureau, *Le problem militaire en Algérie* (Alger, avril 1957), 12, in folder "10e Région Militaire, État-Major Interarmées, 3e Bureau—Le problem militaire en Algérie, avril 1957," Dossier 1H19933 d. 2, Fonds Algérien, SHAT.

30. *Guerrilla Warfare and Airpower in Algeria*, 21; A. H. Peterson, G. C. Reinhardt, and E. E. Conger (eds.), *Symposium on the Role of Airpower in Counterinsurgency and Unconventional Warfare: The Algerian War* (Santa Monica, CA: Rand Corporation, July 1963), 13.

31. Lettre (No. 1135/DEF/EMAT/SH/DEP), Chef du Service Historique de l'Armée de Terre [Général Delmas] au Chef d'État-Major des Armées, Division Organisation-Logistique, Vincennes, 11 février 1983, sujet: Statistiques des effectifs et pertes en A.F.N., Annexe 2 E1 ("Effectif total de L'Armée de Terre [sauf Gendarmerie] ayant sejourné en A.F.N. entre 1 janvier 1952 et le 1e juillet 1962") and Annexe 3 E ("Répartition des effectifs ayant sejourné dans chaque territoire") [cited hereafter as *Statistiques des effectifs et pertes en A.F.N.*]. French ground forces in North Africa totaled 121,500 on 1 January 1954 and 373,700 on 1 January 1962, having reached a peak of 476,300 on 1 January 1957 (see *Statistiques des effectifs et pertes en A.F.N.*, Annexe 1E ["Effectifs de l'Armée de Terre {sauf Gendarmerie} engagés pendant les périodes de combat au 1er janvier de chaque année"]).

32. Waraschitz, 95–96.

33. *Guerrilla Warfare and Airpower in Algeria*, 22–23; Philippe Vial and Pascal Tanchoux, "Les Archives 'Algérie' de l'Armée de l'Air," *Revue Historique des Armées*

187, no. 2 (June 1992), 67. On 1 November 1954 the number of French Air Force personnel in North Africa was 18,300, of whom only eight thousand were assigned directly to Algeria.

34. Patrick Facon, "L'Algérie et la politique générale de l'Armée de l'Air (1954–1958)," *Revue Historique des Armées* 187, no. 2 (June 1992), 77.

35. Jean Kessler, "Témoignage: La surveillance des frontières maritimes de l'Algérie, 1954–1962," *Revue Historique des Armées* 187, no. 2 (June 1992), 94. In June 1955 the Algerian Maritime Command was redesignated the 4th Maritime Prefecture (*4e Préfecture Maritime*; PRÉMAR IV).

36. *Symposium on the Role of Airpower*, 13.

37. Ibid., 7.

38. Edgar O'Ballance, *The Algerian Insurrection, 1954–62* (London: Faber and Faber, 1967), 65.

39. Hamon, 36.

40. *Le problem militaire en Algérie*, 14–15; Hamon, 30 and 36. Some deserters from the Algerian nationalist rebels were also employed. For example, the 5th Company of the 3rd Colonial Parachute Regiment was composed entirely of ALN deserters.

41. Claude Carré, "Aspects opérationnels du conflit Algérien, 1954–1960," *Revue Historique des Armées* 166 (March 1987), 84.

42. Etcheverry, 46.

43. *Guerrilla Warfare and Airpower in Algeria*, 22. The older Type 107 infantry battalions had only three rifle companies.

44. Carré, 84.

45. France, Ministère de l'Armée de Terre, École d'État-Major, École Supérieure de Guerre et École d'État-Major, *Aide-Mémoire pour les Travaux d'État-Major* (Provisional ed. in two volumes; Paris: École Supérieure de Guerre et École d'État-Major, February 1952), I, ¶214b: "Bataillon d'Infanterie Parachutiste—Tableau Numerique (Personnels et Véhicules)."

46. *Aide-Mémoire*, I, ¶213b: "Bataillon d'Infanterie Porté—Tableau Numerique (Personnels et Véhicules)."

47. Waraschitz, 95.

48. Ibid., 91–92.

49. Ibid., 92. For a time the 7th Light Mechanized Division was also included in the General Reserve.

# 3

# French Logistics in Algeria: Supply and Maintenance

The French military command in Algeria benefited from the prior existence of an established and competent logistical support structure. In fact, the adaptation of the logistical organization in Algeria and the rest of North Africa to an active theater of operations antedates the beginning of the Algerian war in 1954. The first steps were taken in 1952, when the French military authorities undertook a study of how to defend French North Africa from an external attack; subsequently they began the conversion of the existing logistical structure to a "rear base."[1] The 1952 study proved prescient in that the four "principles" (really assumptions) enunciated with respect to logistical support indeed came to pass in the period 1954–1962 and became the basis for the organization of the French logistical services during the Algerian conflict. The four key assumptions made by the French planners in 1952 were that: the proposed "rear base" would perform field army-level functions and would thus have to be augmented with additional service units; since the army corps (*Corps d'Armée*) would play the field army role both operationally and logistically, it would assume responsibility for the control and direction of the logistical services and logistical support units; levels of stocks in the theater (Algeria) sufficient to cover any interruption of the flow of resupply from metropolitan France would be needed; and it would be necessary to push advanced depots and repair workshops as far forward toward the supported troop units as possible.

Informed by the accurate assumptions of the 1952 study and backed by the resources of metropolitan France, the autonomous supply services of the French Army in Algeria were organized on a territorial basis corresponding to that of the combat forces, and were operated in accordance with proven principles of logistics. Although plagued by a chronic insufficiency of trained personnel and funds; an aging logistics infrastructure, occasional shortages of key weapons, ammunition, and equipment; and antiquated regulations ill suited to the tasks at hand, the French Army logistical services in Algeria were far superior to anything available to the rebels and thus constituted a major advantage for the French forces.

## OPERATIVE PRINCIPLES OF LOGISTICS

French logistical doctrine as it evolved in Algeria during the period 1954–1962 emphasized many of the same basic principles as its contemporary American counterpart, including command responsibility for logistics, impetus of supply from the rear, and centralized control of decentralized operations.[2] The principal difference was that the autonomous French logistical services in Algeria were organized primarily on a territorial rather than a unit basis and were thus better suited to operations from fixed bases rather than to mobile warfare.

### Supply Service Autonomy

In the main, at the higher staff levels the French logistical system in Algeria very much resembled its American counterpart in the years before World War II, when the General Staff of the United States Army had been limited to operational planning and coordination, and the autonomous supply departments took care of all aspects of operating the supply system, including the selection, training, and management of their own officers and enlisted men. The now-familiar American ''G-staff'' system, in which staff responsibility for each of the four primary functional areas (personnel and administration, intelligence, training and operations, and logistics) is assigned to a principal staff officer (the G-1, G-2, G-3, and G-4, respectively), was derived in fact from the French staff organization of the World War I period. In French usage the four principal G-staff sections were designated as *bureaux* (*1er Bureau, 2e Bureau, 3e Bureau,* and *4e Bureau*, respectively). In Algeria the senior French staffs added a *5e Bureau*, with responsibility for psychological warfare and civil affairs. At military region and army corps area levels the key staff bureau concerned with logistical matters (the *4e Bureau*) was organized in five sections, each responsible for overseeing a particular group of functional areas: *1er Section*—general organization, studies, plans, and statistics; *2e Section*—movements, transport, and fuel; *3e Section*—quartermaster, medical, veterinary, treasury, and postal services; *4e Section*—ordnance service, transport, and petroleum (POL); and *5e Section*—military installations, engineers, and barrier materials.

Although French operational staffs in Algeria were organized on the well known G-staff basis, the various supply services were organized on the more traditional bureau system, wherein each supply service was responsible for all aspects of the management of its assigned commodities or service. Thus the quartermaster service (*Service de l'Intendance*; SI) determined requirements and procured, stored, distributed, maintained, and issued food and clothing, while the ordnance service (*Service du Matériel*; SM) did the same for weapons, ammunition, vehicles, and other equipment. The *Train* handled most motor and inland water transport, and a separate petroleum service (*Service des Essences*; SE) was responsible for all matters having to do with petroleum products. The engineers (*Service du Genie*), signal corps (*Service de Transmissions*), and medical service (*Service de Santé*) each took care of all supply and maintenance activities connected with their particular commodities and functions. The supporting services, which played an important role in the Algerian war, were rounded out by the veterinary service (*Service Veterinaire*), which inspected foodstuffs and maintained the health of the mules, horses, war dogs, and other animals used by the French forces; the Army finance service (*Trésorerie aux Armées*), which handled high-level fiscal matters; and the military postal service (*Poste Militaire*), which provided mail to the troops. The smaller supporting services were organized on a territorial basis like the main elements of the French logistical system in Algeria. Many financial matters, such as troop pay and allowances, were handled by the *Service de l'Intendance* rather than the *Trésorerie aux Armées*.

## Territorial Organization

In general, the French logistical services in Algeria were organized and operated on a territorial basis using both semimobile service units and fixed facilities (*établissements*), some of which were located in installations that dated back to the French conquest of Algeria in the nineteenth century.[3] The structure of the services was aligned with the overall territorial structure of the French military command in Algeria, with logistical elements at the military region, corps area, zone, and sector level operating in accordance with the principle of centralized control and decentralized operations but remaining responsive to the military region, corps area, zone, or sector commander. Technical matters and certain other functions—for example, overall organization of the services, installations and infrastructure, the expenditure of budgetary credits, and the structuring of units—were controlled centrally by the directors of the various services under the commander in chief of the 10th Military Region, while such matters as the distribution of matériel, the resupply of equipment and consumable supplies (rations, POL, ammunition, spare parts, and construction materials, for example), maintenance, evacuation, housing and the maintenance of caserns, ranges, and other fixed facilities, as well as administrative oversight of the logistical support of troop units, was decentralized to the echelon of the zone. While the static territorial support system created long before World War II was adequate for the routine support of fixed garrisons over established lines of communications relatively free from enemy interdiction, the operational requirements of the war against the Algerian rebels—like the war against the Viet Minh in Indochina—required increasing emphasis on the more mobile, unit-oriented support system developed in the European theater in World War II. Consequently, the particular organization of the various elements, their location, and their methods of operation varied somewhat over the course of the war in response to geographical considerations, the nature of the existing infrastructure, and the operational situation.

## Command Responsibility

As in the American system, the commander was ultimately responsible for the logistical support of the forces under his command. The commander determined supply rates and stockage levels, the allocation of service units and of supplies to subordinate units, and the location of depots and supply points, while the supply services were responsible for receiving, storing, managing, and maintaining their assigned commodities and for organizing and operating depots, supply points, and other service facilities. Thus each territorial and unit commander in Algeria played an important role in the overall logistical system. At the various levels command responsibility for logistical support could be, and often was, delegated to subordinate commanders. For example, the Commander in Chief of Forces in Algeria (CCFA) permanently delegated control over all acquisition and distribution of matériel within established budgetary allocations and regulations to the corps area commanders, who were charged with the conditioning, support, resupply, evacuation, movement, and transportation for all forces in their areas.

The responsibilities of the commander of each operational zone were similar and included the control of all formations of the arms and services in the zone. In practice, the zone commander was responsible for such important logistical matters as administrative oversight of all units in his zone, the administration of funds for compensation, the distribution of ''sector material'' (*matériel du secteur*) control of stockage levels in depots and isolated units, and the allocation of certain critical items such as combat rations and ammunition. The zone commander was also responsible for managing and accounting for the use of petroleum products and for maintenance for formations temporarily in his zone, although temporarily attached units retained their usual chain of command for all other matters, notably for statistical reporting. Normally, the authority for the creation of mobile support elements (for example, mobile bakeries or mobile maintenance teams) and for the establishment of temporary supply depots to support planned operations rested with the zone commander. However, for certain formations, namely units of the military region and corps area reserves and training centers, the responsibility of the zone commander was limited, and the corps area commander reserved the right to support certain units directly. Similarly, the commander of each operational or pacification sector was charged with overseeing the execution of logistical directives received from the zone commander, coordinating and controlling the expression and satisfaction of the logistical requirements of subordinate formations, overseeing the maintenance of ''sector material,'' and maintaining direct contact with organizations of the various logistical services providing direct support to his formations. When a series of important operations extended over several territorial zones or sectors, a temporary interzonal authority was designated and empowered to dispose of the logistical means necessary to the operation. In such cases the commander of the operation was responsible for defining the logistical support required and how it was to be used before the operation began, and for the timely notification of the commander of the corps area of the overall logistical requirements of the operation.

The proper functioning of the logistical system required a command structure capable of determining and expressing requirements, adapting available resources to requirements, and controlling the utilization of resources. In general, the commander at each level was assisted in these functions by his logistical staff officers (for example, the *4e Bureau* of the staff of the corps area commander) and by the senior representative of each of the logistical services in his area. As is so often the case, only the commander who foresaw logistical requirements well ahead could expect to receive satisfactory support from his logistical services. The best commanders regulated resupply and transport like they regulated operations—by issuing clear orders to the various logistical services, giving them the resources to carry out those orders, keeping the logisticians fully informed of the commander's intentions, seeking their advice, and delegating to them sufficient authority through directives and instructions to execute their missions.

### Impetus of Supply from the Rear

Just like the contemporary American logistical system, the French supply system in Algeria operated on the basic principle of impetus of supply from the rear toward the front. The key to maintaining impetus of supply from the rear was the existence of preestablished levels of equipment and supplies for each unit, levels that the various supply services sought to maintain in a more or less automatic manner. French units in Algeria were authorized equipment and supplies under three main categories: organic unit equipment and supplies; nonconsumable "sector material"; and consumable material. The theoretical authorizations for basic organic unit equipment and supplies were determined by standard TOE applicable to French Army units everywhere.[4] However, to take into account the special circumstances in Algeria, the Minister of National Defense introduced the concept of the "*dotation practique*," which corresponded to what was known in the United States Army as a "Modified Table of Organization and Equipment (MTOE)." The *dotation practique* authorized supplies and equipment for each unit in Algeria on the basis of the formal theoretical authorization, the actual availability of matériel, and the peculiar needs of the unit. The actual quantities under the *dotation practique* were fixed by the Minister or in some cases by the commander in chief in Algeria, and actual allocations were controlled by the commander of each corps area.

"Sector material," authorized in addition to basic unit *dotations practiques*, consisted of such nonconsumable supplies and equipment as furniture, camp equipment, refrigeration units, generators, and other items required primarily for operation of the territorial infrastructure. Overall types and quantities of "sector material" were determined by the commander in chief, and such equipment was allocated by zone commanders based on allocations approved by the corps area commander. In theory, the "sector material" was included in the overall total of supplies and equipment authorized to the logistical services, the engineers, the signal service, and transportation units in Algeria. "Sector material" was administered in operational sectors by the sector headquarters company (*Compagnie de Commandement de Secteur*; CCS) or some other stable unit designated by the sector commander, and in pacification sectors by the headquarters company of the sector pacification battalion (*Compagnie de Commandement du Bataillon de Secteur de Pacification*; CCBSP). "Sector material" was left in place if a unit was transferred out of a particular zone or corps area.

Consumable material, such as rations, ammunition, and petroleum products, was delivered directly to units by the logistical services without command intervention, except in the case of certain critical items (tents, combat rations, and 9 mm pistol ammunition, for example), the allocation of which was reserved to the corps area commander. The requisitioning and issue of such consumable supplies and equipment was handled under normal resupply procedures.

## Centralized Control and Decentralized Execution

The guiding principle of French logistical operations in Algeria was that of centralized control and decentralized execution. Overall regulation of logistical support rested at the echelon of the commander in chief, particularly the evaluation of requirements and their expression to the civil government in Paris, the request for resources, the allocation of the available resources, and overall management of logistical support. The execution of resupply, maintenance, and evacuation was decentralized to the corps area level, and because of the distances and dispersion of forces in Algeria, the commander of each corps area further decentralized portions of his logistical support resources (depots and service units, for example).

Decentralized supply operations in Algeria were based on a timely and accurate knowledge of the logistical situation of all units, depots, and support establishments. Resupply of units was executed by formations of the various services from depots and infrastructure establishments. For the purposes of resupply operations French units in Algeria were classified into three categories based on their isolation and distance from supporting facilities: isolated units, intermediate units situated far from depots but reliably linked to them, and units stationed in secure areas near the depots. To compensate for the potential interruption of resupply, isolated posts were authorized to keep on hand additional quantities of supplies known as *Lots C* (*lots de sécurité de Compagnie*) or *Lots S* (*lots de sécurité de Secteur*), according to the number of assigned personnel.

Depots for all types of supplies were constituted at zone, and when necessary sector, level, as dictated by factors of distance and security. Some depots were classified as "operational" and were operated by personnel of the logistical services. Others were classified as "garrison" and were operated by personnel of other arms and services. In both cases the authorized stockage level was fixed by the corps area commander on the recommendation of the zone commander, taking into account directives in force and the recommendations of the directors of the various supply services as to technical matters.

The principle of centralized control and decentralized execution also applied to maintenance operations. Each zone normally divided its organic maintenance resources to support its subordinate sectors, and in certain cases the corps area temporarily reinforced the zone maintenance forces with its reserve means. Evacuation of matériel was normally effected by rail whenever possible. If the evacuation was not considered urgent, it was integrated into the routine movements plan worked out by the Corps Area Bureau of Movements and Transport (*Bureau de Mouvements et Transports*; BMT) in liaison with the responsible Director of Transport.

The principle of centralized control and decentralized execution of logistics placed the primary burden for providing adequate logistical support on the directors of each of the logistical services at the various territorial levels (corps area, zone, and sector). Thus, the director of each of the logistical services at each territorial level was responsible for advising the commander on the technical matters within his purview and for the execution of the commander's orders regarding logistical support. In general, the mission of the various logistical services was to assure the resupply, evacuation, and maintenance of personnel, matériel, and infrastructure, taking into account the particular conditions of operations in Algeria—which required adaptation to a *de facto* state of war with a *de jure* peacetime administrative organization. The overall aim, of course, was to support the fighting forces in a manner that would permit them to live and fight under the best possible conditions. In practice, the logistical services in Algeria carried out two different types of missions: the "traditional" logistical support missions prescribed by the Law of 16 March 1882, and the "particular" missions prescribed as part of the long-term plan for the civil and military development of Algeria.[5]

To carry out their mission and coordinate technical measures, the directors of the various services at each echelon developed hierarchical technical chains parallel to the normal chain of command. Within the technical chain the directors of the services controlled the various establishments and units charged with the execution of logistical support functions. Those establishments and units, both general support and direct support, operated in accordance with general regulations supplemented by special instructions adapted to the current operational situation. At corps area and zone level, the directors of the various services were responsible for enforcing the normal observance of technical regulations, but at the same time they were charged with the execution without restriction of the orders of the commander, even when such orders conflicted with the technical regulations, in which case the commander assumed the responsibility for diverging from established regulations and procedures.

## QUARTERMASTER (*SERVICE DE L'INTENDANCE*) FORCES

The French Army *Service de l'Intendance* (SI), corresponding roughly in functions and responsibilities to the U.S. Army Quartermaster Department, was responsible for the management of all subsistence and clothing matters (determination of requirements, procurement, storage, and distribution) as well as for troop pay and certain other financial disbursements. Unlike its American counterpart, the SI was not responsible for supplying petroleum products, which in the French service were the responsibility of a separate Petroleum Service (*Service des Essences*). Long established in Algeria, the SI was generally well prepared to support French forces in combat operations. The major challenges faced by the SI during the Algerian war lay in the greatly increased number of troops supported and their dispersion throughout Algeria, the increased tempo of operations, and the necessity of adapting to a rapidly changing tactical environment. On the whole, these challenges were met without significant lapses in support.

### Organization of the *Service de l'Intendance* in Algeria

The *Service de l'Intendance* in Algeria was organized on a territorial basis that corresponded to the territorial organization of combat forces, with SI headquarters and direct and general-support units at echelons corresponding to the military region, corps area, zone, sector, and major units.[6] The overall organization of the SI in Algeria is shown in Figure 3.1.

The Director-General of Intendance (*Directeur de l'Intendance Officier Général*; DIOG) of the 10th Military Region advised the commander in chief on Intendance matters and exercised technical supervision of Intendance operations throughout Algeria. His responsibilities also included technical direction and supervision of Intendance operations in the corps areas and the Joint Sahara Command.

The Director of Intendance (*Directeur de l'Intendance*; DI) at corps area level was subordinate to the corps area commander for command and operational matters and to the Director-General of Intendance, 10th Military Region, for technical matters. He too was responsible for advising his commander on Intendance matters and oversaw the operation of the specialized central Intendance services operating throughout the corps area. These services included:

1. A subsistence service (*Intendance S*), responsible for the supply of rations, combustibles, and other subsistence material, refrigeration services, and the operation of a central subsistence headquarters (*Gestion S*), a territorial cold-storage facility, and a varying number of subordinate depots and annexes. In Algeria the *Gestions S* were

**Figure 3.1**
**Organization of the *Service de l'Intendance* in Algeria, ca. 1960**

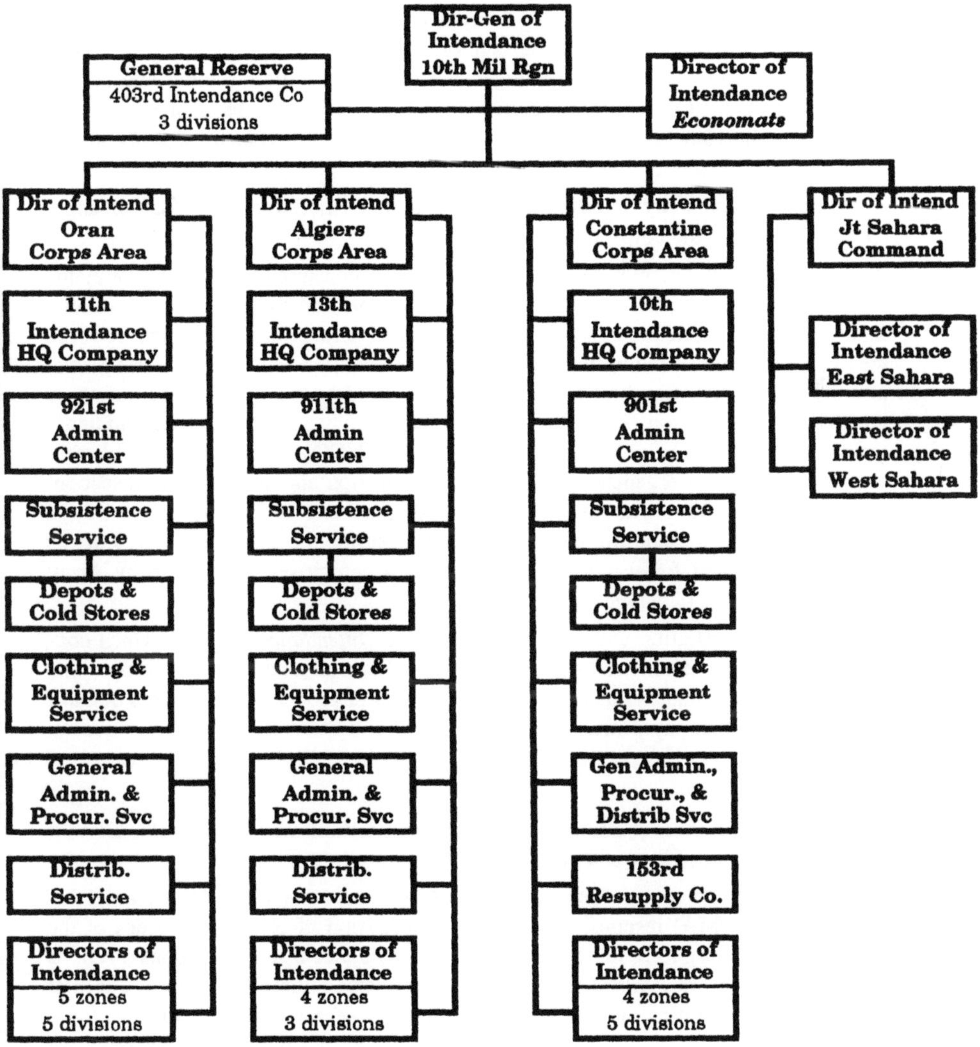


*Source*: Compiled by the author from various sources.

allocated on the basis of one per corps area and one per zone. A standard *Gestion S* was theoretically capable of handling thirty-five thousand rations per day and of storing 100,000–500,000 rations. It was authorized four officers, thirteen NCOs, 113 enlisted men, and twelve vehicles, organized in a headquarters section and an exploitation section.[7] Cold-storage facilities were allocated on the basis of one facility (*atelier*) for the 10th Military Region and for each of the corps areas, and one annex (*sous-atelier*) per zone.

2. A system of *Economats*, retail sales stores similar to the American post exchange, spread throughout the corps area.
3. A clothing and equipment service (*Intendance H*), responsible for the supply of clothing, camp equipment, bedding, furniture, and animals and the operation of various facilities for the fabrication and repair of *Intendance* items, including a central corps area clothing depot, clothing repair and exchange facilities, laundries, shower points, and bakeries. In Algeria the *Gestion H* were allocated on the basis of one per corps area. A standard *Gestion H* was theoretically capable of handling forty thousand transactions per day. It was authorized four officers, ten NCOs, eighty-one enlisted men, and eleven vehicles, organized in a headquarters section and an exploitation section, and it required 1,500 square meters of covered space and an augmentation of three trucks for every thousand transactions.[8] Laundry service for the French forces in Algeria was provided for the most part by civilian enterprises rather than by Army laundry companies (*Compagnies de Blanchisserie*).
4. A general administration and procurement service (*Intendance AG-PR*), which controlled Intendance expenditures proper, oversaw the expenditures of the troop units, and managed such matters as pensions, requisitions of property and services, rentals, and civil reparation payments.
5. A displacements and transport service (*Intendance DT*), which managed the movement of Intendance supplies.
6. A troop administration center (*Corps de Troupes Administration Centrale*; CTAC), which managed troop pay and allowances.
7. A headquarters company (*Compagnie de Commis et Ouvriers Militaires d'Administration*; COMA) responsible for the administration of NCOs and troops of the SI in the corps area except for those assigned to *Intendance* resupply sections (*Sections de Ravitaillement de l'Intendance*; SRI) of major troop units (divisions). The SRI were the basic operating units of the SI at division level and managed all *Intendance* support for the division. They were the main link between the combat units and the supporting SI infrastructure. The SRI in Algeria were equivalent to the earlier *Intendance* resupply companies (*Compagnies de Ravitaillement de l'Intendance*; CRI), which were organic to the divisions and were capable of supporting a thirteen-thousand-man division or fifteen-thousand nondivisional troops in an army zone. The standard CRI was authorized three officers, thirteen NCOs, 103 enlisted men, and twelve vehicles, organized in a headquarters section, a supply section, and a distribution section.[9]

The DI for each corps area was also responsible for technical oversight of the territorial (zone) and divisional Intendants, each of whom had responsibility for a territorial area determined by the corps area commander on recommendation of his DI. The zone and divisional Intendants reported directly on technical matters to the corps area DI, who delegated to them control and administration of certain establishments and annexes belonging to the territorial infrastructure. The territorial (zone) and divisional Intendants also acted as technical advisors to their respective zone or division commanders and oversaw the employment of any divisional SRI in their area.

## *Intendance* Supply Procedures

In theory, the resupply of rations, clothing, and other SI items in Algeria was accomplished automatically based on replenishment of the stocks allocated to each unit by the corps area commander.[10] In practice, the corps area DI received from his commander delegated authority to determine the allocation of *Intendance* matériel, except for certain designated items occasionally in short supply, such as squad tents and bunk beds.[11] Stocks of combat rations were allocated to the various zones in accordance with the desires of the corps area commander and were issued periodically from stocks maintained by the *Gestions S* and subsistence annexes (*Annexes S*).[12] As a rule, rations were supplied on a daily basis by road or rail from the *Intendance* subsistence depots and annexes directly to the SRI and *Economats* serving units in the field, the allocation of rations being determined by the number of men to be supplied. Some isolated units were resupplied by air.

Combustibles and straw were supplied on the same basis as rations and were issued with rations, with a few exceptions. For the resupply of clothing, troop units submitted quarterly statements of requirements through the DI of the zone in which they were located to the supporting *Intendance H* of the corps area. Clothing was then distributed from the supporting clothing depot (*Magasin Militaire d'Habillement*; MMH) by road or by rail either directly to individual units or to the servicing SI clothing annex (*Annexe H*) with instructions for distribution to units. As in the case of rations, some units were supplied clothing by air. Other SI matériel and spare parts were issued to units based on demand data maintained by the *Intendance S* of each corps area. Normally such items were issued on a one-for-one direct-exchange basis whenever mass replacements were required, but closer controls were maintained on critical items.[13]

Because of the distance and transportation time required for the delivery of supplies from France and the difficulties of movement within Algeria, the SI made every effort to make each corps area as logistically autonomous as possible. This was accomplished in part by setting suitable stockage levels and controlling consumption. Stocks of rations and clothing maintained in each corps area were determined by the commander and changed slightly from time to time. The normal stockage level of rations was ninety days of supply (DOS) in the *Gestion S* of the corps area, another ninety DOS in the *Annexes S* (or *SH*), and fifteen DOS of normal rations plus two of fresh rations at unit level. As exceptions to the usual stockage levels authorized, in the Algiers Corps Area, for

**Table 3.1**
**Weight of the French Ration**

| Unit | Normal Rations No. 1 | Individual Combat Rations | Reserve Rations |
|---|---|---|---|
| Infantry Division | 50 tons | 36 tons | 28 tons |
| Armored Division | 46 tons | 33 tons | 25.5 tons |
| Corps Troops | 30 tons | 21.5 tons | 17 tons |
| Per 10,000 Men | 26 tons | 19 tons | 15 tons |

*Source*: France, Ministère de l'Armée de Terre, École d'État-Major, École Supérieure de Guerre et École d'État-Major, *Aide-Mémoire pour les Travaux d'État-Major*, provisional ed. (Paris: École Supérieure de Guerre et École d'État-Major, February 1952), II, ¶634. The weight given includes packaging.

example, the authorized stock for roasted coffee was only fifteen DOS, and forty-five DOS of wheat, pasta, and couscous were maintained from 15 June through 15 October of each year.[14] Stockage levels for hay, straw, and firewood varied according to seasonal availability, and those for frozen meat and wine were determined by the available storage capacity. The allocation of combat rations was somewhat different. In the Oran Corps Area, for example, the authorized allocations were ninety DOS held at corps area level, thirty DOS held at zone level, and only two in the units. As a rule, 120–180 DOS of clothing were stocked by the corps area clothing depot and the units maintained ninety DOS, while no clothing was stocked at zone level.

Rations were the major items for which the SI was responsible. Several different types of rations were used in Algeria, and there were significant differences between the rations issued to European troops and those for indigenous Moslem forces. For example, the Moslem troops required fresh mutton (usually issued live) and were not issued pork products or wine. In general the normal French ration for Europeans during the period 1954–1962 was slightly heavier than the ration for indigenous personnel; it consisted of 500 grams of bread, 350 grams of meat, 250 grams of wine, 500 grams of "small rations" (condiments, sweets, cigarettes, etc.), and 700 grams of fruits and vegetables (*denrées d'ordinaire*), for a total of 2.3 kilograms per man per day.[15] The normal ration was about 3,280 calories. The standard European combat ration, which included either 420 grams of hardtack or 500 grams of fresh bread, weighed about 1.9 kilograms and contained four thousand calories.[16]

The storage and distribution of rations created significant demands on both storage facilities and transportation. As shown in Table 3.1, one day's rations for an infantry division amounted to some fifty tons. Its movement required eight to eleven rail cars, or twenty to twenty-eight trucks.[17] At prevailing scales, thirty days of rations for a hundred thousand men amounted to six thousand tons of normal rations, 5,700 tons of combat rations, and 1,500 tons of wine; it required 17,700 square meters of covered storage area and 60,000 square meters of dispersed, uncovered storage; and it needed forty-five to sixty rail cars or 120–160 trucks for its movement.[18]

### *Intendance* Problem Areas

The *Service de l'Intendance* in Algeria experienced few major problems beyond chronic shortages of funds and of qualified technical personnel. In part the shortage of funds was occasioned by the continued application of peacetime rates and the wide dispersion of French units in Algeria.[19] There were occasional temporary shortages of certain key items of matériel, such as tentage, bedding, and refrigeration equipment. Overall, the challenges of servicing numerous and widely dispersed forces were taken in stride without major modifications to the overall organization and operating procedures of the SI.

## ORDNANCE (*SERVICE DU MATÉRIEL*) FORCES

The French Army ordnance service (*Service du Matériel*; SM) in Algeria was responsible for determining requirements, requisitioning, receiving, storing, distributing, and maintaining weapons, tank-automotive vehicles and equipment, ammunition, optical equipment, fixed and rotary-wing light (Army) aircraft, parachutes and related equipment, animal harness, office machines, and related tools and repair parts.[20] In addition, the SM in the 10th Military Region was responsible for the evacuation and salvage of damaged equipment and for explosive ordnance disposal throughout Algeria. The SM also played a role in devising and applying technical modifications and improvements to existing equipment.

### Organization of the *Service du Matériel* in Algeria

The *Service du Matériel* in Algeria was organized for the most part on a territorial basis (as shown in Figure 3.2) and operated in accordance with the principle of centralized control and decentralized execution. Responsibility and authority for the technical organization and operations of all SM units in Algeria were vested in the Regional Director of Matériel (*Directeur Régional du Service du Matériel*; DRM) of the 10th Military Region, with headquarters in Algiers. The successive regional directors of the *Service du Matériel* in Algeria during the war were Général de Brigade René Roy (9 April 1949–28 February 1955); Colonel, then Général de Brigade Pierre Lautier (1 March 1955–15 December 1959); Ingénieur Général de 2e Classe Jacques Mouliner (16 December 1959–30 September 1961); and Ingénieur Général de 2e Classe Pierre Picquet (1 October 1961–30 October 1962).[21] The Regional Director of Matériel exercised his responsibilities through directors of matériel (*Directeurs du Service du Matériel*; DSM) in each of the three corps areas and the Joint Sahara Command. They in turn provided technical direction for the directors of matériel (*Directeurs du Matériel*; DIRMAT) assigned to each of the operational zones and major

**Figure 3.2**
**Organization of the *Service du Matériel* in Algeria, ca. 1960**

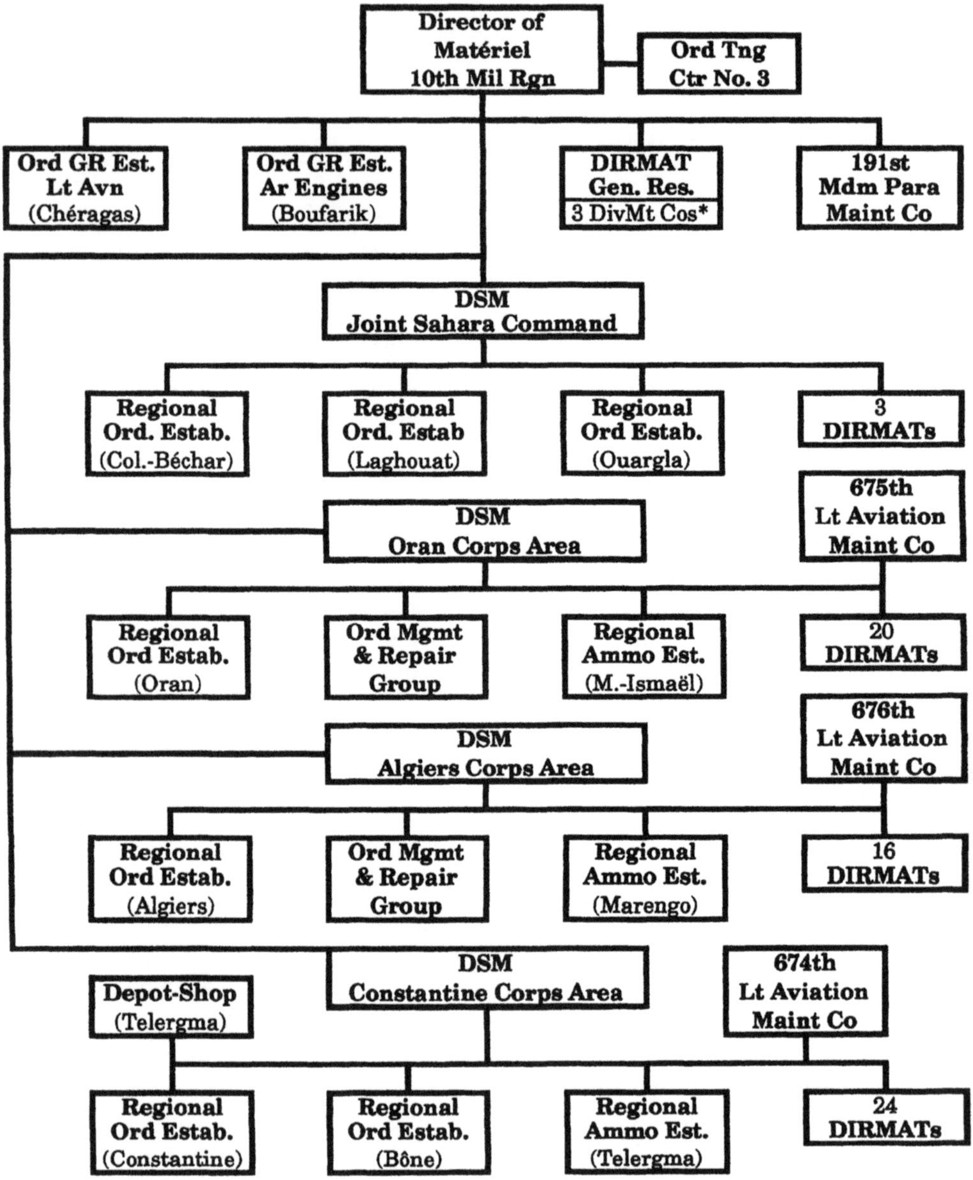


*Usually attached to the Corps Area DSM.

*Source*: Compiled by the author from various sources.

troop units (divisions). When a given zone was occupied by more than one major unit, the DIRMAT of one of the major units was designated as the zone DIRMAT and supervised the others. Such was the case in the East Constantine Zone, where the DIRMAT of the 7th Light Mechanized Division was subordinated to that of the East Constantine Zone/2nd Motorized Infantry Division.

The Regional Director of Matériel retained a number of general support and specialized ordnance units under his direct control. Of course, the assignment of specialized, general support, and direct support units varied over time. As of 1960 the units directly controlled by the DRM, 10th Military Region, included: the SM general reserve establishment for support of Army aviation (*Établissement de la Réserve Générale du Matériel-ALAT*; ERGM-ALAT) at Chéragas; the SM general reserve establishment for support of engines of armored vehicles (*Établissement de la Réserve Générale du Matériel–Engins Blindés*; ERGM-EB) at Boufarik; a medium parachute repair company (the *191e Compagnie Moyenne de Réparation de Parachutes*; 191e CMRP); and an SM training center (*Centre d'Instruction du Service du Matériel No. 3*).

The depots, workshops, and other facilities of the SM infrastructure in Algeria, as well as the relatively few general support units, were normally attached to and controlled by the Director of Matériel (DSM) of the corps area in which they were. The DSM-Oran and DSM-Algiers were both organized with a regional ordnance establishment (*Établissement Régionale du Matériel*; ERM), consisting of an ordnance personnel company (*Compagnie d'Ouvriers du Service du Matériel*; COSM) and a depot company (*Compagnie Magasin*; Cie Mag), which provided management and storage of ordnance equipment repair parts in general support of the SM establishments in the corps area. Each also had an ordnance management and repair group (*Groupement de Gestion-Réparation*; GGR), consisting of an automotive park management company (*Compagnie de Gestion du Parc Auto*; Cie Parc) and a general support heavy maintenance company (*Compagnie Lourde de Réparation Auto*; CLRA). The Oran GGR also included a heavy marine maintenance company (*Compagnie Lourde de Réparation Auto-Marine*; CLRA-Marine), and the Algiers GGR included a medium ordnance repair company (*Compagnie Moyenne de Réparation du Matériel*; CMRM). The GGRs at Oran and Algiers were created in August 1959 and absorbed the automotive parts depots (*Dépots de Rechanges Automobiles*; DRA) and end-item depots (*Dépots de Matériels Complets*; DMC) that had developed around the ERMs of Oran and Algiers.[22]

The DSM-Constantine was organized somewhat differently, with two ERM. The ERM-Constantine consisted of an ordnance personnel company, a heavy maintenance company, and a depot company. The ERM-Bône consisted of a general reserve ordnance battalion headquarters (*Bataillon du Matériel de la Réserve Générale*; BMRG), an automotive park management company, a heavy maintenance company, and a transit unit. The port of Bône was equipped to receive heavy material, and the ERM-Bône was oriented toward the reception and storage of complete vehicles and weapons systems. The transit unit managed the flow of such material through the port and into the SM depot system. In addition, the DSM-Constantine controlled a depot workshop (*Magasin-Atelier*) at Telergma operated by the 611th Depot Company (611e Cie Mag), which stocked and supplied repair parts. In the Constantine Corps Area the functions of the former DRA and DMC were absorbed into the operations of the depot-workshop at Telergma and the ERM-Bône, which were established in 1959.

All three corps area Directors of Matériel (DSM) also oversaw the operations of ammunition supply and light (Army) aircraft maintenance units. The DSM in each corps area exercised his responsibility for munitions management through a regional munitions establishment (*Établissement Régionale des Munitions*; ERMu), consisting of a headquarters element, an ammunition company (*Compagnie de Munitions*; Cie Mu), and a number of depots sited throughout the corps area. The maintenance of Army light fixed-wing aircraft and helicopters was performed by the direct support Army light aviation repair company (*Compagnie de Réparation de l'Aviation Légère de l'Armée de Terre*; CRALAT) assigned to each corps area. A subdetachment of the *Inspection du Matériel* was attached to each DSM.[23] These subdetachments were part of the metropolitan SM establishment and were involved in routine inspection and experimenting with and testing new equipment and procedures.

Ordnance supply and maintenance infrastructure and operations in the Sahara region were much different from those in the rest of Algeria. The SM infrastructure in the Sahara remained limited; it developed slowly and late in comparison with that in "useful Algeria."[24] By 1960 the existing depot workshops at Laghouat, Ouargla, and Colomb-Béchar had been converted into regional ordnance establishments, and three DIRMATs had been established (at Ain-Sefra, Touggourt, and Laghouat).[25] The functions of the Saharan regional ordnance establishments were limited, and they depended more on the Regional Director of Matériel in Algiers than did those of the corps areas. Ordnance supply support was provided by the DSM-Oran, and both direct and general support maintenance was performed by the specialized Saharan ordnance companies (*Compagnies Sahariennes du Matériel*; CSM).[26] The development of SM ammunition service in the Sahara region was even slower. As late as November 1957 there were still piles of uncovered ammunition in the courtyard of the regional ordnance establishment at Colomb-Béchar, but in 1958 groups of ammunition handlers were added to the Saharan ordnance companies, and steps were taken to create a network of operational depots and special depots for airborne troops in the Sahara region.[27]

Direct support maintenance in the Joint Sahara Command was provided until 1954 by several depot workshops operated by two repair platoons of the Foreign Legion (*Péletons de Réparation de la Légion Étrangère*; PRLE). The PRLE disappeared in 1954, with the creation of the first of three specialized Saharan ordnance companies, the 10e CSM.[28] The CSMs were in effect divisional maintenance companies (*Compagnies de Réparation Divisionnaire*; CRD) specially adapted to desert conditions and able to operate sections far from the main unit.[29]

Since the regional ordnance establishments in the Sahara were limited to performing depot ordnance supply functions, the Saharan ordnance companies performed both direct support and general support maintenance.

## Ordnance Personnel

The chief difficulty faced by the *Service du Matériel* in Algeria stemmed from the lack of trained technical personnel, particularly maintenance specialists. The accepted standards of the time for an "active theater," such as the 10th Military Region, called for a complement of some twenty-six thousand ordnance personnel to support a force of four hundred thousand men.[30] However, the number of ordnance personnel in Algeria never approached the standard of 6.5 percent. In fact, despite a sixfold increase in SM personnel between 1954 and 1959, the proportion of ordnance personnel actually fell, from 3.6 to 3.3 percent, dipping to 2.8 percent in 1955 with the heavy influx of combat units.[31] Moreover, a good part of the direct support and general support maintenance capability of the French Army in Algeria was provided by recalled reservists. The productivity of these recalled personnel was substantially lower than either the expected standard or the actual productivity of regular SM troops.[32]

The absolute lack of trained specialists and the low productivity of reserve SM personnel was compounded by the high percentage of temporary workers and the low percentage of European technicians available. In the representative Constantine Corps Area in 1959, for example, nearly half of the employees of the *Service du Matériel* were temporary workers, and approximately two-thirds were Algerian Moslems.[33] The ratio of European technicians to Moslem employees was even lower, there being only one European logistical technician for every eighteen Moslems.[34] The Moslem workers were generally less well trained and less productive; thus both the quality and quantity of their repair work was below theoretical norms.[35]

## Supply of Vehicles and Other Equipment

The *Service du Matériel* was responsible for the requisitioning, receipt, storage, and issue to units of vehicles, weapons, Army aircraft, and other equipment and spare parts. As the number of units in the 10th Military Region increased, the ordnance supply workload grew proportionately. Units deployed to Algeria normally brought with them their normal allocation of vehicles, weapons, and other equipment, but heavy usage and poor maintenance combined to create heavy demands for spare parts and replacement weapons and vehicles. For example, the tank-automotive inventory of the 5th Armored Division (*5e Division Blindé*; 5e DB) turned over every eighteen months.[36] The usual allocation of selected weapons for various types of tactical units is shown in Table 3.2.

The standard planning factors for the resupply of repair parts and complete end items used by the French Army at the time of the Algerian war are shown

**Table 3.2**
**Weapons Authorized for Various Tactical Units**

| Weapon | Infantry Regiment | Tank Regiment | Fld Arty Battalion | AAA Battalion |
|---|---|---|---|---|
| Submachinegun | 990 | 329 | 61 | 60 |
| Automatic Rifle | 84 | 36 | 19 | 12 |
| Light Machine-Gun | 24 | 110 | | |
| Heavy Machine-Gun | 38 | 67 | 24 | 11 |
| 60 mm Mortar | 18 | 54 | | |
| 81 mm Mortar | 12 | 54 | | |
| Antitank Rocket Launcher | 92 | 54 | 33 | 41 |
| Antitank Gun, Towed/Recoilless Rifle | 28 | | | |
| Antiaircraft Gun (under 40 mm) | 16 | | 6 | 33 |
| 105 mm Howitzer, Towed, M–3 | 6 | 6 | | |
| 105 mm Howitzer, Towed, HM2 | | | 18 | |
| Tank Gun, 76 or 105 mm | | 53 | | |

*Source*: France, Ministère de l'Armée de Terre, École d'État-Major, École Supérieure de Guerre et École d'État-Major, *Aide-Mémoire pour les Travaux d'État-Major*, provisional ed. (Paris: École Supérieure de Guerre et École d'État-Major, February 1952), II, ¶415.

in Table 3.3. The standard replacement rate for ordnance matériel averaged about two kilograms per man per day, but the actual replacement rates in Algeria probably exceeded the planning factor.

Some parts and rebuilt weapons and vehicles were obtained locally, but most of the ordnance matériel used in Algeria had to be shipped from metropolitan France. For example, 75 percent of all repair parts were obtained from metropolitan France, the remainder being purchased locally in Algeria.[37] Replacement stocks maintained in Algeria varied from corps area to corps area. In March 1960 the DSM-Oran was maintaining three months of normal consumption of tank-automotive parts and assemblies in general support establishments and at least one additional month in the direct support units.[38] The *Service du Matériel* in metropolitan France managed to supply much of what was needed, but deliveries were slow. In general, the order-ship time for complete end-items was about nine months, despite the relative nearness of metropolitan France.[39] The time required to replace vehicles being rebuilt in Algeria was considerably shorter but still excessive. Again, the experience in the Constantine Corps Area was typical: in February 1960 the delay was two–three months.[40]

The diversity and age of most of the armament and vehicles used by the French forces in Algeria had a major impact on the ordnance supply workload. For example, a significant problem was created by the need to obtain and stock a very large number of spare parts in order to meet the needs of an extremely heterogeneous collection of arms and vehicles. At the end of World War II most of the equipment of the French Army had been of American manufacture provided under Lend-Lease and other wartime programs to rearm the "fighting French" forces. The recovery of France's own armaments industry from the Second World War was slow, principally from the lack of funding available for military production. Thus, the United States continued to supply the French with

**Table 3.3**
**Monthly Provisioning Rate for Ordnance Matériel (Kilograms/Man/Day)**

| | Parts | | Complete End Items | | Total (rounded) | |
|---|---|---|---|---|---|---|
| Category | Cbt Arms Units | Service Units | Cbt Arms Units | Service Units | Cbt Arms Units | Service Units |
| Weapons | .84 | .84 | .36 | .36 | 1.2 | 1.2 |
| Arty; AAA | .60 | .45 | 3.00 | 2.55 | 3.6 | 3.0 |
| Vehicles | 14.50 | 12.30 | 43.50 | 37.20 | 58.0 | 49.5 |
| **Total** | **15.90** | **13.59** | **46.86** | **40.11** | **62.8** | **53.7** |

Source: France, Ministère de l'Armée de Terre, École d'État-Major, École Supérieure de Guerre et École d'État-Major, *Aide-Mémoire pour les Travaux d'État-Major*, provisional ed. (Paris: École Supérieure de Guerre et École d'État-Major, February 1952), II, ¶611.

military equipment, under the Mutual Defense Assistance Program (MDAP) and other arrangements.[41] As a consequence, the French Army in Algeria was equipped with a variety of types and models of small arms, artillery, trucks, and armored vehicles, most of which were of American provenance and all of which had seen long and heavy use under difficult conditions. Again, the situation in the Constantine Corps Area was typical: in 1959 the DSM-Constantine was responsible for a pool of some twenty-one thousand weapons of all calibers, of which 42 percent were of American origin.[42] There were three models of pistols, three models of submachine guns, three models of automatic rifles, eight models of rifles, and six models of machine guns in use. While the mobile field artillery was equipped throughout with the U.S. 105 mm HM2 howitzer, the position artillery included French, German, and U.S. types.

During the Algerian war aging American weapons and vehicles slowly began to be replaced by new French models. In January 1957 a full 80 percent of all tank-automotive equipment being used by the French in Algeria was of American manufacture, but by January 1961 the proportion of American types had fallen to about 60 percent.[43] Some improvement in the ordnance workload was achieved by a reduction in the variety of weapons and vehicles in use, and the replacement of worn-out equipment with items fresh off the production line was also a positive factor. However, many of new vehicles supplied by French industry left much to be desired in the way of durability; consequently they were a maintenance liability almost from the moment they were put into service. Both the new thirteen-ton AMX tank and the new Panhard EBR 75 armored scout car proved to have heavy maintenance requirements, as did the new French 105 mm TF and 155 mm ABS howitzers and the Simca-Ford 4×4 three-ton trucks.[44] The 75 mm gun of the AMX had a short barrel life, repair parts for the new artillery pieces were in short supply, and the chassis of the Simca-Ford truck generally failed after only about twenty thousand kilometers. In general, and despite their age, the American trucks, ambulances, and armored vehicles were preferred for their robustness in the Algerian environment. The new French Berliet Type GLR 8 diesel trucks, however, were favored for work in the Sahara, due to their use of diesel fuel, mechanical robustness, and capacity (equal to three GMC trucks).

## Maintenance Services

The maintenance of arms, vehicles, and other ordnance equipment was a major responsibility of the Regional Director of Matériel of the 10th Military Region, and in theory all five echelons of maintenance were performed in Algeria.[45] In practice the *Service du Matériel* in Algeria focused on third-echelon repairs and some fourth-echelon work; most fourth- and fifth-echelon work was passed to maintenance facilities in metropolitan France or performed by civilian firms in Algeria.[46] The effectiveness of SM maintenance services during the Algerian war was a function of the maintenance workload, in terms of the number of type items supported, and the availability of direct and general support maintenance units and of trained personnel. The SM struggled throughout the war to meet a comparatively high and constantly increasing maintenance workload with inadequate direct and general support resources, a problem that directly affected the mobility of the highly mechanized French forces in Algeria.

### *Maintenance Workload*

The maintenance workload of the *Service du Matériel* in Algeria was influenced by the number of weapons, vehicles, and other ordnance items supported; the diversity, age, and sturdiness of the items supported; the tempo of tactical operations; the terrain and climate of the area of operations, and the effectiveness of maintenance training for both operators and technical personnel. As noted, the SM in Algeria was responsible for maintaining a large, very diverse, and mostly aged collection of armaments and vehicles operating under difficult topographic and climatic conditions at a rapid tempo. Moreover, the quality of both operators and unit maintenance was chronically poor. The result was a high level of maintenance demand. For example, the third-echelon monthly repair rate in the key Constantine Corps Area ranged between 12 and 16 percent of the entire tank-automotive park, about twice the established norm.[47] At a rate of 16 percent every vehicle would enter third-echelon maintenance twice a year. The maintenance workload of the 7th Light Mechanized Division (*7e Division Mécanique Rapide*; 7e DMR), stationed in the Constantine Corps Area, was typical; in 1958 almost three thousand tank-automotive maintenance actions were needed, requiring some 181,000 man-hours and costing 385 million FF, or about 8 percent of the value of the tank-automotive matériel of the division.[48]

Both the number of French troops in Algeria and the number of weapons and vehicles supported increased rapidly in the first two years of the war, and they continued to grow thereafter. The growth in the number of items supported by the *Service du Matériel* from 1954 to 1956 is shown in Table 3.4. Expressed in fiscal terms, the maintenance workload of the SM in Algeria increased from 371

**Table 3.4**
**Growth in Number of Items Supported by the SM in Algeria, 1954–1956**

| Category | Number of Items Supported | | |
|---|---|---|---|
| | 1954 | 1955 | 1956 (est.) |
| Tank-Automotive | 8,000 | 26,000 | 60,000 |
| Arms (Small Caliber) | 241,000 | 300,000 | 400,000 |
| Arms (Large Caliber) | 690 | 753 | 800 |
| Airborne Items | 8,500 | 21,500 | 25,000 |
| Munitions Items | 15,000 | 19,300 | 30,000 |

*Source*: 10e Région Militaire, État-Major, 4e Bureau, *Dossier 4e Bureau*, 24-5-56 (Alger, 24 mai 1956), Dossier VIII (Organisation et Fonctionnement des Services: Matériel—"Fonctionnement" [AI/CQ, 10RM/EM/4e, Alger, 24 mai 1956]), page a, in folder "Aide-Major Général en AFN—Organisation du commandement et coordination des services et moyens logistiques, 1946–57," Dossier 1H2665 d. 1, Fonds Algérien, SHAT. Weapons ceded to the Algerian civil government are not included.

million FF in 1954 to 4,777 million FF in 1956.[49] By 1959 the number of vehicles supported had increased to 85,000, and by 1961 the vehicle inventory of the 10th Military Region had reached 95,659 vehicles.[50]

The Constantine Corps Area was the key operational area in the later stages of the war, and in 1959 the DSM-Constantine supported a pool of some 21,000 weapons of all types as well as 28,846 wheeled and tracked vehicles and trailers.[51] By 1 February 1962 the DSM-Constantine was supporting a force that included over one vehicle for every seven men plus some 170,000 individual and sixteen thousand crew-served weapons.[52] Additional maintenance burdens were created by the rapid operational tempo, heavy use of old, worn vehicles on poor roads and difficult terrain, overloading, poor operator practices, and lack of attention to unit (second-echelon) maintenance.[53] Operator and unit maintenance were notoriously bad throughout the French forces in Algeria. As a 1955 report noted, "Maintenance was a luxury in which the operational forces were little interested" (*"L'entretien est en général un luxe auquel les cadres ne s'intéressent que peu"*).[54]

### *Direct Support Maintenance Units*

Maintenance activities at the level of the zone and major unit was the responsibility of the DIRMAT assigned to each zone or division headquarters. The actual direct support maintenance work was accomplished for the most part by the third-echelon direct support units, or the sections thereof, assigned or attached to the zones, major troop units, or sectors for operational purposes.[55] These units reported up the SM chain only on technical matters. The basic direct support maintenance unit in Algeria was the divisional maintenance company (*Compagnie de Réparation Divisionnaire*; CRD). Organic to the division, the CRD was designed to provide all third-echelon maintenance and repairs for the division in peacetime and 54 percent of such maintenance and repairs in wartime.[56] The number of such units grew rapidly with the influx of units into Algeria in 1955–1956, but many of the newly arrived CRD lacked personnel and equipment and had to be adapted to the Algerian operational environment. For example, the CRD of the newly motorized divisions (the 52e CRD of the 2nd Motorized Infantry Division and the 64e CRD of the 4th Motorized Infantry Division) were reorganized for quick inspection and evacuation to the rear of damaged or worn-out equipment and thus required considerably increased third-echelon repair capability. This was provided out of the feeble resources of the 10th Military Region, which also had to provide shop vehicles and tools for many of the other arriving divisional maintenance companies.[57]

The divisional repair companies in Algeria were initially organized under the then-standard TOE known as *MAT 501*, with a headquarters section, a depot section, two armament-optics repair sections, two tank-automotive repair sections, and two recovery-emergency repair sections.[58] The "classic" CRD, designed for the support of a division in conventional operations, proved poorly adapted to the operational environment in Algeria, where it was called upon to support a diverse and much larger collection of forces in a large operational zone, and where the normal practice was to support as far "forward" as possible and to minimize the distances over which vehicles and equipment had to be evacuated for repair.[59] The principal result was that in Algeria the divisional maintenance company was seldom employed as an integral unit. Instead, the various sections were parceled out over the operational zone and had to operate independently. A CRD section operating independently was authorized one junior officer, two or three NCOs, and about twenty-five mechanics and was theoretically able to support six hundred vehicles; however, the internal organization of the normal CRD, its technical means, and the distribution of personnel by specialty were not adapted to independent operation of the sections.[60] Moreover, dispersion of the unit by section complicated the problems of unit administration, workload control, and supply. Nevertheless, almost all repair of small-caliber weapon and optical equipment in Algeria was accomplished by mobile teams dispatched from the supporting divisional maintenance companies.[61]

Studies were conducted to find the optimum organization to meet the problem of limited ordnance direct support maintenance resources dispersed over a large area; as a result the Regional Director of Matériel of the 10th Military Region obtained permission to organize the CRD in Algeria on the basis of TOE *MAT 605*, which was derived from one used in Indochina (TOE *FTEO 93.00*).[62] At the same time permission was obtained to organize the 62e CRD under an experimental Type AFN TOE, also derived from that Indochina table. The CRD Type AFN (shown in Figure 3.3) differed from the CRD Type 605 in the internal organization of its sections, which had a simplified structure and autonomous repair parts supply capability. The four or five mobile tank-automotive repair sections allowed the application of 47.2 percent of the unit's personnel directly

**Figure 3.3**
**Divisional Maintenance Company—Type AFN**

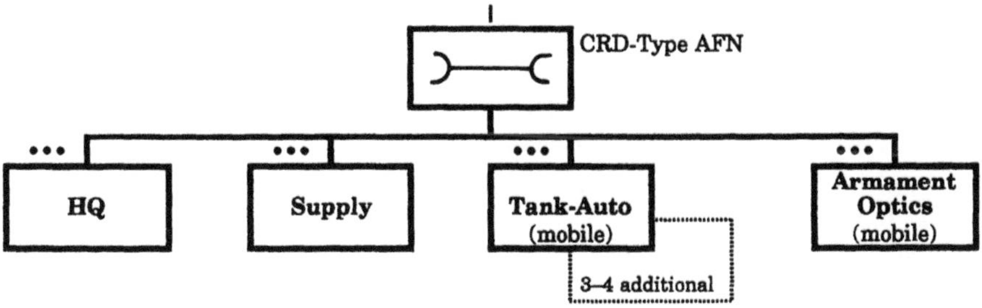


*Source*: Luc Bourgeois, *Le Matériel pendant la guerre d'Algérie* (Paris: Inspection du Matériel de l'Armée de Terre, September 1987), I, Part 1, 10.

to the repair of tank-automotive equipment, versus only 37 percent in the classic (Type 501) CRD.[63]

Even with a modified organization, the direct support maintenance units in Algeria did not always match up well with their areas and the customers they were required to support. Optimum correspondence between mission and means was seldom achieved: either a section had too much or too little capability. Direct support maintenance units in Algeria supported from 1,170 to 3,800 vehicles and repaired two to four hundred vehicles per month on average, but the workload was by no means equally distributed.[64] For example, the 428e CRD supported some 3,800 mostly new French vehicles operating on good roads on level terrain, while the 407e CRD supported only around 1,600 vehicles, but they were old American models operating in the *djebel.* As a result the concept of the "work cell" (*cellule de travail*) emerged as a more accurate measure for matching resources to workload. The optimum "work cell" was fifty men for a supported unit of five hundred vehicles; this standard was used to calculate the required direct support maintenance forces and personnel specialities, and authorizations of tools, equipment, and repair parts.[65]

***General Support Maintenance Units***

The number of general support maintenance units in Algeria was generally inadequate for the size of the tank-automotive inventory supported, especially since in the absence of a sufficient number of direct support maintenance units they were required to perform third-echelon work for troop units, generally nondivisional units, in their vicinity.[66] Unlike the direct support maintenance units, however, the general support units were normally not dispersed but operated as a unit as part of the regional ordnance establishments (ERM) and ordnance management and repair groups (GGR) in the corps areas. The automotive park management company (Cie Parc) assigned to each GGR handled the receipt, storage, and issue of complete new and rebuilt wheeled and tracked vehicles, while the supply of repair parts and accessories was the responsibility of the depot company (Cie Mag) assigned to the ERM in each corps area. The standard Cie Mag was authorized five officers, twenty-five NCOs, 150 enlisted men, and ten vehicles, organized in a headquarters section, a reception and expediting section, a tank-automotive depot section, and a section that handled weapons, instruments, and optical equipment; a Cie Mag was theoretically capable of managing a depot with fifteen days of supply supporting thirty to four thousand men.[67] A heavy automotive maintenance company (*Compagnie Lourde de Réparation Auto*; CLRA) was assigned to the GGR in each corps area to handle fourth-echelon repairs for wheeled vehicles. With five officers, twenty-four NCOs, 172 enlisted men, and twenty-five vehicles organized in a headquarters section, a depot and general mechanical section, and two automotive repair sections, the standard CLRA was theoretically capable of supporting a park of 2,500 vehicles with 180–270 vehicular-repair and 450–600 armament-repair jobs per month.[68] The CLRA was also responsible for the recovery and evacuation of equipment to fifth-echelon maintenance facilities. The heavy maintenance companies in Algeria were tasked with supporting a significantly greater number of vehicles, and their productivity was substantially lower than the theoretical norms, the capacity of a CLRA in Algeria being only about sixty vehicles per month at full operation. In 1959 the direct support maintenance units in the Constantine Corps Area, for example, turned in about 425 vehicles per month for repair by the two CLRAs in the corps area. Since the two companies had, at best, a capacity of only 120 vehicles per month, the excess workload had to be sent to France, but in fact the monthly quota was only a hundred vehicles. The shortfall in fourth-echelon maintenance capacity resulted in a delay of about six months for vehicles repaired in-country and eighteen months for those sent to France for repair.[69]

Backup third-echelon support for the direct support maintenance units in Algeria was provided by the general support medium ordnance maintenance companies (*Compagnies Moyennes de Réparation du Matériel*; CMRM) and medium automotive maintenance companies (*Compagnies Moyennes de Réparation Auto*; CMRA) assigned to the ordnance establishments in the various corps areas. The standard CMRM was authorized six officers, twenty-nine NCOs, 160 enlisted men, and fifty-two vehicles organized in a headquarters section, a depot and general mechanical section, an artillery section, an armament-optics section, a tank-automotive section, and a recovery section; it was theoretically capable of providing third-echelon recovery, repair, supply, and evacuation of tank-automotive, armament, and optical equipment for an infantry division in combat, with 150–200 vehicular repair jobs, 450–600 armament repair jobs, and fifty artillery repair jobs per month.[70] With four officers, fifteen NCOs, 101 enlisted men, and twenty vehicles organized in a headquarters section, a depot and armament section, and two automotive repair sections, the standard CMRA was theoretically capable of providing third-echelon repair of weapons and general-usage vehicles for a park of 1,200 vehicles, with three hundred vehicular and four to six hundred light weapons repair jobs per month.[71] The CMRA was also capable of providing recovery and evacuation services on 100–150 kilometers of fully utilized lines of communications. Most of the CMRM and CMRA in Algeria were converted to divisional maintenance companies (CRD) in 1955–1956, but there were still one CMRM in the Algiers Corps Area and two CMRA in the Constantine Corps Area as late as 1960.[72]

***Aircraft Maintenance***

French air power, and particularly the helicopter, played a major role in the Algerian war.[73] In 1955 there were only eighteen Army fixed-wing aircraft and four Army helicopters in Algeria when the French government purchased one hundred Vertol H-21 transport helicopters for use there.[74] An in-depth study of helicopter operations, supply, and maintenance was undertaken for the French Ministry of National Defense by the Vertol Aircraft Corporation in late 1956 following the delivery of its H-21s in the summer of that year.[75] The Vertol study did much to determine the future tactical and logistical use of helicopters in Algeria as well as the organization and methods of the supporting maintenance activities.

Delivery of the H-21s, along with a number of H-34 helicopters, began in June 1956, and thereafter the number of both fixed and rotary-wing Army aircraft in Algeria escalated rapidly. By 1959 the French Army in Algeria was operating some 140 helicopters (sixty-four Vertol H-21s, nine Sikorsky H-19s, thirty-eight Bell H-13s, and twenty-nine Allouettes) and 120 fixed-wing aircraft.[76] At the same time the French Navy had thirty-six helicopters (twenty H-21s plus H-5s, H-19s, and H-34s) in Algeria, and the French Air Force had 122 (about eighty H-34s and twenty-five Allouettes plus H-13s and H-19s). The number of Army divisional aviation platoons increased from thirteen at the beginning of 1956 to thirty-two by the end of 1960; by the end of the war in 1962 there were over six hundred helicopters in use by the French forces in Algeria.[77]

The substantial growth in the number of Army light aircraft and helicopters in use in Algeria between 1954 and 1962 was matched by the development of maintenance facilities for their support. Many of the organizational structures and technical methods for the maintenance of Army aviation in the field were developed by the French forces in Algeria, and in general the facilities, equipment, and technical personnel were excellent. All aircraft maintenance was controlled at the top by the French Air Force, which managed all fifth-echelon and most fourth-echelon aircraft maintenance and controlled, stocked, and issued all aircraft spare parts, thereby controlling lower echelons of aircraft maintenance as well.[78] No modifications could be made to Army helicopters without approval through Air Force technical channels. Although the French Navy normally provided for all of its own logistical support, the three squadrons of Navy helicopters in Algeria were supported by Air Force and Army aircraft maintenance organizations.[79]

Initially, recovery and limited maintenance services for Army aircraft, both fixed and rotary-wing, were provided by the divisional repair companies; the

52e, 64e, and 71e CRD were particularly involved, and the latter even had a Sikorsky helicopter for recovery operations. Even after the new Army light aviation maintenance companies (*Compagnies de Réparation de l'Aviation Légère de l'Armée de Terre*; CRALATs) assumed responsibility for third-echelon aircraft maintenance in 1957, some of the CRDs continued to recover aircraft for some time. Increasing aircraft maintenance requirements led to the establishment of a support facility at Sétif in 1955. The facility at Sétif suffered initially from a shortage of trained aircraft maintenance specialists; it was subsequently somewhat relieved by the creation of an Army aviation school at Bourges in metropolitan France.[80]

Crews, squadron maintenance personnel, and the three service squadrons of the 2nd Helicopter Group at Sétif provided first and second-echelon service and repairs for Army aircraft, but the primary role in Army aircraft direct support maintenance in Algeria after 1956 was taken by the Army CRALAT, which was developed independently in Algeria and metropolitan France at the same time.[81] All third-echelon and some fourth-echelon direct support maintenance for Army fixed and rotary-wing aircraft was the responsibility of the three CRALATs assigned to the 10th Military Region: the 675e at Sidi-bel-Abbès (Oran Corps Area), the 676e at Algiers (Algiers Corps Area), and the 674e at Ain-Arnat near Sétif (Constantine Corps Area).[82] The organization of the 674e CRALAT at Ain-Arnat, shown in Figure 3.4, was typical of the others as well.

General support fourth-echelon aircraft maintenance and repair parts supply was provided by the General Reserve Army Aviation Maintenance Establishment (*Établissement de Réserve Général du Matériel d'Aviation Légère de l'Armée de Terre*; ERGM-ALAT) at Chéragas, which opened in 1957 and also supported the civilian Société Héli-Service and the civilian airports of Maison-Blanche and Blida.[83] Additional fourth-echelon aircraft maintenance in Algeria was provided by contractors. For example, H-21 airframe repairs were done by Société Héli-Service and Air France repaired H-21 engines. The Vertol Aircraft Corporation provided technical representatives to advise on the maintenance of the H-21.[84] Army aircraft requiring extensive fourth-echelon repairs or fifth-echelon rebuilding were evacuated to facilities in metropolitan France. Since Army aviation maintenance support was decentralized, substantial stocks of repair parts and assemblies were required in Algeria, most of them delivered by air. After October 1957 there was a weekly transport flight to Sétif (Ain-Arnat) and Sidi-bel-Abbès from the Montauban ERGM-ALAT near Toulouse, from Orléans, or from the repair center at Versailles via Le Bourget airfield near Paris.[85]

Aircraft, and helicopters in particular, are maintenance intensive, and the climatic and operational conditions in Algeria substantially increased the workload. By 1959 the French experience in Algeria indicated that the requirement for hours of first-through partial fourth-echelon maintenance per hour of flight was 9.1 hours for the Vertol H-21, 3.5 hours for the Bell H-13, and 2.7 hours for the Allouette.[86] The principal maintenance problem encountered was a high rate

**Figure 3.4**
**Organization of the 674th Army Light Aviation Maintenance Company**

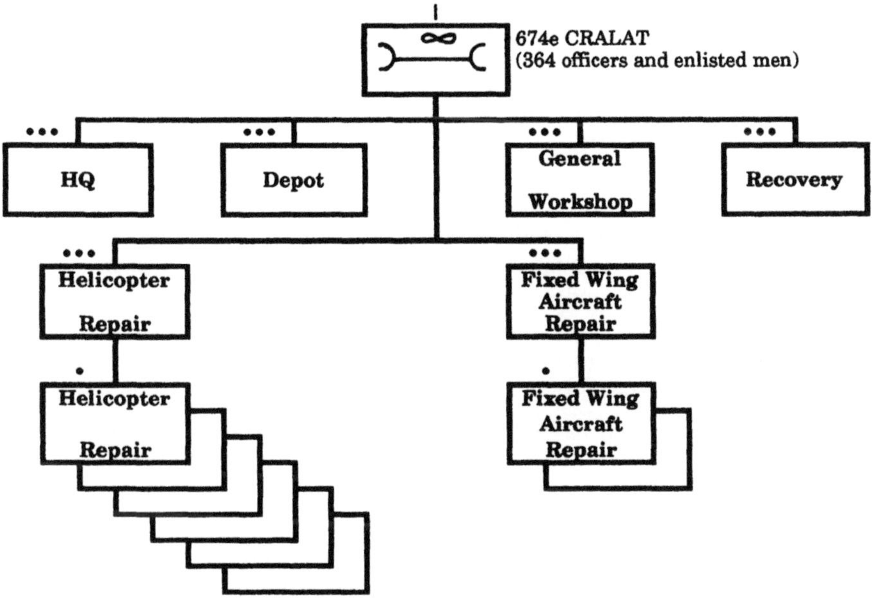


*Source*: Luc Bourgeois, *Le Matériel pendant la guerre d'Algérie* (Paris: Inspection du Matériel de l'Armée de Terre, September 1987), I, Part 1, 11.

of repairs to reciprocating engines due to excessive oil consumption. Other problems included clutch and clutch actuator failures; engine cowling, filter, and ducting failures; carburetor malfunctions; the shearing of landing gear; leaking hydraulic servos; and sand erosion of rotor blades.[87] Few insurmountable obstacles were encountered in the aircraft maintenance field, however, and the overall excellence and effectiveness of the French effort in Algeria is reflected in an availability rate for all types of helicopters in combat of almost 80 percent and a combat utilization rate of over sixty flying hours per month per helicopter.[88]

## Ammunition Service

The Regional Director of Matériel of the 10th Military Region was responsible for the management and technical oversight of ammunition units and operations in Algeria.[89] In the framework of decisions of the commander in chief regarding the general location of ammunition depots, stockage levels, and the allocation of ammunition to the various combat units, the Regional Director oversaw the deployment of ammunition supply units, the establishment of stockage levels, monthly review of actual consumption, and quarterly requests for replenishment of stocks from metropolitan France. Similarly, the corps area directors of matériel (DSM), acting on the decisions of their commanders, supervised the organization, operation, and security of ammunition depots and supply points in their corps areas and prepared monthly reports of depot balances to 10th Military Region headquarters. Commanders at zone and sector level were responsible for the protection of depots and movements of ammunition in their area as well as for the preparation of monthly reports of consumption and stockage for the next higher level.

***Ammunition Units***

The storage, maintenance, and issue of ammunition and related ordnance items was accomplished through a network of operational and garrison depots and smaller ammunition supply points (ASP) scattered throughout Algeria. The DSM in each corps area carried out his responsibilities for ammunition supply through a regional ammunition establishment (*Établissement Régionale des Munitions*; ERMu). The three ERMu became operational in the fall of 1957. Before that time ammunition supply and management had been a function of the *Service Munitions*, a part of the regional ordnance establishment in each corps area. Each ERMu included one or more central ammunition depots and an attached ammunition company (*Compagnie de Munitions*; Cie Mu), which was authorized seven officers, nineteen NCOs, 161 enlisted men, and twenty-two vehicles organized in a headquarters section and six ammunition detachments; the Cie Mu was capable of operating a depot, ASP, or transfer point of one to two thousand tons twelve hours per day.[90] The corps area DSM were also responsible for explosive ordnance disposal (EOD). The lack of qualified specialists left most EOD work to military ammunition handlers assigned to the ammunition companies or in some cases to military personnel from the regular direct support maintenance units. The 311th Medium Automotive Maintenance Company and the 62nd and 402nd divisional maintenance companies were particularly involved in such work. The danger of EOD activities was underlined by the loss of five NCOs and one officer during the Algerian war.[91]

***Stockage Levels and Storage of Ammunition***

The central depots in each corps area (at Moulay-Ismaël for the Oran Corps Area, at Marengo for the Algiers Corps Area, and at Telergma and Bône for the Constantine Corps Area) supplied the operational and garrison depots and provided ammunition general support maintenance and technical support. The principle of "throughput" from the ports to depots in the zones was followed as much as possible. For example, in the Constantine Corps Area the regional ammunition establishment did not play a centralizing role; ammunition was shipped by rail directly from the ports of Bône and Philippeville to the zones. The central depot at Bône did, however, handle resupply of all East Constantine Zone ammunition depots and those depots served by coastal shipping (Bougie, Djidjelli, and Philippeville), while the central depot at Telergma resupplied other depots in the corps area.[92] In general, the goal was to create sufficient capacity in the operational and garrison depots in the zones to limit the frequency of resupply from the two central depots to twice a month.

Operational depots in the various zones were operated by ordnance personnel from the regional ammunition establishment's ammunition company. The widespread antiguerrilla war and necessary dispersion of combat units required that the operational depots, of three to five hundred tons capacity, be no more than a hundred kilometers from the areas of operations of the units they supported.[93] The operational depot network in the Constantine Corps Area was typical. The ERMu at Telergma controlled main depots at Telergma and Bône and eleven operational depots (three each in the East, North, and South Constantine Zones and two in the West Constantine Zone), all operated by the 758th Ammunition Company. The capacity of the more important of these depots was: Telergma (main depot), ten thousand tons; Duzerville, three thousand tons; Tebessa, one thousand tons; Batna and Khenchela, nine hundred tons each; Philippeville, six hundred tons; and Sétif and Bougie, 250 tons each.[94] Smaller operational stocks were maintained in the garrison depots and ammunition supply points operated by personnel of the other arms and services, under general SM supervision.

Normally, the ammunition stocks maintained in Algeria averaged from five to ten months of consumption plus the critical items of another five months of consumption, the non-critical items of which were stored in metropolitan France.[95] In the later stages of the war in Algeria the 10th Military Region stockage level was based on eight months of supply, with four months plus a one-month "safety level" in Algeria and three months on call in metropolitan France. The ammunition stocks maintained by the ERMu central depot in each of the three corps areas consisted of various authorizations, including part of the so-called "ministerial reserve"; the regional reserve controlled by the Regional Director of Matériel of the 10th Military Region; and part of the corps area stocks directly under the authority of the corps area DSM. Since there was no general reserve ammunition establishment (*Établissement de Réserve Générale des Munitions*; ERGMu) in Algeria, the major part of the ministerial reserve was maintained by the ERGMu at Miramas in metropolitan France.

Each tactical unit in Algeria also kept on hand a "safety level" of ammunition equivalent to the initial authorization (basic load) that could be carried on the unit's organic transport. An additional amount of ammunition, the equivalent of a "maintenance float," to ensure that the unit stockage level was never less than 70 percent of the safety level, completed the unit authorization. Of course, additional stocks were issued to units for specific operations.

Until the completion of infrastructure improvements in the late 1950s, the physical condition of most of the main ammunition depots in Algeria was poor. Crowding and a lack of adequate covered storage were common, as were shortages of materials-handling equipment. For example, the ERMu at Moulay-Ismaël handled about three thousand tons of ammunition per month but had no rail connection to the port of Oran and no forklifts.[96] The principal storage problems occurred in the garrison depots and ASPs operated by the tactical units.

Standards of storage, handling, segregation, and maintenance of ammunition stocks were frequently not met due to a lack of command interest and to incompetent personnel. Of 165 units in the North Constantine Zone inspected in 1958, fifty-two were found to be storing ammunition in unsatisfactory conditions.[97] Inspections in the Algiers Corps Area found in 10 percent of the units ammunition stored in locations occupied by troops, in one case mortar rounds under the beds of the night crew; mixing of classes of ammunition in 90 percent of the units; problems in compatibility and imperfect lot segregation in 80 percent; and lack of adequate firefighting controls and equipment in 25 percent.[98]

***Ammunition Consumption***

Ammunition storage and supply operations in Algeria were characterized by the invariable urgency of resupply operations, fluctuations in consumption corresponding to periods of high or low operational activity, heavy security requirements, and the general inadequacy of the lines of communications. The irregularity of consumption, and weakness of the lines of communications in particular, required careful distribution of ammunition stocks so as to meet all potential operational requirements while minimizing replenishment operations. Accounting for consumption was an especially heavy burden, compounded by the tendency of combat unit commanders to overstate their consumption figures in order to augment their ready reserves of ammunition, particularly of items in short supply.[99]

The usual method for calculating ammunition consumption and supply requirements was the Unit of Fire (UF). The UF was an arbitrary unit of measurement, fixed by the commander and subject to variation, representing the consumption of munitions of all types for a given unit (battalion, regiment, division, etc.) during a given period of combat. At the end of World War II the UF represented the consumption of an average day of combat in offensive warfare. The UF was divided into "tranches" (slices) representing the various types of ammunition that might be used by a unit. Both the UF and the "tranches" were usually expressed in either the number of rounds or weight.[100] At higher staff levels the concept of the Day of Supply (DOS) was substituted for the UF.[101] Table 3.5 shows the weight in tons of a standard UF for various type units.

French staff manuals in use during the Algerian war provided standard planning factors for ammunition consumption under various combat conditions. The factors for monthly consumption are shown in Table 3.6. (It should be noted that the planning factors account for days without firing.)

The nature of combat operations in Algeria varied over time and from area to area, and ammunition consumption fluctuated accordingly. In general, the consumption rates of artillery, tank, antitank, and antiaircraft artillery ammunition was lower in Algeria than in the classic European warfare on which the planning factors were based. However, artillery ammunition consumption on the eastern and western barrages in 1958–1960 probably exceeded the standard fac-

**Table 3.5**
**Weight of Standard Units of Fire for Various Units**

| Type Unit | Tranche in Tons | | | | | | | | Total Tons |
|---|---|---|---|---|---|---|---|---|---|
| | A | B | C | D | E | F | G | H | |
| Infantry Battalion | 1.7 | 3.8 | 3.9 | – | 0.5 | 4.5 | 0.3 | – | 14.7 |
| Infantry Regiment | 6.3 | 11.8 | 20.8 | 17.5 | 2.0 | 18.0 | 1.1 | – | 77.5 |
| Mdm Tank Regiment | 6.6 | 1.1 | 36.2 | – | 0.5 | 9.0 | 0.3 | – | 53.7 |
| FA Bn (Gp) (105 mm) | 0.9 | 0.1 | 1.6 | 82.0 | 0.7 | 0.9 | 0.1 | – | 86.3 |
| FA Bn (Gp) (155 mm) | 0.9 | 0.1 | 1.6 | 167.0 | 0.7 | 0.9 | 0.1 | – | 171.3 |
| Lt AAA Bn (Group) | 1.3 | 0.1 | 2.4 | – | 18.6 | 0.9 | 0.1 | – | 23.4 |
| Cbt Engineer Bn | 2.4 | 2.8 | 2.5 | – | – | 5.3 | 0.1 | 6.0 | 19.1 |
| Signal Battalion | 3.0 | 0.1 | 3.9 | – | – | – | – | – | 7.0 |
| Transport Battalion | 2.2 | – | 1.8 | – | – | – | – | – | 4.0 |
| Infantry Division | 47.8 | 41.8 | 159.6 | 458.5 | 29.0 | 88.6 | 4.6 | 6.0 | 842.1 |
| Armor Division | 71.7 | 70.9 | 253.5 | 246.0 | 25.6 | 93.4 | 3.3 | 6.2 | 772.6 |

*Source*: France, Ministère de l'Armée de Terre, École d'État-Major, École Supérieure de Guerre et École d'État-Major, *Aide-Mémoire pour les Travaux d'État-Major*, provisional ed. (Paris: École Supérieure de Guerre et École d'État-Major, February 1952), II, ¶617.

tor for "Defense—1st Day" during certain periods. In more precise terms, a field artillery battalion of eighteen 105 mm howitzers might fire as many as 3,240 rounds (1.1 UF) in a period of only three hours.[102] Land mines were another important component of the French Army ammunition requirements in Algeria, particularly on the eastern and western barrages. The standard planning factors for the allocation of antitank and antipersonnel mines is shown in Table 3.7.

The rapid expansion of the Algerian war in 1955–1956 quickly led to an enormous surge in the consumption of ammunition. Between 1 November 1954 and 1 June 1955 French forces in Algeria fired only fourteen thousand cartridges and dropped only 2.3 tons of bombs, but in the following six months they fired fifty-five thousand cartridges and dropped sixteen tons of bombs.[103] Expressed in monetary terms, the consumption of ammunition in Algeria increased from some millions of FF monthly in 1955 to an average of 300 million FF per month in 1956 and two billion FF per month in 1958.[104] The increased consumption of munitions in Algeria sharply reduced ammunition reserves in metropolitan France, but increased expenditures for ammunition were politically unacceptable. Consequently, restrictions on consumption were applied in 1958–1959. Despite restrictions and continuing review of ammunition consumption in Algeria, expenditures in 1959 exceeded 21.5 billion FF for current ammunition and another 3.2 billion FF for mines for the barrages.[105] In 1960–1961 the monthly consumption rate increased to 34 million FF, and the first drawdowns of ammunition stocks in Algeria came only at the very end of the war, in 1962.[106]

The production of munitions and other explosives was a government monopoly in France; production was controlled by the Powder Service (*Service des Poudres*), which operated twelve government-owned factories.[107] The production of munitions in France in the 1950s was limited by the incomplete recovery

**Table 3.6**
**Standard Planning Factors for Monthly Ammunition Consumption by Arm**

| Nature of Operations | Infantry | | Armor | | Artillery | | AA Arty | |
|---|---|---|---|---|---|---|---|---|
| | Days | UF | Days | UF | Days | UF | Days | UF |
| Periods of Calm | 10 | 2.0 | 4 | 1.0 | 15 | 1.5 | 15 | 1.5 |
| Offense - 1st Day | 1 | 1.0 | 1 | 1.0 | 1.5 | 1.5 | 1 | 1.0 |
| Defense - 1st Day | 1 | 1.5 | 1 | 1.5 | 1.5 | 2.0 | 1 | 2.0 |
| Following Days (Off/Def) | 3 | 1.5 | 5 | 2.5 | 3 | 1.0 | 5 | 1.5 |
| Days without Firing | 15 | – | 19 | – | 9 | – | 8 | – |
| **Total** | **30** | **6** | **30** | **6** | **30** | **6.0** | **30** | **6** |

*Source*: France, Ministère de l'Armée de Terre, École d'État-Major, École Supérieure de Guerre et École d'État-Major, *Aide-Mémoire pour les Travaux d'État-Major*, provisional ed. (Paris: École Supérieure de Guerre et École d'État-Major, February 1952), II, ¶618b.

from World War II, antiquated equipment, and lack of raw materials. As a consequence, a major portion of the ammunition used in Algeria was provided by the United States, under the Mutual Defense Assistance Program. By February 1956 the United States had already delivered to France under the MDAP some 65,628,279 rifle (.30 caliber) cartridges, 547,000 75 mm recoilless rifle rounds, 1,345,600 high-explosive howitzer (105 mm) shells, 709,600 M-6 high-explosive antitank (2.36 in) rockets, and 569,500 M-6 and M-15 series high-explosive antitank mines, among many other items.[108]

### *Ammunition Problem Areas*

The supply of ammunition posed few problems for the French forces in Algeria beyond the obvious difficulties of delivery to widely dispersed combat units, and fluctuating consumption levels. Nevertheless, the French found it necessary, primarily for economic reasons, to reduce ammunition consumption, particularly the consumption of artillery ammunition.[109] Although certain items, such as colored-smoke grenades and 75 mm recoilless rifle rounds, were sometimes on the critical item list, shortages of ammunition were seldom a problem, except temporarily at such points as Souk-Ahras and Tebessa on the eastern barrage, where consumption was relatively high and storage facilities were limited.[110] Actual restrictions on the monthly consumption of ammunition were rarely imposed, except in the case of 9 mm pistol ammunition and 105 mm howitzer shells; an example occurred in the Constantine Corps Area in 1959–1960, when it was necessary to reduce the level of monthly consumption of 105 mm HM2 howitzer ammunition from 17,600 rounds to 14,400.[111]

## ***Service du Matériel* Problem Areas**

The physical and operational environment of Algeria had a profound impact on the *Service du Matériel*, which had perhaps the most difficult tasks of all the French logistical services in Algeria. The great distances and operational dis-

**Table 3.7**
**Requirements for Land Mines**

| Type Mine Field | Density Per Meter | Number of Mines | Tons |
|---|---|---|---|
| **Anti-tank Mines** | | | |
| Hasty - 1 km | 1.5 | 1,500 | 15 |
| Medium - 1 km | 2.0 | 2,000 | 20 |
| Strong - 1 km | 4.0 | 4,000 | 40 |
| Obstruction of a Key Point | – | 100 | 1 |
| Obstruction in Depth of 10 km of Road | – | 800 | 8 |
| **Anti-Personnel Mines** | | | |
| Booby-Trap - 1 hectare | – | 40 | 0.2 |

*Source*: France, Ministère de l'Armée de Terre, École d'État-Major, École Supérieure de Guerre et École d'État-Major, *Aide-Mémoire pour les Travaux d'État-Major*, provisional ed. (Paris: École Supérieure de Guerre et École d'État-Major, February 1952), II, ¶618. An average minefield depth of one hundred meters is assumed.

persion of units, coupled with the uncertainty of rail communications, intensified the use of motor vehicles. These in turn, due to the climate, difficult terrain, and generally poor roads, experienced a higher than normal rate of accidents and wearing out.[112] Because the enemy did not possess aircraft, artillery, or many antitank weapons, the rates of loss to enemy action and associated fourth-echelon repair were much lower in Algeria than in a "classic" war, but the third-echelon repair workload due to accidents and rapid wear-out of vehicles and other equipment, compounded by poor unit maintenance, was much higher.

Operator and unit maintenance was a particular problem; French combat units in Algeria were notoriously lax in accounting for and maintaining their vehicles, weapons, and other equipment. Thus, second-echelon maintenance was the weak point in the French maintenance system in Algeria, one that required the joint effort of unit commanders and the SM to resolve. Among the corrective measures adopted were frequent inspection of vehicles and unit workshops by unit commanders, rewarding good performance and punishing bad, and the assignment in each company-size unit of a "maintenance/material expediter" (*correspondant du matériel*) to coordinate problems with the SM and facilitate informal inspection and assistance visits by SM teams.[113]

The principle of supporting combat troops as far forward as possible led to piecemeal employment of direct support maintenance units, by section or even smaller increments; their dispersion made control difficult and decreased the efficiency of repair operations.[114] The need of these detachments to provide for their own security and administrative functions in isolated areas, as well as a requirement that personnel of the ordnance units participate actively with the combat units in counter-guerrilla operations (patrols, searches, etc.), significantly reduced the number of personnel available for maintenance. Moreover, in a dispersed company, administration, security, and transport required twice as many personnel as for one grouped in one place. For example, in 1961 the 407th Divisional Maintenance Company could put 65.4 percent of its personnel to work in the shops, but the 80th Divisional Maintenance Company, whose 3rd Section was two hundred kilometers away from the main body, could muster only 40.4 percent of its personnel for primary mission activities.[115] The relative immobility of most direct support maintenance units, their lack of some key equipment (such as portable lighting, maintenance tents, and radios), and the extraordinarily high stockage level they were obliged to maintain restricted their ability to shift quickly to support mobile combat forces, which further decreased their effectiveness.[116]

The general support and specialized ordnance units, operating from aged and generally inadequate fixed facilities in the urban areas along the Mediterranean coast, also experienced difficulty in operating at their full capacity. Moreover, due to the shortage of direct support maintenance units, the general support and specialized units were obliged to take over the direct support of nondivisional units in and around the urban areas. The resulting loss of fourth- and fifth-echelon capability meant that a great deal of reliance had to be placed on the evacuation to metropolitan France of most equipment requiring those levels of repair.

During the period 1955–1958 the *Service du Matériel* in Algeria was preoccupied with the improvement of its infrastructure, the reception and employment of reinforcing units, and the adaptation to the Algerian environment of its direct and general support maintenance units. The sudden influx of combat units into Algeria in 1955–1957 led to a temporary deficit in small arms and vehicles. At the same time, additional SM units were arriving at a rapid pace, and the new units required some time to fill out their complements of personnel and equipment and become operational. By 1957 most of the difficult problems associated with this "start-up" activity had been resolved, and the SM infrastructure in Algeria had made the necessary adjustment to a more intense and more mobile operational situation.[117] Policies were adopted to relieve the overworked direct support and general support units in Algeria by accelerated delivery of repair parts and assemblies, rapid evacuation to metropolitan France of vehicles requiring fourth-echelon maintenance, the use of troop unit maintenance workshops to perform some third-echelon repair work under SM supervision, and most importantly, emphasis on the responsibility of unit commanders for second-echelon maintenance.[118]

By 1959 the *Service du Matériel* in Algeria had reached its apogee. The mistakes of the early years of the conflict were behind it, and the necessary measures had been taken to avoid their repetition. However, many problems remained: replacement of the aging infrastructure; assignment of sufficient numbers of qualified technical personnel; the overall shortfall in direct support maintenance units, and their dispersion; and the general and continuing adjustment to a wartime environment. However, solutions to all of these problems were in train, and the lessons already learned in Algeria were being internalized.

Among those lessons were the need to: simplify and articulate in elementary cells the organic means of provisioning and repair, in order to give SM units the greatest flexibility of employment and produce best results; educate the Army (especially users) on the complexity of the problems posed by mechanization and on the imperative need for preventive maintenance; plan for the imminent replacement of aging and worn U.S. vehicles and armor; reform the methods of recruitment of SM personnel in order to ensure full employment in appropriate positions of draftees with civilian specialist skills; assign absolute priority to the SM technical mission and relieve SM units from the participation in operations for maintaining order (patrols, etc.); achieve an equilibrium between "missions" and "means" by providing the SM with the quantity and quality of supervisory personnel and labor necessary to its work; simplify accountability procedures imposed on the SM by adopting standards normally utilized in time of war; and orient French industry toward the problems of the maintainer in the field and his desire for standardization.[119]

## PETROLEUM SUPPLY (*SERVICE DES ESSENCES*) FORCES

Overall control and supervision of French military petroleum supply and storage in Algeria was vested in a Director of Petroleum Service in Algeria (*Directeur du Service des Essences en Algérie*; DirSEA), colocated with the headquarters of the 10th Military Region in Algiers. Following the usual pattern, the commander in chief determined the allocation of petroleum products to the various corps areas, the general location of petroleum (POL) depots, the stockage levels to be maintained, and the supplementary manpower and transportation required. The DirSEA was responsible for the receipt, storage, and distribution of POL, the organization and operation of depots, and the manning of necessary POL service installations.

Like the *Service de l'Intendance* and *Service du Matériel*, the *Service des Essences* (SE) in Algeria was organized on a territorial basis, the overall structure of which, around 1960, is depicted in Figure 3.5. The DirSEA was represented in each of the corps areas by a territorial delegate (*Délégué du Service des Essences*; DSEA), who oversaw the operation of varying numbers of petroleum importation centers, supply units, and resupply centers. The DSEA also supervised the operations of the SE representative known as the operational delegate (*Délégué Opérationnel*; DO) attached to each of the operational zones and major units. The DOs in turn managed the network of fixed and mobile operational depots, distribution centers, and civilian petroleum depots under contract in their zones of responsibility. The supply of petroleum products for French forces in the Sahara was provided through the territorial delegate in Oran, with direct management exercised by operational delegates at Colomb-Béchar (West Sahara Zone) and Beni-Ounif (East Sahara Zone).

**Figure 3.5**
**Organization of the *Service des Essences* in Algeria, ca. 1960**

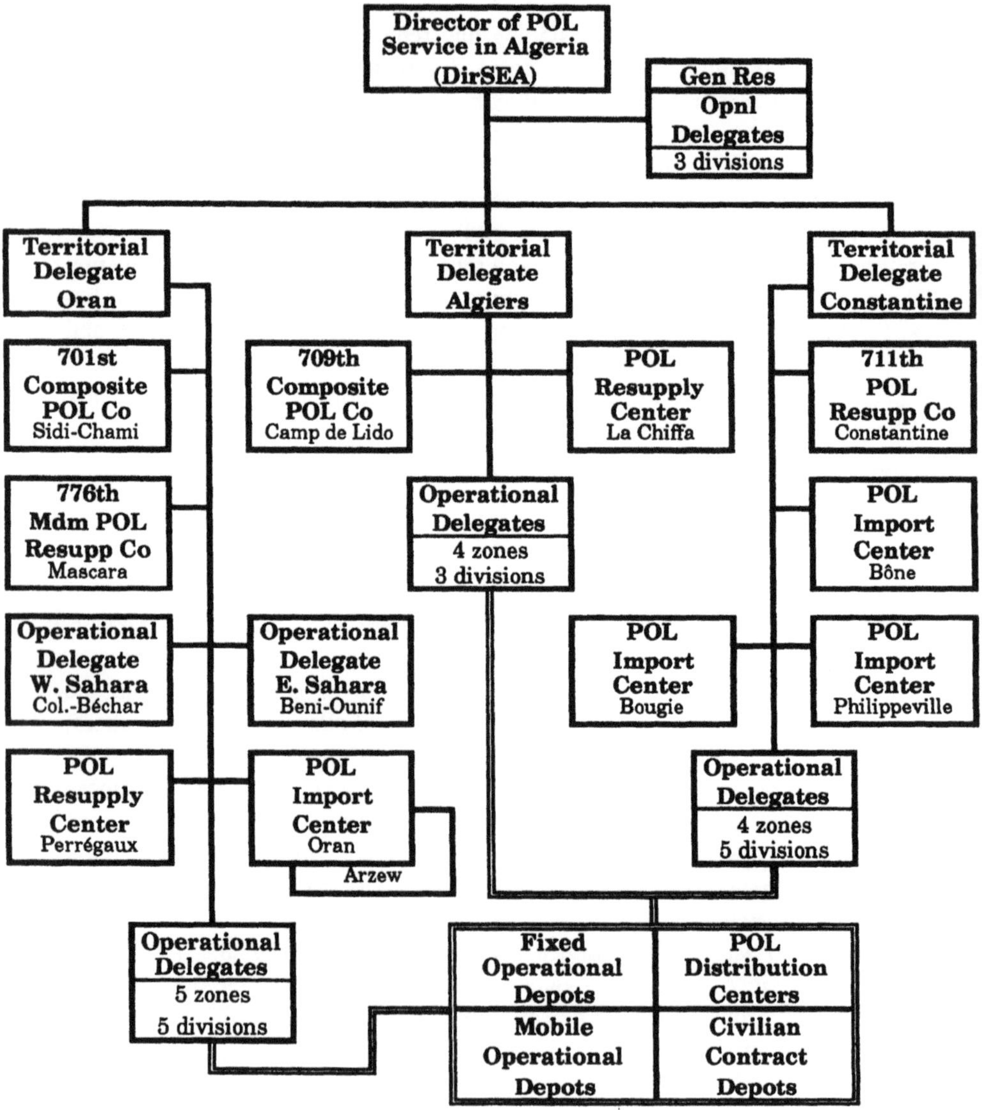


*Source*: Compiled by the author from various sources.

### POL Units

Before 1 November 1954 the means available to the Director of Petroleum Service in Algeria were limited to one bulk POL transport company (the *776e Compagnie de Transport "Gros Porteurs"*; 776e CTGP) at Mascara; one POL resupply and distribution company (the *709e Compagnie de Ravitaillement et de Distribution en Essences*; 709e CRDE), stationed near Algiers; a number of petroleum depots under contract with the civilian firms of Esso Standard, Socony, and Stettine; and a small network of POL distribution centers (*corps répartiteurs*; CR).[120] The 776e CTGP, later reorganized as a medium POL carrier company (*Compagnie de Ravitaillement en Essences Moyenne Porteur*; CREMP), was authorized four officers, fourteen NCOs, 186 enlisted men, and sixty-five vehicles organized in a headquarters section, a maintenance section, and a transport section with three squads of ten POL tankers and two trucks each.[121]

The expansion of combat and support forces in Algeria once the war began was accompanied by an impressive growth of the military petroleum distribution system throughout Algeria, with the creation of a number of both fixed and mobile operational depots, the opening of additional POL distribution centers and special depots for artillery and aviation units, and significant personnel reinforcements.[122] Petroleum importation centers were established at Oran and Arzew in the Oran Corps Area and at Bône, Bougie, and Philippeville in the Constantine Corps Area. POL resupply centers were created at Perrégaux (Oran Corps Area) and La Chiffa (Algiers Corps Area). The 709e CRDE, reorganized as a composite POL company (*Compagnie Mixte des Essences*; CME) and stationed at Camp de Lido (at Oued-el-Alleug near Algiers), was reinforced by the 701e CME at Sidi-Chami in the Oran Corps Area and by the 711th POL Resupply Company (*711e Compagnie de Ravitaillement et d'Exploitation en Essences*; 711e CREE) in the Constantine Corps Area. The 711e CREE was authorized four officers, sixteen NCOs, 172 enlisted men and fifty vehicles organized in a headquarters section, a transport section of two squads of ten tanker trucks each and one squad of nine cargo trucks, and an exploitation section of three squads.[123] In theory, the POL resupply company (CREE) was capable of transporting 10,900 U.S. gallons of POL products and of operating one principal POL depot and conditioning center and two secondary depots. The composite POL company (CME) had similar capabilities.

### POL Consumption

The consumption of all types of automotive and aviation fuels increased enormously during the first two years of the Algerian war and continued to rise thereafter. In the Constantine Corps Area POL consumption increased fifteenfold in the first year alone.[124] Consumption of fuel for ground vehicles rose from 4,544,700 U.S. gallons in 1954 to 12,523,200 in 1955 and reached 24,642,400

**Table 3.8**
**Average Monthly Consumption (in U.S. Gallons) of POL in Algeria, November 1954–October 1955**

| Date | Type Fuel | Oran Corps Area | Algiers Corps Area | Constantine Corps Area |
|---|---|---|---|---|
| Before 1 November 1954 | Ground | 129,700 | 139,600 | 43,400 |
| | Aviation | 288,200 | 128,900 | 12,700 |
| November–December 1954 | Ground | 163,000 | 305,400 | 372,000 |
| | Aviation | 434,900 | 117,000 | 44,400 |
| 1st Quarter, 1955 | Ground | 252,000 | 148,300 | 209,300 |
| | Aviation | 756,200 | 147,900 | 60,000 |
| 2nd Quarter, 1955 | Ground | 130,900 | 348,800 | 376,700 |
| | Aviation | 567,400 | 188,500 | 59,200 |
| 3rd Quarter, 1955 | Ground | 128,000 | 315,500 | 610,600 |
| | Aviation | 475,000 | 218,100 | 124,800 |

*Source*: Ministère de la Défense Nationale et des Forces Armées, Service des Essences des Armées, Direction des Essences en A.F.N., *Fiche sur le Service des Essences en A.F.N.* (Alger, 14 octobre 1955), 2, in folder "Organisation du Service des Essences, 1945–61," Dossier 1H2692 d. 1, Fonds Algérien, SHAT. Figures converted from cubic meters to U.S. gallons (1 cubic meter = 201.987 U.S. gallons) and rounded to nearest hundred gallons.

in 1956.[125] During the same period the consumption of aviation fuel rose from 3,938,700 U.S. gallons in 1954 to 9,847,400 in 1956.[126] Table 3.8 provides a detailed breakdown of the growth in consumption of both ground and aviation fuel in the first year of the war. Note that consumption of both ground and aviation fuels declined somewhat in the Oran Corps Area after the first quarter of 1955 and that of ground fuels declined in the Algiers Corps Area after the second quarter of 1955, while consumption of both types in the Constantine Corps Area continued to rise. This was due to the shifting focus of rebel activity and corresponding French counter-guerrilla operations in the first year of the war.

The significant distances to be covered in Algeria and the tempo of search-and-clear operations and later of defensive operations on the eastern and western barrages, as well as the increased use of helicopters, resulted in POL consumption at least equal to that prescribed by the standard planning factors in use at the time for a "classic" European-type war. The usual unit of measurement of POL consumption planning prescribed by 1950s French staff manuals was the Unit of Fuel (*Unité d'Essence*; UE). UE were calculated in two ways, per vehicle and per unit. With respect to a given type of vehicle the UE equaled the amount of fuel consumed traveling one hundred kilometers.[127] For example, the UE for a jeep was 5.3–9.3 U.S. gallons, for a GMC 2.5-ton truck 11.9–13.2 U.S. gallons, and for a medium tank 79.3–105.7.[128]

In an operational context the UE equaled the amount of fuel consumed by a unit in a day of active operations: 32,742 U.S. gallons for an infantry division or 63,929 for an armored division, as shown in Table 3.9. In periods of calm each type of division was expected to consume only about one-half UE per day (ca. 16,361 U.S. gallons for an infantry division or ca. 31,914 U.S. gallons for

**Table 3.9**
**Standard Units of Fuel (UE) in U.S. Gallons for Various Units**

| Type Unit | UE (US Gals) |
|---|---|
| Infantry Regiment | 2,121 |
| Light Tank Regiment | 6,666 |
| Medium Tank Regiment | 6,686 |
| Field Artillery Battalion (105 mm HM2) | 1,111 |
| Field Artillery Battalion (155 mm HM1) | 1,111 |
| Light Antiaircraft Artillery Battalion | 1,818 |
| Combat Engineer Battalion (Infantry Division) | 1,818 |
| Signal Battalion (Infantry Division) | 1,616 |
| Medical Battalion (Infantry Division) | 1,050 |
| Headquarters Company (Infantry Division) | 1,273 |
| Traffic Regulation Company (Infantry Division) | 303 |
| Transport Battalion (Infantry Division) | 1,818 |
| Intendance Section/Company (Infantry Division) | 101 |
| Maintenance Company (Infantry Division) | 1,050 |
| **Infantry Division** | **32,742** |
| **Armor Division** | **63,929** |

*Source*: France, Ministère de l'Armée de Terre, École d'État-Major, École Supérieure de Guerre et École d'État-Major, *Aide-Mémoire pour les Travaux d'État-Major*, provisional ed. (Paris: École Supérieure de Guerre et École d'État-Major, February 1952), II, ¶643. The table shows calculations of unit UE at corps level based on 10.6 U.S. gallons per wheeled vehicle, 23.8 U.S. gallons per half-tracked vehicle, 58.1 U.S. gallons per light tank, and 105.7 U.S. gallons per medium or heavy tank.

an armored division), and in exploitation operations each type of division was expected to consume 1.5 UE (ca. 28,884 U.S. gallons for an infantry division or 95,944 U.S. gallons for an armored division).[129]

### *Service des Essences* Problem Areas

The very high tempo of operations in eastern Algeria from late 1957 to mid-1960 resulted in consumption rates for POL that exceeded the available allocations, particularly in the key Constantine Corps Area. The battles on the eastern barrage and the subsequent Challe offensive created a minor fuel crisis, especially with respect to diesel fuel, in the Constantine Corps Area in the early months of 1960, at which time the available quarterly allocation covered only 2.5 months of consumption.[130] The allocation of ground fuels for the Constantine Corps Area for the first quarter of 1960 included only 5,454,000 U.S. gallons of V-70 automotive fuel, 525,200 U.S. gallons of MT-80 armored vehicle fuel, and 171,700 of diesel fuel, against expected consumption of 6,060,000 U.S.

gallons of V-70, 606,000 of MT-80, and 242,400 of diesel.[131] The fuel crisis in early 1960 was compounded by the great increase in requirements due to intensive engineer work on the barrage, pacification roads, and the oil pipeline from the Sahara, all at the same time.

Considering the high demand for scarce fuels and the usual shortages of trained personnel and specialized equipment, the *Service des Essences* was generally successful in supporting the French forces in Algeria. The existence of a civilian petroleum distribution network helped considerably, and the techniques for forward support of mobile combat forces in the field had been well worked out in Indochina. New problems, such as the increased requirement for the resupply of petroleum products to Army aviation units, were handled effectively; it seldom became necessary to reduce the operational tempo for lack of fuel, although strict conservation measures had to be enforced and allocations shifted from one area to another.

## FRENCH LOGISTICAL DEFICIENCIES

Despite its many strengths, the French logistical system in Algeria suffered from several deficiencies that lessened its effectiveness and increased its requirements for resources. While the efficiency and effectiveness of the French logistical services far surpassed anything available to the rebels, there was a continuous gap between requirements and resources arising from the rapid expansion of the pacification (*quadrillage*) program, which required substantial numbers of men and amounts of equipment; from the construction, maintenance, and operation of the barrages on the Tunisian and Moroccan borders to stifle the delivery of men, heavy weapons, and other crucial supplies to the rebels; and from the development in 1959 of active offensive operations, which brought increased consumption of rations, ammunition, POL, and transport. The solutions to these problems lay in eliciting the political decisions necessary to establish a definitive program of priorities, provide additional means, and improve the infrastructure. For the most part these decisions were obtained, but even at the end of the war in 1962 much remained to be done.

The fundamental weakness, from which most of the other deficiencies derived, was that the financial resources dedicated by the French government to the prosecution of the war in Algeria were never abundant; resources of all types were therefore perpetually in short supply. Thus constant attention was required to economy, achieving maximum productivity with minimal resources. The most striking defect of the French logistical system in Algeria occasioned by insufficient funding was the chronic shortage of logistical personnel, particularly trained technical specialists. Another major deficiency arising in part from the shortage of funds was the inadequacy of the military logistical infrastructure at the beginning of the conflict in 1954. Most storage areas and workshops were old, small, and inefficient.

Not all of the problems faced by the French supply services in Algeria were the result of insufficient funding. Poor troop supply discipline, inadequate training of draftees and reservists, and the very nature of the physical environment in Algeria also lessened the effectiveness and efficiency of French logistical support. The general dispersion of combat units over the enormous territory of Algeria, and the consequent dispersion of support units, increased logistical requirements and interfered with the full employment of available personnel and matériel resources, as well as with the judicious use of funds. The continued application until 1958–1959 of peacetime regulations little adapted to the missions assigned or to the operational environment in Algeria also served to reduce the effectiveness of logistical support.

## NOTES

1. Luc Bourgeois, *Le Matériel pendant la guerre d'Algérie* (Paris: Inspection du Matériel de l'Armée de Terre, September 1987), I, Part 1, 9.

2. The principles, organization, and procedures of the French Army supply services in Algeria are described in three main documents; 10e Région Militaire, Commandement de Région Territoriale et Corps d'Armée d'Oran, État-Major, 4e Bureau, *Instruction Provisoire sur l'organisation et le fonctionnement de la logistique sur le Territoire de la Région Territoriale et du Corps d'Armée d'Oran*, No. 100/CAO.4.CRG (Oran, 1 mars 1960), in folder "Corps d'Armée d'Oran, État-Major, 4e Bureau—Organisation logistique des forces en Algérie et au Sahara, 1959–62," Dossier 1H3215 d. 1, Fonds Algérien, SHAT [cited hereafter as *Instruction Provisoire*]; Commandement de Région Territoriale et Corps d'Armée de Constantine, État-Major, 4e Bureau, *Directive Logistique*, No. 1383/CAC.4.ETG, (Constantine, 16 octobre 1957), 1, in folder "Corps d'Armée de Constantine, État-Major, 4e Bureau—Organisation et directives logistiques, 1957–62," Dossier 1H2989 d. 1, Fonds Algérien, SHAT; and Commandement de Région Territoriale et Corps d'Armée de Constantine, État-Major, 4e Bureau, *La Logistique dans la Région Territoriale et le Corps d'Armée de Constantine—Principes d'Organisation et de Fonctionnement*, No. 973/RT-CAC/4/ETG (Constantine, 1 février 1960), 4–5, in folder "Corps d'Armée de Constantine, État-Major, 4e Bureau—Organisation et directives logistiques, 1957–62," Dossier 1H2989 d. 1, Fonds Algérien, SHAT [cited hereafter as *La Logistique*]. Although the documents named pertain specifically to either the Oran or the Constantine corps area, the principles and procedures enunciated apply generally to the other corps areas as well.

3. *La Logistique*, 4–5.

4. The TOE and Tables of Allowances (TA) for the various types of units of the French Army during the war in Algeria can be found in France, Ministère de l'Armée de Terre, École d'État-Major, École Supérieure de Guerre et École d'État-Major, *Aide-Mémoire pour les Travaux d'État-Major*, provisional ed. (Paris: École Supérieure de Guerre et École d'État-Major, February 1952) [cited hereafter as *Aide-Memoire*].

5. Commandement de Région Territoriale et Corps d'Armée de Constantine, État-Major, 4e Bureau, *Le Service de l'Intendance dans la Région Territoriale et Corps d'Armée de Constantine* (Constantine, 8 mars 1960), 1, in folder "Corps d'Armée de Constantine, État-Major, 4e Bureau—Organisation et directives logistiques, 1957–62," Dossier 1H2989 d. 1, Fonds Algérien, SHAT [cited hereafter as *Le Service de l'Intendance*].

6. The organization and functions of the SI in Algeria are described in *Instruction Provisoire*, Annexe III, 2e Partie: Service de l'Intendance; *Directive Logistique*, Title III, Chapter 2; *Le Service de l'Intendance*; and Commandement Supérieur des Forces en Algérie, Région Territoriale et Corps d'Armée d'Alger, État-Major, 4e Bureau, *Instructions sur le soutien logistique du corps d'armée d'Alger*, No. 260/CAA/4/ETG (Alger, 5 fevrier 1962), in folder "Corps d'Armée d'Alger, État-Major, 4e Bureau—Instructions sur le soutien logistique de CAA, 1957–62," Dossier 1H2762 d. 3, Fonds Algérien, SHAT [cited hereafter as *Instructions sur le soutien logistique*].

7. *Aide-Mémoire*, I, ¶332, and II, ¶633.

8. Ibid.

9. Ibid., I, ¶331, and II, ¶633.

10. Ibid., II, ¶631.

11. *Instructions sur le soutien logistique*, 13. The procedures were essentially the same in the other corps areas.

12. *Instructions sur le soutien logistique*, 12–13.

13. Ibid., 13.

14. Ibid., 12.

15. *Aide-Mémoire*, II, ¶634.

16. Ibid.

17. Ibid., ¶635. The average rail car could carry from ten to fifteen tons. The 2.5-ton Dodge or GMC truck could carry from two to five tons, depending on road conditions.

18. Ibid. The standard was one metric ton of normal rations (two metric tons of combat rations) per square meter, in 2.4-meter stacks.

19. Commandement de Région Territoriale et Corps d'Armée de Constantine, État-Major, 4e Bureau, *Fiche—Logistique du Corps d'Armée de Constantine* (Constantine, 24 decembre 1958), 4–5; *Fiche—Logistique du Corps d'Armée de Constantine*, No. 36/RT.CAC/4 (Constantine, 10 mars 1960), 3; and *Fiche—Logistique du Corps d'Armée de Constantine*, No. 26/RT.CAC/4.ETG.F (Constantine, 26 février 1962), 3, all in folder "Corps d'Armée de Constantine, État-Major, 4e Bureau—Organisation et directives logistiques, 1957–62," Dossier 1H2989 d. 1, Fonds Algérien, SHAT.

20. The principal sources for the organization and operation of the SM in Algeria include: Bourgeois, I; *Instruction Provisoire*, Annexe IV, Partie 1; 10e Région Militaire, Direction du Matériel, Section Organisation, *Synthèse sur la Campagne d'Algérie: Les Unités du Service du Matériel Soutien Logistique des Troupes Engagées dans le Maintien de l'Ordre* (Alger, 1956), in folder "10e Région Militaire, État-Major Interarmées, 3e Bureau—Rapports et synthèses des services, 1956," Dossier 1H1905 d. 2, Fonds Algérien, SHAT; and Commandement de Région Territoriale et Corps d'Armée de Constantine, État-Major, 4e Bureau, *Le Service du Matériel dans la Région Territoriale et Corps d'Armée de Constantine* (Constantine, 15 decembre 1959), in folder "Corps d'Armée de Constantine, État-Major, 4e Bureau—Organisation et directives logistiques, 1957–62," Dossier 1H2989 d. 1, Fonds Algérien, SHAT.

21. Bourgeois, I, Part 2, passim.

22. Ibid., Part 1, 13 and 32. The former DRAs ordered and stocked automotive repair parts, tools, and various controlled items, and the former DMCs received, delivered, and accepted turn-ins of vehicles and complete equipment sets.

23. Gaston Gouineau, "Trente-cinq ans d'inspection du Matériel de l'armée de Terre," *Revue Historique des Armées*, Special Issue (No. 3 for 1980) (1980), 131.

24. Bourgeois, I, Part 1, 13.

25. Ibid., 21 and 50.

26. Ibid., 16–17

27. Ibid., 67.

28. Ibid., 10 and 18. The 11e and 12e CSMs were established subsequently to reinforce the 10e CSM.

29. Ibid., 10. A very high proportion of the personnel of the CSM (64 percent) were dedicated to tank-automotive repairs.

30. *Synthèse sur la Campagne d'Algérie*, 5.

31. Bourgeois, I, Part 1, 54 (Tableau XII). Between 1954 and 1959 the number of SM personnel assigned to the 10th Military Region rose from 2,500 to 15,000, but overall 10th Military Region strength rose from 70,000 to 450,000.

32. *Le Service du Matériel*, 14; *Synthèse sur la Campagne d'Algérie*, 11–12.

33. *La Logistique*, 13–14.

34. Commandement de Région Territoriale et Corps d'Armée de Constantine, État-Major, 4e Bureau, untitled document containing notes on logistics in the Corps d'Armée de Constantine in 1959 [Constantine, 1960?], 2–3, in folder "Corps d'Armée de Constantine, État-Major, 4e Bureau—Organisation et directives logistiques, 1957–62," Dossier 1H2989 d. 1, Fonds Algérien, SHAT.

35. *Directive Logistique*, Title III, Chapter 3, page 7.

36. Bourgeois, I, Part 1, 53.

37. Ibid., 55.

38. *Instruction Provisoire*, 3. Armament stockage levels were about 2 percent of the total number of weapons in the Oran Corps Area.

39. Bourgeois, I, Part 1, 55.

40. Commandement de Région Territoriale et Corps d'Armée de Constantine, État-Major, 4e Bureau, *Problemes Logistiques Essentiels du Corps d'Armée de Constantine* (Constantine, 8 février 1960), 4–5, in folder "Corps d'Armée de Constantine, État-Major, 4e Bureau—Organisation et directives logistiques, 1957–62," Dossier 1H2989 d. 1, Fonds Algérien, SHAT.

41. United States Department of the Army, Office of the Assistant Chief of Staff, Intelligence, *The French Motor Vehicle and Ground Munitions Industry*, Intelligence Research Project No. 10,027 (Washington, DC: Office of the Assistant Chief of Staff, Intelligence, Headquarters, Department of the Army, 1 August 1957), 2. As of 31 January 1956 the United States had provided to France under the MDAP some 1,167 M-24 light tanks, 1,254 M-4 Sherman medium tanks, 9,035 jeeps, 16,763 2.5-ton 6×6 cargo trucks, 672 M2A1 105 mm howitzers, 232,499 M-1 .30 caliber rifles, 981 75 mm recoilless rifles, and 10,008 3.5-in. M-20 rocket launchers, among other items (see *The French Motor Vehicle and Ground Munitions Industry*, 31 and 49–50).

42. *Le Service du Matériel*, 9.

43. Bourgeois, I, Part 1, 52 (Tableau XI).

44. *Fiche—Logistique du Corps d'Armée de Constantine* (Constantine, 24 decembre 1958), 4; *Fiche—Logistique du Corps d'Armée de Constantine*, No. 36/RT.CAC/4 (Constantine, 10 mars 1960), 5; *Fiche—Logistique du Corps d'Armée de Constantine*, No. 26/RT.CAC/4.ETG.F (Constantine, 26 février 1962), 2; Bourgeois, 52; *Synthèse sur la Campagne d'Algérie*, 6; Commandement de Région Territoriale et Corps d'Armée de Constantine, État-Major, 4e Bureau, *Fiche—Logistique du Corps d'Armée de Constantine (draft)*, No. 49/RT.CAC/4/ETG/F (Constantine, 16 juin 1961), 2–3, in folder "Corps

d'Armée de Constantine, État-Major, 4e Bureau—Organisation et directives logistiques, 1957–62,'' Dossier 1H2989 d. 1, Fonds Algérien, SHAT.

45. First-echelon maintenance involved inspection, cleaning, and minor adjustments carried out by the equipment operator. Second-echelon maintenance was performed at the unit level and involved recovery, evacuation, inspection, troubleshooting, and some replacement of parts and assemblies. Third-echelon maintenance was carried out by the direct support units and involved recovery, evacuation, inspection, replacement of major parts and assemblies, and the repair of some assemblies. Third-echelon maintenance was carried out in fixed maintenance facilities or by mobile maintenance teams; items submitted for third-echelon maintenance were generally returned to the using unit after they were repaired. Fourth-echelon maintenance involved the systematic repair and rebuilding of equipment in general support of the third-echelon maintenance units and was generally carried out in fixed facilities by heavy maintenance units; items repaired or rebuilt at the fourth-echelon were returned to general stocks rather than to the original using unit. Fifth-echelon maintenance, also carried out by designated heavy maintenance units in fixed facilities, involved the complete rebuilding of entire vehicles and the renovation of major assemblies (such as motors or transmissions) for return to general stocks.

46. Bourgeois, I, Part 1, 58. The rebuilding of equipment accounted for about one-fifth of the total fifth-echelon maintenance workload in Algeria. In March 1958, for example, of 1,485 vehicles requiring rebuilding, only 210 were rebuilt in Algeria; the remainder were evacuated to France. The most important rebuild line was at Oran, where some 1,650 GMC 2.5-ton trucks were rebuilt in 1959.

47. Bourgeois, I, Part 1, 57; *La Logistique*, 6; Commandement de Région Territoriale et Corps d'Armée de Constantine, État-Major, 4e Bureau, *Fiche—Logistique du Corps d'Armée de Constantine* (Constantine, 24 decembre 1958), 3; *Fiche—Logistique du Corps d'Armée de Constantine*, No. 26/RT.CAC/4.ETG.F (Constantine, 26 février 1962), 2.

48. Bourgeois, I, Part 1, 56.

49. 10e Région Militaire, État-Major, 4e Bureau, *Dossier 4e Bureau, 24-5-56* (Alger, 24 mai 1956), Dossier VIII—''Organisation et Fonctionnement des Services: Matériel—'Fonctionnement' (AI/CQ, 10RM/EM/4e, Alger, 24 mai 1956),'' page b, in folder ''Aide-Major Général en AFN—Organisation du commandement et coordination des services et moyens logistiques, 1946–57,'' Dossier 1H2665 d. 1, Fonds Algérien, SHAT [cited hereafter as *Dossier 4e Bureau, 24-5-56*].

50. Bourgeois, I, Part 1, 54 (Tableau XII).

51. *Le Service du Matériel*, 9; *Fiche—Logistique du Corps d'Armée de Constantine*, No. 36/RT.CAC/4 (Constantine, 10 mars 1960), 5; Commandement de Région Territoriale et Corps d'Armée de Constantine, État-Major, 4e Bureau, *Fiche—Documentation statistique relative au 4e Bureau* (Constantine, avril 1959), 5, in folder ''Corps d'Armée de Constantine, État-Major, 4e Bureau—Organisation et directives logistiques, 1957–62,'' Dossier 1H2989 d. 1, Fonds Algérien, SHAT.

52. *Fiche—Logistique du Corps d'Armée de Constantine*, No. 26/RT.CAC/4.ETG.F (Constantine, 26 février 1962), 2.

53. *Le Service du Matériel*, 7–8; Bourgeois, I, Part 1, 53.

54. Cited by Bourgeois, I, Part 1, 56. Bourgeois notes that the Foreign Legion and cavalry units were exceptions to the rule of poor operator and unit maintenance practices.

55. *Le Service du Matériel*, 2. Each of the four divisions assigned to the 10th Military

Region General Reserve had its own DIRMAT and a supporting divisional maintenance unit that was usually attached to the DSM of the corps area in which the reserve division was based.

56. Bourgeois, I, Part 1, 10; *Aide-Mémoire*, I, ¶¶311–312.

57. *Synthèse sur la Campagne d'Algérie*, 8.

58. Bourgeois, I, Part 1, 10; *Aide-Mémoire*, I, ¶¶311–313. The standard infantry CRD was authorized ten officers, forty-five NCOs, and 233 enlisted men. The standard armored division maintenance company was considerably larger, with an authorization of thirty officers, 127 NCOs, and 567 enlisted men. The two airborne divisions in Algeria had a maintenance company of the infantry division type in lieu of the standard airborne division's airborne equipment maintenance company (*Compagnie de Réparation de Matériel Aéroportée*), with seven officers, thirty-one NCOs, and 154 enlisted men, and the parachute packing and maintenance company (*Compagnie d'Entretien, de Pliage, et de Réparation de Parachutes*), with three officers, twenty-seven NCOs, and 234 enlisted men.

59. *Synthèse sur la Campagne d'Algérie*, 2.

60. *Le Service du Matériel*, 3; untitled document containing notes on logistics in the Constantine Corps Area in 1959 [Constantine, 1960?], 4.

61. Bourgeois, I, Part 1, 61. Such items were either repaired within twenty-four hours or replaced on the spot. The 64e CRD (of the 14th Infantry Division in the North Constantine Zone) repaired some thirty thousand weapons and twenty thousand pieces of other equipment between November 1958 and November 1960—a very mobile contact team on the eastern barrage, covering some forty thousand kilometers in only two months.

62. Ibid., 10.

63. Ibid. The formula of one mobile armament-optics repair section and four-five mobile tank-automotive repair sections for each divisional maintenance company was generally adopted in the 1959 reorganization of divisions in metropolitan France as being better suited to both nuclear and counter-guerrilla warfare scenarios.

64. Ibid., 54.

65. *Synthèse sur la Campagne d'Algérie*, 3. The previous workload standard was six hundred vehicles per section, and the actual workload in the Constantine Corps Area in 1959, for example, was about 550 vehicles per section (see *Le Service du Matériel*, 5). The concept was also applied to general support units.

66. *Le Service du Matériel*, 14; *Synthèse sur la Campagne d'Algérie*, 3–4; Bourgeois, I, Part 1, 10. As a consequence, the general support maintenance units in Algeria generally performed the same functions as the divisional direct support maintenance companies, but they were a bit more specialized. Normally general support maintenance units supported a depot workshop or reinforced the divisional maintenance companies in their area.

67. *Aide-Mémoire*, I, ¶314.

68. Ibid.

69. *Le Service du Matériel*, 6.

70. *Aide-Mémoire*, I, ¶314.

71. Ibid.

72. Bourgeois, I, Part 1, 10, n. 11.

73. Helicopter unit organization, operations, and maintenance are discussed in ibid., 10–11, 20–21, and 62–66; Vertol Aircraft Corporation (T. R. Pierpont, Director, Euro-

pean Operations), *French Army Helicopter Operations in Algeria, June 1956–September 1959*, Report No. SM-406 (Morton, PA: Vertol Aircraft Corporation, 1 November 1959); and United States Air Force, Air University, Aerospace Studies Institute, Concepts Division, *Guerrilla Warfare and Airpower in Algeria, 1954–1960*, Project No, AU-411-62-ASI (Maxwell AFB, AL: Concepts Division, Aerospace Studies Institute, Air University, March 1965).

74. Bourgeois, I, Part 1, 62.

75. France, National Defense Committee for Scientific Action, Operations Research Group, *Report of the Operations Research Mission on H-21 Helicopter Operations in Algeria* (Morton, PA: Vertol Aircraft Corporation, April 1957).

76. *French Army Helicopter Operations*, 6 (Figure 1); *Guerrilla Warfare and Airpower in Algeria*, 55. The characteristics of the various helicopters used in Algeria are listed in Appendix B.

77. Bourgeois, I, Part 1, 10; Bernard B. Fall, *Street without Joy*, 3rd revised ed. (Harrisburg, PA: Stackpole, 1963), 261.

78. *French Army Helicopter Operations*, 5–9 passim.

79. Ibid., 8 (Figure 2).

80. Bourgeois, I, Part 1, 21.

81. Ibid., 20; *French Army Helicopter Operations*, 11. The function of the service squadrons—one for each major type of helicopter in use (H-21, H-19, and Allouette)—was to perform seventy-five-hour and other major periodic inspections.

82. *French Army Helicopter Operations*, 8 (Figure 2); Bourgeois, I, Part 1, 63.

83. Bourgeois, I, Part 1, 21 and 63.

84. *French Army Helicopter Operations*, 8 (Figure 2).

85. Bourgeois, I, Part 1, 63.

86. *French Army Helicopter Operations*, 25. The figures include combat damage repairs.

87. Ibid., 90.

88. Ibid., 1 and 91–92.

89. The organization and operation of the ammunition service in Algeria are described in Bourgeois, I, Part 1, 66–69.

90. *Aide-Mémoire*, I, ¶314.

91. Bourgeois, I, Part 1, 69.

92. Ibid., 68; *Directive Logistique*, Title II, chap. 3, 9.

93. *Dossier 4e Bureau, 24-5-56*, Dossier IV—"Service du Matériel: Munitions—Importance des Stocks à Constituer (Alger, 24 mai 1956)."

94. *Le Service du Matériel*, 3.

95. Ammunition stockage levels in Algeria are discussed in Bourgeois, I, Part 1, 66–67.

96. Bourgeois, I, Part 1, 69.

97. Ibid., 68.

98. Ibid.

99. Ibid., 69. The administrative paperwork involved was enormous. The *Service du Matériel* in Algeria was required to submit annually some seventy-three status reports, the preparation of which required about three hundred man-days of work.

100. *Aide-Mémoire*, II, ¶614. The standard tranches were: A—small-arms cartridges up to 12.7 mm; B—explosive infantry munitions (grenades, antitank rockets, mortar shells); C—tank and antitank gun munitions; D—artillery munitions; E—antiaircraft ar-

tillery munitions; F—mines; G—pyrotechnics (flares, colored smoke grenades, etc.); H—explosives (TNT, C-4, etc.); and I—flamethrower fuel.

101. *Aide-Mémoire*, II. ¶614. In most cases the DOS was equal to from one-tenth of the standard UF for most ammunition of Tranche A to one-fifth of the standard UF for other tranches, except hand and rifle grenades (included in Tranche B), for which the DOS was equal to the UF.

102. Ibid., ¶451.

103. Aide-Major Général en Afrique du Nord, *Charte: "Evolution Consummation Mensuelle Munitions"* (Alger, n.d.), in folder "Aide-Major Général en AFN—Organisation du commandement et coordination des services et moyens logistiques, 1946–57," Dossier 1H2665 d. 1, Fonds Algérien, SHAT.

104. Bourgeois, I, Part 1, 66.

105. Ibid., n. 91.

106. Ibid., 69.

107. *French Motor Vehicle and Ground Munitions Industry*, 36.

108. Ibid., 50 (Table XIV: "FRANCE—US Weapons and Ammunition Delivered to France under the MDAP [to February 1956]").

109. *Fiche—Logistique du Corps d'Armée de Constantine*, No. 36/RT.CAC/4 (Constantine, 10 mars 1960), 6.

110. *Fiche—Documentation statistique relative au 4e Bureau* (Constantine, avril 1959), 5; *Fiche—Logistique du Corps d'Armée de Constantine (draft)*, No. 49/RT.CAC/4/ETG/F (Constantine, 16 juin 1961), 4.

111. *Fiche—Logistique du Corps d'Armée de Constantine*, No. 26/RT.CAC/4.ETG.F (Constantine, 26 février 1962), 3; *Problems Logistiques Essentiels du Corps d'Armée de Constantine*, 5.

112. *Directive Logistique*, Title III, chap. 3, 7; *Le Service du Matériel*, 2; untitled document containing notes on logistics in the Corps d'Armée de Constantine in 1959 [Constantine, 1960?], 3.

113. *Synthèse sur la Campagne d'Algérie*, 9.

114. Bourgeois, I, Part 1, 12. Although the dispersion of maintenance units facilitated direct support of combat units, it made control of the units themselves difficult, because the commanders of such units had no radio equipment.

115. Bourgeois, I, Part 1, 12, n. 12.

116. *Synthèse sur la Campagne d'Algérie*, 3–5 passim.

117. Bourgeois, I, Part 1, 22.

118. *Directive Logistique*, Title III, chap. 3, 7–8.

119. *Synthèse sur la Campagne d'Algérie*, 9.

120. Ministère de la Défense Nationale et des Forces Armées, Service des Essences des Armées, Direction des Essences en A.F.N., *Fiche sur le Service des Essences en A.F.N.* (Algiers, 14 octobre 1955), 1, in folder "Organisation du Service des Essences, 1945–61," Dossier 1H2692 d. 1, Fonds Algérien, SHAT.

121. *Aide-Mémoire*, I, ¶341.

122. *Fiche sur le Service des Essences en A.F.N.*, 1.

123. *Aide-Mémoire*, I, ¶341, and II, 642.

124. *Fiche sur le Service des Essences en A.F.N.*, 2.

125. *Dossier 4e Bureau, 24-5-56*, Dossier IX—"Services Communs: Chart—'Consommations de Carburants en M3—Comparaison Terre-Air' (Alger, 24 mai 1956)." In monetary terms the consumption of all types of fuels increased from 2,738,123,000 FF

in 1954 to 4,878,219,000 FF in 1955 and over 8.5 billion FF in 1956 (see *Dossier 4e Bureau, 24-5-56*, Dossier IX—''Services Communs: Chart—'Dépensés de Carburants en Deniers' (Alger, 24 mai 1956).''

126. Ibid., ''Services Communs: Chart—'Consommations de Carburants en M3—Comparaison Terre-Air' (Alger, 24 mai 1956).''

127. *Aide-Mémoire*, II, ¶641.

128. Ibid., ¶643. Figures converted from liters (1 liter = .2642 U.S. gallons) and rounded to the nearest tenth.

129. *Aide-Mémoire*, II, ¶642.

130. *Fiche—Logistique du Corps d'Armée de Constantine*, No. 36/RT.CAC/4 (Constantine, 10 mars 1960), 2.

131. *Problems Logistiques Essentiels du Corps d'Armée de Constantine*, 4.

# 4

# French Logistics in Algeria: Transportation

The disparity between the French forces in Algeria and their rebel opponents was nowhere more striking than in those essential to logistical movements and the mobility of combat forces. Controlling the existing transportation infrastructure of Algeria, the French soon reinforced—and proceeded to use with maximum effectiveness—their superior ground, sea, and air transport. Special benefits were derived from the existence of French air forces, which provided both close air support and the air transport needed to overcome the vast Algerian distances, as well as the aerial observation and attack capabilities required to hamper rebel mobility. For the first time helicopters were used extensively for the movement of cargo and the transport of combat troops, thus adding another whole dimension to the war.

The problems encountered in providing adequate transportation services to the French forces were the usual ones of a large area of operations, difficult terrain and climate, widely dispersed forces, an inadequate logistical infrastructure, a high operational tempo, and too few units, vehicles, and trained personnel to meet requirements. To overcome these problems, particularly the weaknesses of the rail net and the limitations of motor transport, the French commanders in Algeria turned to centralized control and decentralized execution, combined with careful economy of the means available. The railroad was used whenever possible, in order to preserve motor transport for operational movements, and air transport was used only for urgent requirements and the regular resupply of isolated posts.[1] On the whole, the French forces in Algeria enjoyed superior transportation support, which gave them a decided advantage over their Algerian nationalist opponents, who lacked all but the most basic transport.

## GENERAL ORGANIZATION OF TRANSPORT IN ALGERIA

Ground and limited inland water transport, movement control, and traffic regulation in Algeria were the responsibility of the Train of the French Army. The Train in Algeria underwent massive reinforcement and tremendous organizational change in the first three years of the Algerian war, in response to the increased demand for transportation services. On 1 November 1954 the Train elements present in Algeria were limited to a regional support company in Algiers to provide administrative support for the 10th Military Region; three squadrons (at Oran, Algiers, and Constantine); a motor transport group at Beni-Messous; three Saharan motor transport companies; one traffic control company; and one aerial delivery company.[2] Reinforcement and reorganization proceeded rapidly after 1 November 1954, and by the end of 1955 one-fourth of all Train formations in the French Army were in Algeria.

The management of all movements and command of Army ground and water transport units of the Train in Algeria were the responsibility of the Commandant of the Train and Director of Transport in Algeria (*Commandant du Train et Directeur des Transports en Algérie*; CTDT) of the 10th Military Region, commonly known as the Director of Transport-Algeria. That title was assumed by the former Regional Director of Transports (*Directeur Régional des Transports*) of the 10th Military Region in 1957, when the functions of Train commander and Regional Director of Transport were combined. He was assisted by a staff composed of a command section, a courier bureau, and three staff sections: First Section (organization and personnel management); Second Section (plans, operations, security, studies, regulations, and documentation); and Third Section (matériel). The commander in chief assigned to the Director of Transport-Algeria the general missions of planning, organizing, executing, supervising, and coordinating all surface movements and transport for all French air, ground, and naval forces in Algeria.[3] The Director of Transport-Algeria was the principal transport advisor to the commander in chief and acted in coordination with the commander of the 5th Air Region as the regional air regulator for air transport and aerial resupply operations in Algeria and the Sahara. He was also responsible for coordinating water transport operations with the commander of French naval forces in Algeria (PRÉMAR IV) and for ensuring the mobile defense of all surface lines of communications. Furthermore, the Director of Transport-Algeria was charged with coordinating all military movements and transport requirements with his civilian counterparts in the Algerian railway organization (*Société Nationale des Chemins de Fer Algérien*; SNCFA), the Department of Public Works and Transport (*Direction des Travaux Publiques et des Transports*; known at lower echelons as *Ponts et Chaussées*), and the commercial trucking industry in Algeria (*Interfédération des Transports Routières*).[4] Similar liaison was maintained at corps area and zone levels.

The size of the operational area in Algeria, the necessary dispersion of forces, and the need to decentralize control of support operations required the concentration of technical responsibility for transport operations, and in December 1956 the existing Direction of Transport annexes and the Train command headquarters in each of the three military territorial divisions were consolidated.[5] With the creation of the corps areas on 1 March 1957 the recently combined transport commands were redesignated as Train Commands/Transport Directions (*Commandement du Train et Direction des Transports*; CTDT).[6] The officer in charge was known as the Commandant of the Train/Director of Transport (*Commandant du Train et Directeur des Transports*; CTDT). The functions of the Commandants of the Train/Director of Transport at corps area level were similar to the functions performed by the Director of Transport-Algeria: to command units of the Train and to prepare, organize, execute, supervise, and coordinate all movements and transport operations in their areas of interest in coordination with the 4e Bureau of the corps area staff.[7] The corps area CTDTs were also responsible for the security of ground lines of communication and acted as the Principal Air Regulator (*Principal Régulateur Aérien*) and Military Commissioner of the Algerian railway subdivision (*arrondissement d'exploitation SNCFA*) in their area.

Decentralization was later extended to the echelons subordinate to the corps area, and a CTDT was created for each operational zone and its corresponding tactical division. The Commandant of the Train and Director of Transport at zone level performed the same functions as the Commandant of the Train/Director of Transport at higher levels.[8] Figure 4.1 depicts the resulting territorial organization of the Train in Algeria.

To provide central control and coordination of all transportation requirements and services a Bureau of Movements and Transport (*Bureau de Mouvements et Transports*; BMT) was created in the staff of the commander in chief to establish priorities, promulgate regulations, and coordinate operational and logistical movements and transport requirements.[9] BMTs were also established in the staff of commanders at lower echelons (Corps Area and zone). At all levels the BMT was not an organic part of the commander's staff but was rather an ad hoc organization composed of an officer representative from the 4e Bureau (in principle the chief of the Second Section), the 3e Bureau, and the CTDT. The essential distinction between the CTDT and the BMT was that the former was an operating agency, while the latter was a staff coordinating agency. At sector level a single officer of the Train filled the role of the BMT for the sector commander.

## OPERATIONAL UNITS OF THE TRAIN

To each of the Train Commands/Transport Directions, down to zone level, was assigned a number of operational ground transport units of the Train. These units were of various types and included training, headquarters support, traffic control, motor transport, pack mule, and aerial resupply units as well as Train units organized and employed as infantry. Table 4.1 shows the Train units by type in Algeria in January 1960.

Figure 4.1
Train Headquarters in Algeria, January 1960

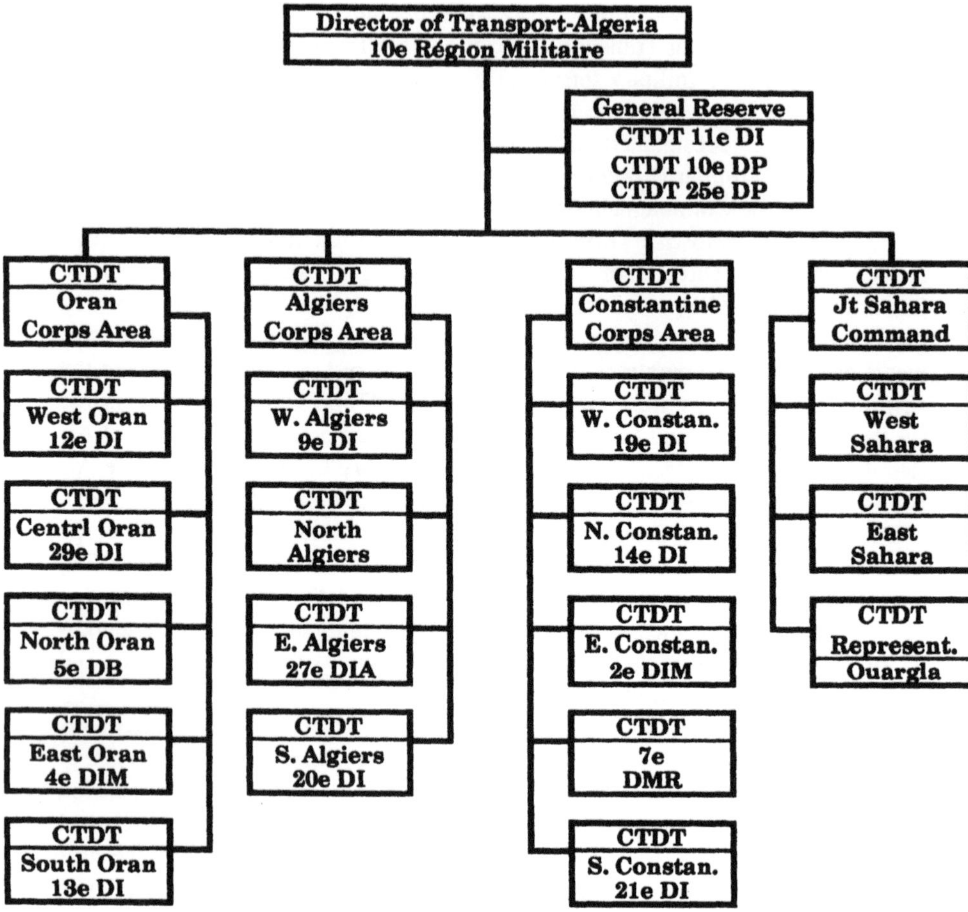


**Key:**

| | |
|---|---|
| CTDT | Train Command/Commandant—Transport Direction/Director |
| DB | Division Blindée (Armored Division) |
| DI | Division d'Infanterie (Infantry Division) |
| DIA | Division d'Infanterie Alpine (Mountain Infantry Division) |
| DIM | Division d'Infanterie Motorisée (Motorized Division) |
| DMR | Division Mécanique Rapide (Mechanized Infantry Division) |
| DP | Division Parachutiste (Parachute Division) |

*Source*: Commandement en Chef des Forces en Algérie, Commandement du Train et Direction des Transports, *L'Arme du Train Algérie-Sahara*, No. 94/CTDT.AL/1/ORG, (Alger, janvier 1960), Charte ("Les CTDT en Algérie-Sahara, 1-1-60."), in folder "10e Région Militaire, État-Major Interarmées, 3e Bureau, Commandement du Train et Direction des Transports—Le Train en Algérie et au Sahara, 1960–61," Dossier 1H1917 d. 6, Fonds Algérien, SHAT.

**Table 4.1**
**Units of the Train in Algeria, January 1960**

| Type | 10th Military Region Reserve | | Oran Corps Area | | Algiers Corps Area | | Constantine Corps Area | | Joint Sahara Command | | |
|---|---|---|---|---|---|---|---|---|---|---|---|
| Unit | Non-Div'l | Divisional | Non-Div'l | Divisional | Non-Div'l | Divisional | Non-Div'l | Divisional | ZOS | ZES | Total |
| CTDT | 1 | 1 | 1 | 5 | 2 | 3 | 1 | 5 | 2 | 1 | 22 |
| CIT | 1 | | | | | | | | | | 1 |
| CRT | 1 | | | | | | | | | | 1 |
| CTB | 1 | | | | | | | | | | 1 |
| CQG | 1 | 3 | | 5 | 1 | 3 | 3 | 5 | 2 | 1 | 21 |
| ET | | | 1 | | 1 | | 1 | | | | 3 |
| CCS | | | | | | | 2 | | | | 2 |
| CSQP | | | | | | | 1 | | | | 1 |
| CM | | | 1 | | 1 | | 1 | | | | 3 |
| CA | | | 2 | | 2 | | 2 | | | | 6 |
| CCR | 1 | | 1 | 5 | 1 | 3 | 1 | 5 | | 1 | 18 |
| GT | 2 | 3 | 2 | 4 | 2 | 3 | 3 | 3 | | | 22 |
| CT | | | 2 | 1 | 2 | | 2 | 2 | | | 9 |
| GST | | | | | | | | | 1 | 1 | 2 |
| CAST | | | | | | | | | | 1 | 1 |
| Cie Mule | | | 1 | | | | | | | | 1 |
| Ptn Mule | | | | 5 | 1 | 3 | 2 | | | | 11 |
| CLA | 2 | | | | | | | | | | 2 |
| BT | | | 1 | | 5 | | | | | | 6 |

*NB*: See Glossary for type unit abbreviations.

*Source*: Commandement en Chef des Forces en Algérie, Commandement du Train et Direction des Transports, *L'Arme du Train Algérie-Sahara*, No. 94/CTDT.AL/1/ORG. (Alger, janvier 1960), Table 21 (''Tableau Recapitulatif des Formations du Train en Algérie-Sahara''), in folder ''10e Région Militaire, État-Major Interarmées, 3e Bureau, Commandement du Train et Direction des Transports—Le Train en Algérie et au Sahara, 1960–61,'' Dossier 1H1917 d. 6, Fonds Algérien, SHAT.

The Director of Transport-Algeria retained a number of operational units as a regional reserve (*Réserve Régionale*), but the bulk of the Train units were integral to the combat divisions that occupied and administered the operational zones. In January 1960 the units constituting the regional reserve included the 160th Transport Training Center (*Centre d'Instruction du Train 160*; CIT 160), the 210th Traffic Control Company (*210e Compagnie de Circulation Routière*; 210e CCR), the 2nd and 3rd Aerial Delivery companies (*2e and 3e Compagnies de Livraison par Air*; 2e and 3e CLA), the 312th Transfer Company (*312e Compagnie de Transbordement*; 312e CTB), and the 514th and 520th Motor Transport Groups (*514e and 520e Groupements de Transport*; 514e and 520e GT). The 160th Transport Training Center, located at Camp Basset in Beni-Messous, eight kilometers south of Algiers, occupied a post of some hundred hectares and was charged with the *quadrillage* of a subsector that provided a training ground for new recruits. The 312th Transfer Company was a water transport unit stationed in Arzew, and the 514th and 520th Transport groups were used to reinforce the motor transport assets of the corps areas. In addition, the Director of Transport-Algeria directly controlled the 110th Headquarters Support Company (*110e Compagnie de Quartier-Générale*; 110e GQG) and the 10th Regional Transport Company (*10e Compagnie Régionale de Transport*; 10e CRT), both of which supported the 10th Military Region headquarters in Algiers. A number of smaller detachments completed the list of units of the regional reserve.

In each of the corps areas the territorial transportation reserves were grouped under control of a Train squadron (*Escadron du Train*; ET), which provided corps area–wide transportation services and, when necessary, reinforced the traffic control and motor transport units assigned to the operational zones and combat divisions. The 28e ET was assigned to the Oran Corps Area; the 27e ET to the Algiers Corps Area; and the 25e ET to the Constantine Corps Area. There was no Train squadron assigned to the Joint Sahara Command, transport units in the Sahara region being directly controlled by the two zone headquarters (West Sahara [ZOS] and East Sahara [ZES]). Operating under the direct control of the corps area CTDT, each of the corps area's ETs was organized somewhat differently, but in general each included a headquarters support company, a traffic control company, a pack mule element, a composite transport company (*Compagnie Mixte*; CM), two transportation car companies (*Compagnie Auto*; CA), and two or more motor transport groups.

Among the more unusual Train organizations in Algeria were the Train infantry battalions (*Bataillons de Marche du Train*; BT).[10] The need for additional combat formations to carry out the defensive *quadrillage* program in Algeria and the lack of sufficient numbers of infantry officers and NCOs led in March 1956 to the creation of the first of the Train infantry battalions. The 504th Train Infantry Battalion, created from the 57th Rifle Battalion (*57e Bataillon de Tirailleurs*), was joined in November 1956 by five additional BTs (the 584e, 585e, 586e, 587e, and 588e). By the end of 1956, the Train infantry battalions in

Algeria included 140 officers, 660 NCOs, and 4,100 enlisted men, or about 8 percent of the total strength of the Train in the French Army.[11] The 588e BT was disbanded in December 1958 and the 586e BT in September 1961. The remaining four battalions, plus the 519e BT, which arrived in Algeria from Morocco in January 1958, operated in Algeria until the end of the war; they were subsequently disbanded in metropolitan France between April and August 1962.[12]

The Train infantry battalions—led by officers and NCOs drawn from the Train, many of whom had combat experience in Indochina, and filled with draftees who had received infantry training in France or at CIT 160 in Beni-Messou—were assigned to the Train squadrons for administrative control. In January 1960 the 28th Train Squadron in Oran administered the 519e BT, and the 27th Train Squadron in Algiers administered the 504e, 584e, 585e, 586e, and 587e BTs. There were no BTs assigned to the Constantine Corps Area. Operational control of the Train infantry battalions was exercised by the tactical commander to whose area they were attached for operations.

Organized as infantry units under TOE *Inf 107*, each BT had a complement of around seven hundred men organized in a headquarters and service company and four 150-man combat companies.[13] The BTs differed somewhat in organization. For example, the 519e BT had six combat companies, one of which—KIMONON 22—was organized as a *commando de chasse*. Most Train infantry battalions were reinforced, as were the regular infantry units, with armored vehicles, Moslem *harkis* (some mounted), dog teams, heavy weapons, and other supporting personnel and equipment. Scattered in the various zones and sectors, primarily in the Algiers Corps Area, and often occupying isolated posts (*bordjs*) resembling the forts of the American Old West, the Train infantry battalions performed all of the normal pacification functions of infantry units in Algeria: defense of facilities, patrols, search-and-clear operations, the pursuit and destruction of rebel units, and civic action operations designed to pacify the Moslem population and retain their support of the French government of Algeria. The isolation of many of the Train infantry battalions made their logistical support difficult; several had to be supplied by air, especially during the winter months.

Some of the Train infantry battalions, notably the 519e and 585e BTs, amassed combat records equal to the best French infantry units; one French general officer offered the opinion that with adequate helicopter support a Train infantry battalion was equal to a battalion of "paras."[14] Concrete proof of their participation in active operations is provided by the casualties incurred by the Train infantry battalions in Algeria: 244 killed and over four hundred wounded in action.[15] The 587e BT, assigned the dangerous specialized mission of providing protection for railroads in the Algiers Corps Area, deployed some five hundred men per day to operate armored trains and *draisines* (railroad-sabotage detection devices) and to escort passenger and freight trains. In six years the 587e BT lost forty-four officers and men killed in action.[16] Although the creation

**Table 4.2**
**Personnel Status of the Train in Algeria, January 1960**

| Grade | Status | Total Authorized | Total Actual | North Africans (included in Total) Authorized | North Africans (included in Total) Actual |
|---|---|---|---|---|---|
| Officer | Regular | 567 | 550 | None | |
| | Reserve | 248 | 362 | | |
| | **Total** | **815** | **912** | | |
| NCO | Regular | 2,261 | 2,290 | 224 | 186 |
| | Reserve | 1,333 | 1,496 | 98 | 20 |
| | **Total** | **3,594** | **3,786** | **322** | **206** |
| Enlisted | Regular | 3,528 | 1,782 | 1,368 | 1,017 |
| | Reserve | 20,373 | 22,919 | 2,511 | 2,389 |
| | **Total** | **23,901** | **24,701** | **3,879** | **3,406** |
| **Total Train Personnel** | | **28,310** | **29,399** | **4,201** | **3,612** |

*Source*: Commandement en Chef des Forces en Algérie, Commandement du Train et Direction des Transports, *L'Arme du Train Algérie-Sahara*, No. 94/CTDT.AL/1/ORG. (Alger, janvier 1960), 8 (Charte 31: "Situation des effectifs du Train en Algérie-Sahara"), in folder "10e Région Militaire, État-Major Interarmées, 3e Bureau, Commandement du Train et Direction des Transports—Le Train en Algérie et au Sahara, 1960–61," Dossier 1H1917 d. 6, Fonds Algérien, SHAT.

of the Train infantry battalions was a necessary response to the operational situation in Algeria and the units performed with a high degree of morale and effectiveness, the diversion of experienced Train officers and NCOs had a negative effect on the efficiency of normal transportation operations in Algeria, by depriving the transportation staffs and operating units of their capable leadership.

## STRENGTH OF THE TRAIN IN ALGERIA

In the early years of the Algerian war the Train, like the other logistical services, suffered from a severe shortage of qualified personnel and from the fact that personnel authorizations for most Train units were based on peacetime rather than wartime allowances. However, as the war progressed the recall of reservists, the application of the greater part of each year's draftees to the Algerian war, and recruitment among the loyal Moslem population of Algeria brought some amelioration of the personnel situation for the Train in Algeria, as shown in Table 4.2.

## WATER, RAIL, AND ANIMAL TRANSPORT OPERATIONS

Transoceanic and coastal water transport in support of the French forces in Algeria was provided by French and foreign-flag steamship lines and, in extraordinary circumstances, by French naval vessels. The availability of ocean and coastal shipping was never a problem, and the ports of Algeria were well developed, although delays were experienced in the discharge of such bulky items as aircraft, tanks, and wheeled vehicles, which required special equipment or handling.

The lack of navigable inland waterways made inland water transport both impractical and unnecessary. Limited inland and coastal water movement support for the French combat forces was provided by the 312th Transfer Company (312e CTB), stationed at Fort de la Pointe in Arzew and controlled by the 28th Train Squadron. The 312e CTB was authorized five officers, thirty-four NCOs, 155 enlisted men, thirteen vehicles, eight landing craft, mechanized (LCM), and twenty LVT-4 tracked amphibians organized in a headquarters and service platoon, a platoon of LCMs, a platoon of LVT-4s, and a salvage team.[17] The principal vehicles—LCMs and the LVT-4—had a speed, range, and capacity, respectively, of eight knots, 128 miles, and sixty troops or sixty tons of cargo, and four knots, forty-fifty miles, and twenty-five troops or 2.5 tons.[18]

Logistical transport (the carriage of supplies as opposed to tactical troop movements) in Algeria was accomplished primarily by rail. The Algerian railway system (*Société Nationale des Chemins de Fer Algérien*; SNCFA) was limited, both with respect to modern right-of-way and to locomotives and rolling stock, but in 1959 over forty-five thousand tons of cargo per month plus a large number of troop movements were being carried by the SNCFA.[19] An average of 5,500 rail cars were required for the movement of matériel each month, and between 1 January and 1 November 1959 some eighty-five special trains were dispatched. Although efforts to extend and improve the 4,390 kilometers of SNCFA right-of-way and to acquire additional motive power and rolling stock continued throughout the war, the road found it more and more difficult to meet its military and civilian requirements, in part because of increased demand but also due to the fact that rebel actions further reduced the rolling stock, lowered the speed and daily circulation, and all but eliminated night operations on certain sections of line.[20] Various active and passive defensive measures to protect vulnerable rail lines and facilities were adopted, with varying degrees of success, but even the near elimination of rebel bands operating inside Algeria by mid-1960 did not relieve the pressure altogether, and greater use had to be made of civilian air and motor transport for logistical movements.[21] Security of rail lines and railway movements within their area of operations was a responsibility of the various tactical (zone and division) commanders. It was accomplished by patrols, armed escorts, overflights by light aircraft and helicopters, improved radio networks, and various technical means (*draisines* and other sweeping mechanisms).

In general, the use of animals for transport was rare in the Algerian war, but French mountain infantry units were supported by a small number of pack mules. As of 1 November 1959 the 128th Pack Mule Company (*128e Compagnie Muletière*; 128e Cie Mule) was attached to the 28th Train Squadron in the Oran Corps Area, one pack mule platoon (*Peloton Muletière*) was attached to the 27th Train Squadron in the Algiers Corps Area, and a detachment of two platoons was attached to the 25th Train Squadron in the Constantine Corps

**Table 4.3**
**Operational vs. Logistical Motor Transport Movements, July–September 1959**

| | % Operational | % Logistical |
|---|---|---|
| Train Units, 10th Military Region Reserve | 41 | 59 |
| Train Units, Oran Corps Area | 54 | 46 |
| Train Units, Algiers Corps Area | 65 | 35 |
| Train Units, Constantine Corps Area | 71 | 29 |
| Train Units, Joint Sahara Command | 14 | 86 |

*Source*: Commandement en Chef des Forces en Algérie, Commandement du Train et Direction des Transports, *L'Arme du Train Algérie-Sahara*, No. 94/CTDT.AL/1/ORG. (Alger, janvier 1960), 26, in folder "10e Région Militaire, État-Major Interarmées, 3e Bureau, Commandement du Train et Direction des Transports—Le Train en Algérie et au Sahara, 1960–61," Dossier 1H1917 d. 6, Fonds Algérien, SHAT.

Area.[22] Five divisions in the Oran Corps Area and three in the Algiers Corps Area also had one platoon of pack mules each. A mule could carry eighty kilograms eighty kilometers in one day, but the French experience in Indochina had demonstrated that mule trains were extremely vulnerable and themselves required significant support (forage, veterinary service, etc.). That experience had led to the conclusion that the movement by pack train of heavy weapons was impractical and the suggestion that mule trains ought to be replaced by aerial resupply. The greater availability of the helicopter in Algeria significantly reduced the need for pack mule units. Although combat troops were frequently obliged to carry their supplies and equipment on their backs, porters were never used extensively in Algeria, as they had been in Indochina.

## MOTOR TRANSPORT OPERATIONS

The bulk of French motor transport assets in Algeria were concentrated in the combat divisions that administered the various operational zones and sectors. These divisional assets were largely dedicated to operational (tactical troop) movements and the necessary housekeeping. Logistical (bulk supply) movements were effected by rail, civilian trucking firms, and the motor transport groups that formed the 10th Military Region and corps area reserves.[23] The proportions of logistical to operational movements in the third quarter of 1959 are shown in Table 4.3.

In January 1960 the total military motor transport assets available in 10th Military Region (outlined in Table 4.2) amounted to three train squadrons (one per corps area), twenty-two motor transport groups, thirteen independent transport companies, two Saharan transport groups, and one Saharan transport company, for a combined total theoretical lift of 12,880 tons. The 10th Military Region regional reserve consisted of one independent company and two transport groups equipped with ten-ton Willème tractors, ten-ton semitrailers, and Simca five-ton 4×2 trucks and accounted for 1,580 tons of lift. The motor transport assigned to the corps areas and combat divisions included three Train squadrons, twenty transport groups, and twelve independent transport companies equipped for the most part with GMC or Simca 4×4 trucks; it accounted for another 9,300 tons of lift. The remaining two thousand tons was provided by the two transport groups and one independent transport company in the Sahara equipped with Berliet GBO-15 6×4, GBC 6×6 diesel, and GLR-8 diesel trucks.

On a day-to-day basis the available military motor transport was augmented by the assets of the civilian trucking companies operating in Algeria, particularly for logistical movements on the routes south into the Sahara. Two civilian trucking companies operating under contract had exclusive rights to operate in the Sahara; they provided for the movement of POL and the delivery of urgent and fragile cargo from Algiers and Colomb-Béchar. Civilian tanker trucks were also used for the movement of wine for the *Service de l'Intendance*. Under emergency conditions the military authorities in Algeria could also activate the so-called "Park of National Interest" (*Parc d'Intérêt National*; PIN). The PIN consisted of vehicles, drivers, and supervisory personnel levied from the civil departments and private industry under previously negotiated agreements, the personnel being Army reservists. The PIN was organized in eight groups (three in the Oran Corps Area, three in Algiers Corps Area, and two in the Constantine Corps Area), each composed of a captain, six lieutenants, eighteen NCOs, and sixty-six enlisted men ready for operations on forty-eight hours' notice and capable of moving around four hundred tons per day, for a combined augmentation of 3,200 tons. In January 1960 the PIN included some 506 buses (21,867 troops seats), 585 flatbed trucks (5,945 tons capacity), 656 flatbed trailers (8,256 tons capacity), and a number of tanker trucks and trailers with a capacity of over 1.3 million U.S. gallons.[24]

The key organization providing traffic control, motor transport, and other transportation services to the combat units in the field in Algeria was the Divisional Train, which in most cases consisted of a Train headquarters, a headquarters support company (CQG) to provide services to the parent division's headquarters, a traffic control company (CCR), and one or more motor transport groups (GT). The organization, personnel authorizations, and vehicle allowances of the standard infantry division Train are shown in Figure 4.2. The allowances for armored division Trains and parachute division Trains were somewhat lower.

The traffic control company (CCR) of the divisional Train provided traffic regulation services in the division's area of operations, including route control in coordination with the Gendarmerie, infrastructure security, night ambushes, so-called "Rat Patrols" (motorized patrols, usually in jeeps armed with machine guns), and escort and guide services.[25] The personnel authorizations of the original TOE, *TRN 984*, under which the traffic control companies in Algeria were organized included five officers, thirty-six NCOs, 139 enlisted men, thirty-four motorcycles, twenty-one jeeps, and sixteen trucks, but this was insufficient for the operational environment, and authorizations were subsequently increased.[26]

**Figure 4.2**
**Organization of an Infantry Division Train**

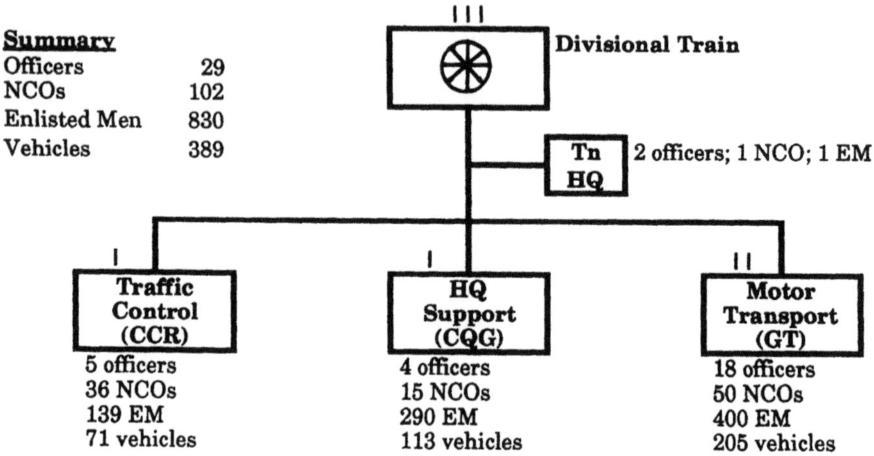


*Source*: France, Ministère de l'Armée de Terre, École d'État-Major, École Supérieure de Guerre et École d'État-Major, *Aide-Mémoire pour les Travaux d'État-Major*, provisional ed. (Paris: École Supérieure de Guerre et École d'État-Major, February 1952), I, ¶¶122b and 271.

Efforts were also made to reequip the CCRs in Algeria with additional radio equipment and armored scout cars or other protected vehicles in lieu of motorcycles, jeeps, and trucks.[27] Convoy discipline was never very good among the French forces in Algeria, and the personnel of the CCRs and the Gendarmerie experienced some difficulty in performing their mission due to the ill will of drivers and vehicle commanders.[28]

The workhorse of the French motor transport forces in Algeria was the motor transport group (GT). Motor transport groups were to be found as part of the regional and corps area reserves. One or more motor transport groups were assigned to each of the combat divisions, except for the 4th Motorized Infantry Division (4e DIM) in the Oran Corps Area and the 2nd Motorized Infantry Division (2e DIM) and 7th Light Mechanized Infantry Division (7e DMR) in the Constantine Corps Area, which had a separate motor transport company (CT) as part of their organic service battalion. The motor transport groups in Algeria were organized under TOE *TRN 019*, which was designed for a peacetime garrison setting and did not provide sufficient personnel for operations in a combat environment.[29] The organization and authorizations of the standard infantry division motor transport group are shown in Figure 4.3.

A variety of types of tractors, semitrailers, and trucks were in use by the French forces in Algeria. About 80 percent of the vehicles at first were older, well-worn American models; these were gradually replaced by newer French models as the war progressed. The U.S. Dodge or GMC 2.5-ton 6×6 truck was capable of carrying twenty-five men or up to five tons of cargo on good roads,

**Figure 4.3**
**Organization of a Divisional Motor Transport Group**

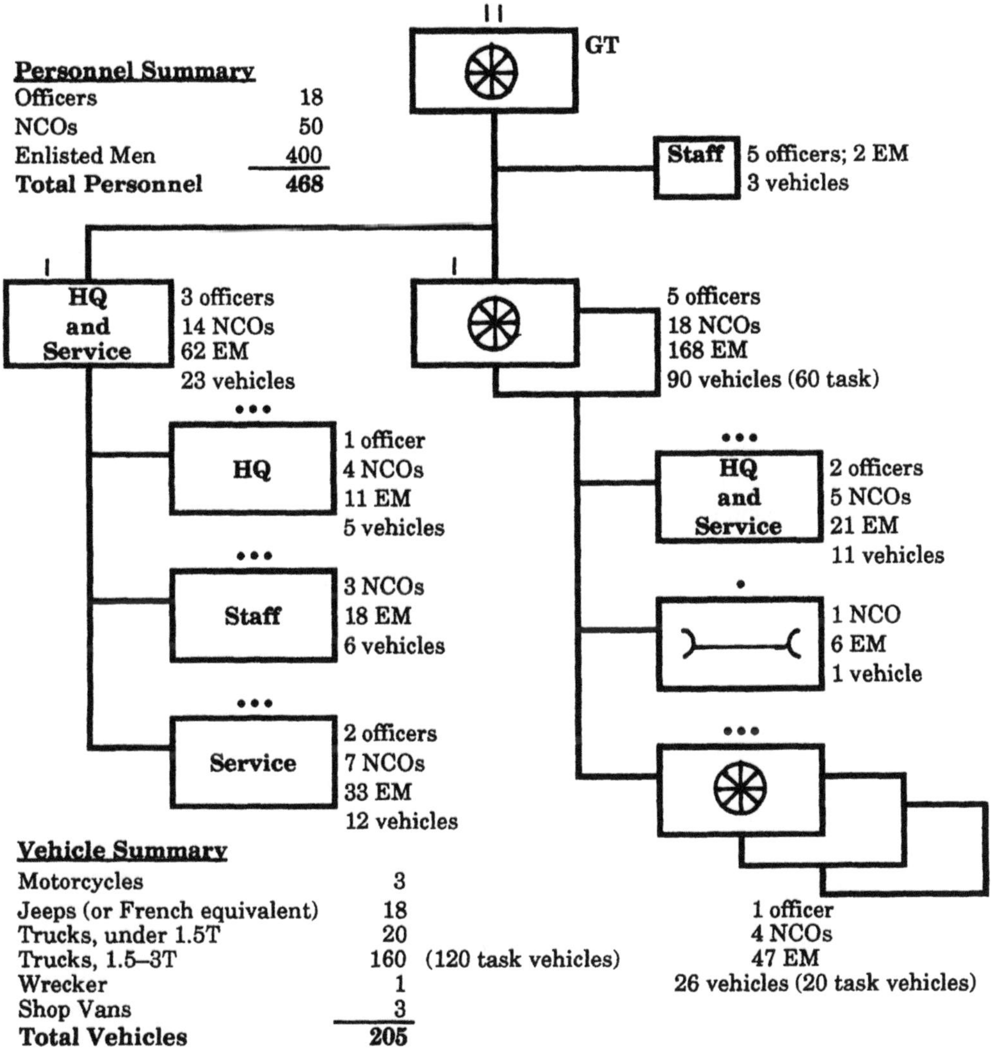


*Source*: France, Ministère de l'Armée de Terre, École d'État-Major, École Supérieure de Guerre et École d'État-Major, *Aide-Mémoire pour les Travaux d'État-Major*, provisional ed. (Paris: École Supérieure de Guerre et École d'État-Major, February 1952), I, ¶274.

although in Algeria the normal load was closer to two tons. The newer French vehicles in use by 1960 generally had a larger capacity. For example, the Berliet GBO-15 could haul about sixteen tons, the Berliet GLR-8 diesel about eight tons.[30]

A number of factors served to reduce the efficiency and thus the overall capacity of the motor transport units available in Algeria. Difficult terrain limited cargo loads, there were shortages of authorized vehicles, vehicles were lost temporarily to maintenance, and the units were dispersed; all of this adversely affected general performance. Shortages of authorized vehicles and the rate of nonavailability due to maintenance were particularly vexing. For example, on 23 January 1960 the availability rate in the 25th Train Squadron in the Constantine Corps Area ranged from 68 percent for the 101e CT to 91 percent for the 102e CT.[31]

Most of the motor transport groups assigned to the corps areas and combat divisions were employed piecemeal, which reduced their efficiency and increased the difficulties of maintenance and administration. For example, the 515e GT had its headquarters at Méchéria, one company at Géryville, and one company at Colomb-Béchar, 640 kilometers from Géryville.[32] The two companies were divided into some twelve detachments of from one to thirty-two vehicles each and spread along some 1,600 kilometers of trails from Géryville to Tindouf. Similarly, the 543e GT had its headquarters and one company at Djelfa and the other company at Touggourt some 530 kilometers away, and the 512e GT was divided into thirteen detachments ranging in size from one vehicle to twenty.[33]

The 510th Colonial Transport Group (*510e Groupement Coloniale de Transport*; 510e GCT) assigned to the 21st Infantry Division (21e DI) at Batna in the Constantine South Zone was typical of the divisional transport groups in Algeria during the earlier part of the war.[34] The 510e GCT arrived at Bône from metropolitan France on 8 November 1954, one week after the start of the war, and assembled at Batna, in what was to become the Constantine South Zone, in June 1955. The 510e GCT was assigned four primary missions: motorization of infantry for operations; provision of permanent detachments of vehicles to the sectors of Constantine South Zone; transportation of supplies (logistical transport); and short-haul housekeeping transport (*brouettage*) for the garrisons and other services in its area of operations. Like the other transport groups in Algeria in 1956, 510e GCT was undermanned as a result of a peacetime TOE that did not provide sufficient personnel to perform all of the unit's wartime tasks. Moreover, the group was unable to fill all of its authorized positions, even though 25 percent of its strength was composed of Moslem Algerians. Authorized a total strength of 349 officers, NCOs, and men, on 1 November 1956 the 510e GCT had only 315 personnel assigned, being short five NCOs and thirty-one men but having two extra officers. In January 1956 the group maintained a daily average availability of eighty-six trucks out of 120 theoretically available. The proportion rose to a hundred out of 120 in August and September 1956. The

**Table 4.4**
**510th Colonial Transport Group Operational Statistics, March and July 1956**

| Type Operation | Passengers and Tons | Daily Sorties | % of Total Workload |
|---|---|---|---|
| | **March 1956** | | |
| Infantry Support | 11,063 | 86 | 3.5 |
| Permanent Detachments | passengers | 1,550 | 62.0 |
| Logistical Transport | 3,042 | 628 | 25.2 |
| House-Keeping | tons of cargo | 231 | 9.3 |
| | **July 1956** | | |
| Infantry Support | 22,349 | 834 | 26.3 |
| Permanent Detachments | passengers | 1,852 | 58.5 |
| Logistical Transport | 1,861 | 206 | 6.5 |
| House-Keeping | tons of cargo | 275 | 8.7 |

*Source*: 10e Région Militaire, Division de Constantine, Groupe de Transport de Réserve Générale du Train 510, *Rapport sur la campagne d'Algérie au GTRGT 510*, No. 221/GT510/S (Batna, 20 novembre 1956), 1, in folder "10e Région Militaire, État-Major Interarmées, 3e Bureau—Rapports et synthèses des armes, 1956," Dossier 1H1905 d. 1, Fonds Algérien, SHAT.

operational statistics of 510e GCT for the months of March and July 1956 are shown in Table 4.4.

## AIR TRANSPORT OPERATIONS

Air transport, both fixed-wing and helicopter, played a key role in the Algerian war.[35] The size of the Algerian theater of operations, the difficulty of the terrain, and the poor state of the ground transport infrastructure, as well as the rapid, fluid nature of rebel operations meant that only decentralized air transport could make possible the speedy movement of troops and timely resupply of isolated outposts and tactical units needed to find, fix, fight, and finish the Algerian nationalist rebels. Air transport, including parachute operations, had been used extensively in World War II, Korea, and French Indochina, but helicopter transport was used regularly and extensively for the first time in Algeria, and it added an entirely new dimension to the battlefield. The ability to move men and matériel rapidly by air to any point in the theater of operations gave the French a decisive advantage over the Algerian rebels, who lacked any such capability and who moreover found their own logistical and tactical movements subject to French aerial observation and interdiction.

### Organization of French Air Forces in Algeria

Like the French ground forces, the French air forces in Algeria quickly expanded and reorganized to meet the new requirements of the war in Algeria. In 1954 the 5th Air Region (*5e Région Aérienne*; 5e RA) consisted of only one fighter wing, one transport squadron, a liaison squadron, and several support aircraft operating from only three main airfields. It soon became apparent that centrally controlled airpower could not meet the needs of a fluid counter-guerrilla war, and a more flexible, decentralized system with substantial air forces controlled at lower operational levels emerged. By 1 January 1959 over thirty new operational airfields had been developed, radio and logistics support had been improved, and the 5th Air Region's assets had increased to 35,284 officers and men and over eight hundred aircraft, including 159 jet and propeller fighters, twenty-two B-26 light bombers, sixty-one transports, seventy-three reconnaissance aircraft, fifty-nine liaison aircraft, 306 light aircraft (Broussards and T-6 Texans), and 122 helicopters. It was augmented by the Army's 120 light fixed-wing aircraft and a hundred helicopters and the Navy's ten Neptune surveillance aircraft and thirty-six helicopters.[36] The airfields at Oran, Algiers, and Blida were considerably improved, and three new main bases were developed at Boufarik, Bône, and the old U.S. World War II base at Telergma, near Constantine. In the eastern Sahara alone there were some two hundred fields suitable for light aircraft by the end of 1959.[37]

The 5th Air Region was also reorganized on a decentralized, territorial basis. Decentralized control provided increased flexibility and decreased reaction time and ensured full integration of air and ground operations. The 5th Air Region retained centralized control of air defense and transport elements, but the bulk of the air combat elements were assigned to one of the five tactical air groups (*Groupement Aérien Tactique*; GATAC) linked to the Army's territorial commands, as shown in Figure 2.4. Each GATAC was further subdivided into advanced air commands (AAC), linked to the Army's operational zones and divisions. The GATAC commander acted as the air advisor to the corps area commander and was responsible for coordinating the action of all air units in his area of operations; the AAC commander performed the same functions at zone/division level.

Most of the available Air Force aircraft were assigned to the GATACs and then allocated down to the AAC to give each zone and division its own dedicated reconnaissance and close air support capability, usually one flight of liaison aircraft and one flight of T-6s for light close air support.[38] The 5th Air Region "leased" most of its jet fighters and a few transports to the GATACs, which retained the heavier firepower for use as dictated by the operational situation. The GATAC and AAC assets could be temporarily augmented by aircraft from higher levels or provided on an emergency basis from adjacent GATACs as required by the operational situation. The time required to obtain augmentation from adjacent GATACs varied from fifteen minutes to several hours, depending on flight times.[39]

The management of operational air forces in Algeria was a joint activity, involving personnel and equipment from all three services. By 1959 a system was in place for the continuous coordination of planned, on-call, and quick-reaction air support through joint operations centers that linked the Army territorial commands (corps area and zone/division) with their counterpart Air

**Figure 4.4**
**Integration of Air-Ground Operations in Algeria**

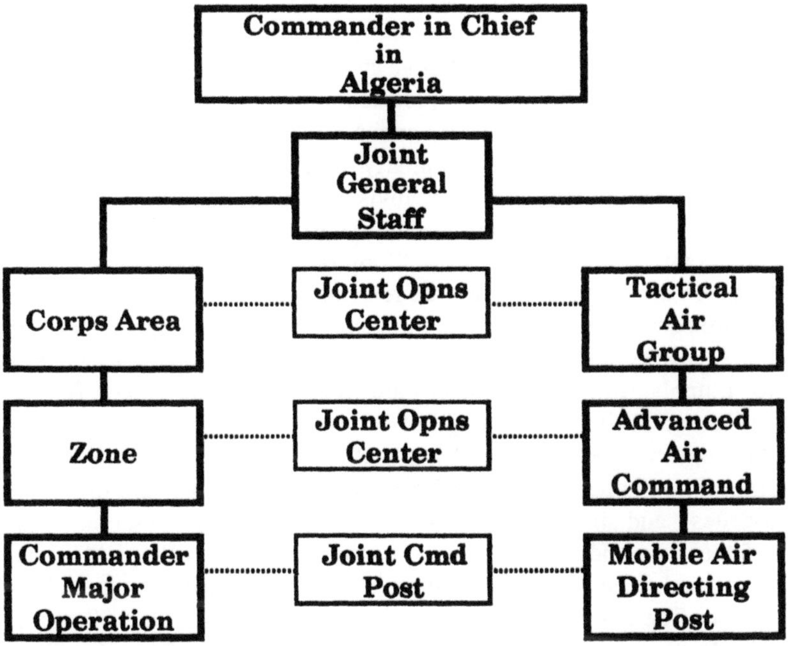


*Source*: France, Ministère des Armées, *Notice on the Air Support in Operations for the Maintenance of Order in Algeria [Manual T.T.A. 172: Notice sur l'appui aérien dans les operations de maintien de l'ordre en Algérie]* (trans. Office of the Assistant Chief of Staff for Intelligence, Headquarters, Department of the Army; Washington, DC: Headquarters, Department of the Army, Office of the Assistant Chief of Staff for Intelligence, 9 December 1960), 4.

Force components (GATAC, AAC), as shown in Figure 4.4. The Army and Air Force headquarters at each level were normally located near each other. The fixed joint operations centers and mobile joint command posts were austere, consisting normally of a single room manned by a few essential operations, intelligence, and communications personnel drawn from both the Army and the Air Force. Mobile joint command posts included truck-mounted and airborne (on board liaison aircraft or helicopters) facilities. Ground combat units in the field were normally accompanied by an air directing post (ADP) to coordinate immediate close air support. Provisions were also in place for the coordination of air-ground operations on an ad hoc basis for major joint tactical operations of limited duration. Naval air forces and Army light aviation assets were integrated into the system as necessary, usually by being placed under the operational control of the Air Force element.

The organization and operation of GATAC 2 in early 1959 in was typical.[40] The commander of GATAC 2 was responsible for the area bounded on the west by the Moroccan border, on the north by the Mediterranean Sea, on the south by the Saharan Atlas mountains, and on the east by a north-south line sixty miles west of Algiers. While most administration was handled by 5th Air Region headquarters, GATAC 2 was responsible for its own operational logistics, and its headquarters, including administrative and maintenance elements, was located at Oran. The joint operations center, located at the air base at La Senia outside Oran, consisted of thirty-eight Army and Air Force personnel supported by an Army photo-interpretation unit. Similarly organized joint operations centers were provided for the six Army zone/divisions in the GATAC 2 area of operations. GATAC 2 assets included some six to eight thousand personnel and 275–300 aircraft. Assigned and "leased" aircraft included one squadron of B-26 light bombers and three to five RB-26 reconnaissance aircraft from the 5th Air Region; two squadrons (fifty aircraft) of Vampire Mistral jet fighters; one squadron of P-47 fighter-bombers; two Noratlas transports from the 5th Air Region, used to resupply Army posts and as flare ships; six flights of T-6 Texan light observation-attack aircraft; one flight of Broussard light observation planes; two Navy helicopters squadrons (ten H-21s each) from the 4th Maritime Prefecture; one wing of Air Force helicopters (two squadrons of H-34s and one squadron of Allouettes); and six flights of Army L-18 and L-19 light observation planes. Most of the T-6s and L-18/19s were reassigned down to the AACs, and the Broussards and helicopters were almost always on "lease" to the AACs and lower levels. The main airfields were located at La Senia (Oran), Tlemcen, and Thiersville with secondary fields at Mostaganem, Tiaret, Aflou, Géryville, Ain-Sefra, Sidi-bel-Abbès, Saida, Méchéria, and El Aricha plus six or seven unoccupied fields and more than eighty helicopter refueling points.

## Fixed-Wing Air Transport

Fixed-wing air transport support for the French forces in Algeria was provided by the Military Air Transport Command (*Groupement des Moyens Militaires Transports Aériens*; GMMTA) based in metropolitan France; the Military Air Transport Sub-Group (*Sous-Groupement des Moyens Militaires Transports Aériens*; S/GMMTA) assigned to the 5th Air Region in Algiers; the GATACs using their assigned and "leased" assets; and civilian airlines under contract to the French government.[41] All types of air transport missions were flown, including tactical troop movement, parachute assaults, air evacuation of casualties, and both routine and emergency aerial resupply of isolated units. Frequently, air transport missions were combined with air reconnaissance and casualty evacuation. The Military Air Transport Command and Air France maintained regularly scheduled and special mission runs between metropolitan France and major bases in Algeria. Using its three squadrons of Noratlas 2501 transports, the 5th Air Region's own Military Air Transport Sub-Group operated twenty-seven scheduled routes serving thirty-four airfields, and the Société Civil "Aérotechnique" (AEROTEC), using DC-3s, operated under contract eight lines serving eighteen airfields, with a weekly capability of nineteen tons.[42] Special, urgent missions were flown by Air Force or Army light aviation as required.

The fixed-wing air transport assigned in Algeria in 1959 amounted to forty-eight Noratlas transports, twenty-three C-47 Dakotas, twenty old Ju-52 Toucans, and thirty-five MD-315 Flamants.[43] Additional capability was provided by the thirty old twin-engine Martinets and sixty-four Broussards—each of which could carry about 1,600 pounds of cargo—assigned to the GATACs.[44] The forty-eight Noratlas transports were grouped in two transport squadrons based at Algiers and one squadron based at Blida, thirty miles southwest. Each squadron was composed of sixteen aircraft and twenty crews, with a total of about four hundred men.[45] The three transport squadrons carried out planned missions as scheduled and maintained about half their aircraft on one-hour alert for emergency missions. French forces in the Sahara were supported by GATAC 4 at Colomb-Béchar (western Sahara) and GATAC 5 at Ouargla (eastern Sahara). Each of the Sahara tactical air groups was supported by a so-called Saharan reconnaissance and transport squadron (*Groupement Reconnaissance Saharienne et Transport*; GRST). The GRST assigned to GATAC 5 in the eastern Sahara in early 1959 had around fifty aircraft, including several old Ju-52s; twelve C-47s; fourteen Broussards; six Dassault MD-310 twin-engine all-purpose aircraft; and about sixteen Army light observation planes (L-18/19s), based at Ouargla, Touggourt, and Laghouat.[46]

Air transport management in Algeria was centralized at 10th Military Region level. Requests for both planned air transport and special air missions generated at any level were passed up through the Train Commandants/Transport Directors of the corps areas, who acted as principal regulators (*Régulateurs Principals*) to the Second Section of the Direction of Transport-Algeria in Algiers, which in turn acted as the regional air regulator (*Régulateur Aérien Régional*). Requests were received by the Second Section, coordinated with the 4e Bureau of the Joint General Staff, and passed to 5th Air Region, where the Air Transport Coordination Office (*Bureau d'Exploitation-Coordination Transports Aériens*; BECTA) approved the requests and passed them to the air transport units for execution.[47] Army liaison detachments at the various airfields furnished Air Force crews with current information on the tactical situation, including enemy activity and the location and radio call signs of friendly troops. The missions were then flown, with enroute diversions or cancellations as necessary.

In the third quarter of 1959 alone military air transport in Algeria carried some 22,100 passengers and 1,550 tons of cargo, while the civilian contractor (AEROTEC) moved another 4,500 passengers and 130 tons of freight.[48] Overall, in 1959 the Military Air Transport Command of metropolitan France flew an average of two thousand hours per month in support of French forces in Algeria, and the Military Air Transport Sub-Group of the 5th Air Region flew some thirty-five thousand hours to transport forty-eight thousand passengers, eight thousand tons of cargo, and eight thousand air evacuation patients.[49] Aircraft assigned to the various GATACs in 1959 flew some 29,400 hours, performing

1,150 paradrop missions, 5,000 air evacuation missions, and 7,500 passenger and cargo missions.[50]

## Aerial Resupply

Initially only one aerial delivery company—the 2nd Aerial Delivery Company (*2e Compagnie de Livraison par Air*; 2e CLA)—was available in Algeria, and it was retained as part of the general reserve directly under the control of the Director of Transport-Algeria acting on behalf of the commander in chief. The headquarters, headquarters platoon, and parachute packing platoon of the 2e CLA were based at the French airborne base at Blida, and the aerial delivery platoons were detached to support the three corps areas and French forces in the Sahara, with one platoon at Blida in support of the Algiers Corps Area; one platoon at La Senia, ten kilometers south of Oran, in support of the Oran Corps Area; one platoon at Telergma, forty kilometers southwest of Constantine, to support the Constantine Corps Area; and one platoon at Colomb-Béchar supporting French forces in the Sahara.[51] By January 1960 a second company, the 3e CLA, had been added at Blida to provide support for heavy drops.

The standard French Army aerial resupply company (*Compagnie de Ravitaillement par Air*; CRA), which had been redesignated as the aerial delivery company (CLA) by the time of the Algerian war, was authorized nine officers, forty-seven NCOs, 366 enlisted men, and some eighty vehicles organized in a company headquarters platoon, a security platoon, a parachute maintenance and packing platoon, a cargo preparation and packing platoon, and three aerial delivery platoons.[52] The organization of the two CLA in Algeria was somewhat different. The 2e CLA had four aerial delivery platoons, and the 3e CLA was organized with a headquarters platoon, one packing platoon, and two aerial delivery platoons.

Aerial resupply was reserved for emergency resupply of forces in the field and the routine resupply of isolated posts and units on the move; stockpiles of ammunition, rations, fuel, and medical supplies were maintained at Blida for emergency resupply operations.[53] For the most part the responsibility for resupplying isolated posts was delegated to the commandant of the Train/Director of Transport in each corps area and in the Joint Sahara Command. In February 1959 the aerial resupply tonnage allocated to the Constantine Corps Area amounted to from ten to twenty tons per day, and that allocated to the Algiers and Oran corps areas accounted for ten tons each.[54] In January 1960 the total aerial resupply tonnage allocation for all of Algeria was 910 tons: 260 tons as an operational reserve retained by 10th Military Region; 300 tons allocated to the Constantine Corps Area; 170 tons to the Algiers Corps Area; and 180 tons to the Oran Corps Area.[55] As late as January 1961, nineteen out of thirty posts in the Constantine Corps Area were being resupplied by air, with some twenty-five tons of supplies per month.[56]

**Helicopter Operations**

The bulk of tactical troop movements and air evacuations as well as a good deal of logistical transport in Algeria was provided by Army, Air Force, and Navy helicopters. In 1955 there were only four Army helicopters in Algeria; by 1959 the French Army in Algeria was operating some 140 helicopters (sixty-four Vertol H-21s; nine Sikorsky H-19s; thirty-eight Bell H-13s; and twenty-nine Allouettes) and 120 fixed-wing aircraft.[57] At the same time the French Air Force had two wings, based at Oran and Boufarik, with 122 helicopters (some eighty H-34s and twenty-five Allouettes plus H-13s and H-19s), and the French Navy had three squadrons (Flotilla 31) with thirty-six helicopters (twenty H-21s plus H-5s, H-19s, and H-34s), operationally controlled by the Air Force in western Algeria.

Most of the Army fixed-wing aircraft and helicopters employed in Algeria were assigned to the 101st Army Light Aviation Group (*Groupement d'Aviation Légère de l'Armée de Terre 101*; GALAT 101) based at Ain-Arnat near Sétif. GALAT 101 was responsible for all helicopter operations in Algeria east of 3° east longitude (about sixty miles west of Algiers). Most of the helicopters controlled by GALAT 101 were assigned to Helicopter Group 2 (*Groupe d'Hélicoptères No. 2*; GH 2) at Ain-Arnat. In 1959 GH 2 controlled some ninety-eight helicopters in five helicopter squadrons and three service squadrons, as shown in Figure 4.5. There were another thirty-five helicopters (thirty-three Bell H-13s and two H-21s) plus eighty-two fixed-wing aircraft in the nine divisional aviation squadrons in eastern Algeria and the GH 2 liaison squadron, all of them also controlled by GALAT 101.[58] Although the Air Force was responsible for all air operations, including troop support helicopter operations, for all forces west of Algiers, a few Army aircraft (about thirty fixed-wing and twenty helicopters in 1959) were assigned to GALAT 102 at Sidi-bel-Abbès in western Algeria and GALAT 105 and GALAT 114 in the Algiers Corps Area; all of these served primarily as training units and administrative and maintenance support centers for Army aircraft assigned to divisional aviation units.[59]

Normally, most helicopters assigned to GH 2 were employed in small groups of seven or eight aircraft and crews with supporting maintenance personnel in what were known as helicopter intervention detachments (*Détachements d'Intervention Hélicoptère*; DIH), attached to combat formations for extended operations.[60] The usual DIH was composed of seven H-21s and one Allouette from one of the GH 2 helicopter squadrons plus an additional Allouette from the Allouette squadron for the use of the tactical unit commander. In combat one H-21 was usually held back as a reserve against combat losses. The DIHs were entirely mobile, including their maintenance personnel and equipment, and were designed to live and operate in the field. The DIHs could be, and often were, shifted between eastern and western Algeria (i.e., between the area controlled by the Air Force and that controlled by the Army).[61]

The principal missions performed by helicopters in Algeria included tactical

**Figure 4.5**
**Organization of the 101st Army Light Aviation Group**

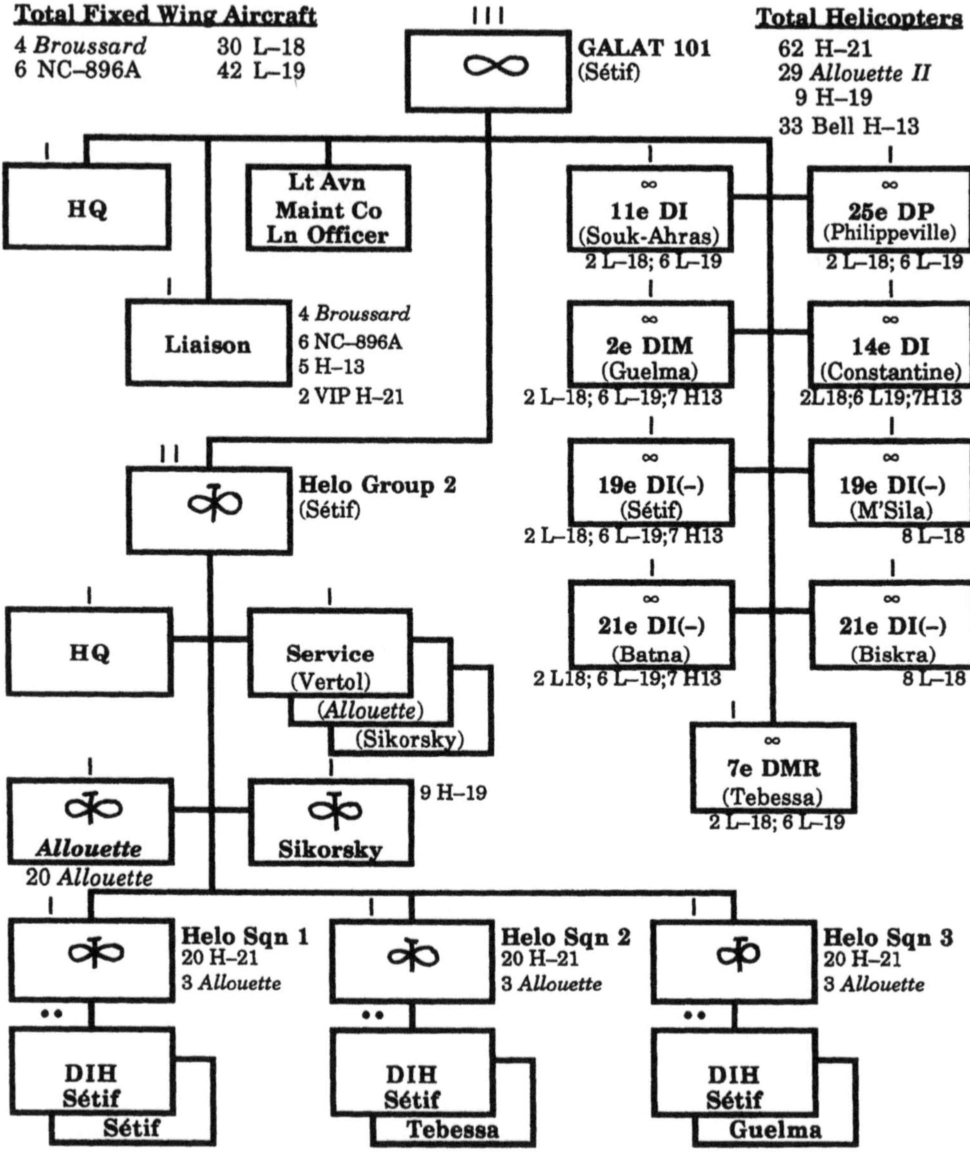


*Source*: Vertol Aircraft Corporation (T. R. Pierpont, Director, European Operations), *French Army Helicopter Operations in Algeria, June 1956–September 1959*, Report No. SM-406 (Morton, PA: Vertol Aircraft Corporation, 1 November 1959), 6 (Figure 1). All data as of 1 October 1959.

**Table 4.5**
**Army H-21 and Air Force H-34 Helicopter Operations, August 1957–July 1958**

| Operational Factor | Type Aircraft | Annual Total | Monthly Average |
|---|---|---|---|
| Average Number of Aircraft Available | Air Force H–34 | – | 40.06 |
| | Army H–21 | – | 31.50 |
| Hours Flown | Air Force H–34 | 12,620 | 1,051 |
| | Army H–21 | 11,014 | 918 |
| Operational Hours | Air Force H–34 | 9,507 | 792 |
| | Army H–21 | 9,575 | 798 |
| Assault Troops Transported | Air Force H–34 | 85,475 | 7,123 |
| | Army H–21 | 134,867 | 11,239 |
| Other Passengers Transported | Air Force H–34 | 8,434 | 703 |
| | Army H–21 | 4,477 | 373 |
| Air Evacuation Patients Transported | Air Force H–34 | 1,995 | 166 |
| | Army H–21 | 1,463 | 121 |

*Source*: Vertol Aircraft Corporation (T. R. Pierpont, Director, European Operations), *French Army Helicopter Operations in Algeria, June 1956–September 1959*, Report No. SM-406 (Morton, PA: Vertol Aircraft Corporation, 1 November 1959), Appendix A.

troop movements, air evacuation, and logistical (resupply) movements. Other missions included reconnaissance, convoy escort, radio relay, leaflet drops, and artillery fire direction. In general, the priority mission for transport helicopters was the movement of combat troops, including helicopter-borne assaults, but helicopters also played a key role in supporting isolated posts and troops on the move in rough country. Air evacuation was performed mainly by the light Bell H-13 helicopters equipped with litters (as seen on the American television program *M*A*S*H*), but on occasion the medium and heavy transport helicopters (H-21s and H-34s) were used.[62]

The development of the armed helicopter was a signal feature of the war in Algeria. Normally the armed helicopters were used to protect flights of troop-carrying helicopters from ground fire in the immediate area of the landing zone, but they were sometimes used to provide close support for ground combat elements during initial deployment, for shows of force, or to cover intelligence operations. Both the Air Force and the Army considered the ideal ratio to be one armed helicopter for every five cargo helicopters.[63]

The operational statistics accumulated by helicopter units in Algeria were impressive. Table 4.5 outlines Army H-21 and Air Force H-34 operations in Algeria for the year ending 31 July 1958. Between June 1956 and September 1959, GALAT 101 helicopters accumulated some 37,000 combat flying hours and transported 393,479 combat troops (an average of nine miles per lift) and 95,600 other passengers (an average of seventy-five miles per lift), 12,820 casualties (an average of nine miles per lift), and 1,400 tons of cargo (an average of fifty miles per lift), with the loss of sixteen helicopters and 121 helicopters damaged and sixteen crew members and seven troops killed.[64] In 1959 alone,

Air Force helicopters handled 7,500 air evacuation patients, 48,000 other passengers including assault troops, and 1,200 tons of cargo.[65]

It is indeed fortunate that the most critical segment of the French mobility system in Algeria should have experienced the fewest problems. On the whole, the air transport elements assigned to the Algerian theater had few major difficulties beyond the obvious ones of a high tempo of operations and a hostile physical environment. Once the initial defects were made good in the first year of the war, both Air Force and Army light aviation manning levels and crew training were maintained at adequate levels. The number of both fixed-wing aircraft and helicopters grew steadily throughout the war, and even the age, primitive fittings, and low speed of some of the aircraft assigned (for example, the old Ju-52 Toucan) proved to be advantages in the counter-guerrilla environment. Maintenance was less complex, and the slow speed and long loiter time of the older aircraft facilitated aerial observation. Coordination and integration of air operations among the Army, Air Force, and Navy were excellent, and the decentralization of air assets was shown to be highly effective. The introduction of numerous helicopters into the battle area and the development of organizations and tactics for their use was likewise a resounding success, providing the French forces in Algeria with a tactical and logistical mobility unmatched by their flightless opponents.

## CONCLUSION

The proximity of metropolitan France and the ready availability of shipping and transport aircraft all but eliminated problems of strategic transportation for the French forces in Algeria. The principal transportation problem encountered in Algeria itself was the difficulty of maintaining a reserve of motor transport for the rapid tactical deployment of units while simultaneously meeting the heavy logistical transport requirements.[66] However, a good deal of the shortfall in tactical motor transport was met by helicopters, the use of which increased tremendously during the course of the war. Indeed, most of the operational transport problems posed by the great distances and difficult terrain of the Algerian area of operations, rebel counter-mobility operations, and the high demand for flexible, rapid movement of troops and supplies were solved by air power to a degree not seen earlier in any active theater of war.

The territorial organization of French forces in Algeria proved well suited to a counter-guerrilla campaign in which the objective was to seal Algeria's international borders, clear populated areas of rebel forces, and protect the population against renewed rebel attack and exploitation. However, it was very costly in terms of the dispersion of forces and the consequent demands on a logistical system that was itself in the process of reorganization and adaptation to a unique operational environment and that, due to restricted budgets, was struggling to attract satisfactory numbers of skilled personnel, specialized equipment, and modern, efficient facilities. That deficiencies in the French logistical structure in

Algeria were not in the end obstacles to military success over the nationalist rebels was due in large part to hard work and the adoption of effective policies of economy and decentralized control.

To overcome the basic scarcity of resources of all types, scarcities that lay at the bottom of all deficiencies in the logistical support of French forces in Algeria, policies were adopted which rested on ten principles: (1) maximum use of maritime and air LOCs to and from metropolitan France; (2) maximum use of "throughput" movement of cargo directly from the Mediterranean ports to units in the field, without intermediate stops; (3) maximum exploitation of the existing logistical infrastructure, including civilian facilities and transport; (4) maximum utilization of the supporting facilities available in metropolitan France; (5) speedy repair of equipment and its prompt return to users; (6) limiting the amount of transport required, by reducing movements from major depots by, in turn, increasing the number of depots in the operational zones, augmenting their capacity, and provisioning them by carefully scheduled deliveries; (7) fixing for each troop unit a sufficient stockage level according to its degree of isolation; (8) concentrating responsibility for all transport operations in the same hands (those of the Train Commandant/Transport Director) at each level; (9) decentralizing control over transport resources; and (10) limiting aerial resupply within Algeria to only the most isolated units and most urgent demands.[67]

Such a policy of economy could be effective only with strict management and control at all levels, discipline on the part of operators, and the full support of commanders. In view of the large territory covered and the frequent insecurity of ground LOCs, the two most effective methods of achieving such economy in the Constantine Corps Area and elsewhere were the pooling of resources and greater decentralization to relieve the corps areas of logistical responsibilities as much as possible and to achieve greater flexibility.[68] In the end, through strict economy and careful management of the existing resources, French commanders and logisticians in Algeria were able to overcome most difficulties.

## NOTES

1. Commandement Supérieur des Forces en Algérie, Région Territoriale et Corps d'Armée d'Alger, État-Major, 4e Bureau, *Instructions sur le soutien logistique du corps d'armée d'Alger*, No. 260/CAA/4/ETG (Alger, 5 février 1962), 6, in folder "Corps d'Armée d'Alger, État-Major, 4e Bureau—Instructions sur le soutien logistique de CAA, 1957–62," Dossier 1H2762 d. 3, Fonds Algérien, SHAT.

2. Pierre Gourmen, "Les bataillons du Train en Algérie," *Revue Historique des Armées*, Special Issue (1978), 173.

3. The missions and responsibilities assigned to the Director of Transport-Algeria are outlined in 10e Région Militaire, Commandement du Train et Direction des Transports, 2e Section, *Notice Technique sur la Commandement du Train et Direction des Transports en Algérie* (Alger, 24 novembre 1958), 4–5, in folder "10e Région Militaire, État-Major Interarmées, 3e Bureau, Commandement du Train et Direction des Trans-

ports—Notice technique sur le Commandement du Train, etc., 1958,'' Dossier 1H1917 d. 1, Fonds Algérien, SHAT [cited hereafter as *Notice Technique*].

4. Ibid., 5; Commandement en Chef des Forces en Algérie, Commandement du Train et Direction des Transports, *L'Arme du Train Algérie-Sahara*, No. 94/CTDT.AL/1/ORG (Alger, janvier 1960), Charte (''Liaisons CTDT et Organismes Civils''), in folder ''10e Région Militaire, État-Major Interarmées, 3e Bureau, Commandement du Train et Direction des Transports—Le Train en Algérie et au Sahara, 1960–61,'' Dossier 1H1917 d. 6, Fonds Algérien, SHAT. To this end the Director of Transport-Algeria was designated Military Commissioner of the SNCFA.

5. Ibid., 1–2.

6. Ibid., *Notice Technique*, 2.

7. 10e Région Militaire, Commandement de Région Territoriale et Corps d'Armée d'Oran, État-Major, 4e Bureau, *Instruction Provisoire sur l'organisation et le fonctionnement de la logistique sur le Territoire de la Région Territoriale et du Corps d'Armée d'Oran*, No. 100/CAO.4.ORG ([Oran], 1 mars 1960), Annexe II, 2e Partie, 1–3, in folder ''Corps d'Armée d'Oran État-Major, 4e Bureau—Organisation logistique des forces en Algérie et au Sahara, 1959–62,'' Dossier 1H3215 d. 1, Fonds Algérien, SHAT.

8. *Notice Technique*, 2; *Instruction Provisoire*, Annexe II, 2e Partie, 2.

9. *Instruction Provisoire*, Annexe II, 2e Partie, 1.

10. The story of the Train infantry battalions in Algeria is told by Gourmen, 172–184.

11. Ibid., 175.

12. Ibid., 174.

13. Ibid., 174–175.

14. Ibid., 184.

15. Ibid.

16. Ibid., 182.

17. *L'Arme du Train*, 44.

18. Ibid., 47; Commandement du Train des Forces Terrestres en Indochine, État-Major, 3e Section, *Mémento Provisoire sur l'Emploi des Pelotons Fluviaux en Indochine* (Saigon: Imprimerie S.P.I. des Forces Aériennes en Extrême-Orient, 11 mai 1953), 13 (table).

19. *L'Arme du Train*, 13. A description of the SNCFA routes and rolling stock is provided in Chapter 1, above.

20. 10e Région Militaire, Commandement du Train et Direction des Transports, *Rapport: Expérience de la campagne d'Algérie en ce qui concerne: le Service des Transports et l'Arme du Train*, No. 1827/CDT (Alger, 3 decembre 1956)], 3, in folder ''10e Région Militaire, État-Major Interarmées, 3e Bureau—Rapports et synthèses des armes, 1956,'' Dossier 1H1905 d. 1, Fonds Algérien, SHAT.

21. Ibid. Normally military motor transport was reserved insofar as possible for tactical troop movements, particularly of distances less than four hundred kilometers. An average train could haul as much as a transport group with a hundred trucks (see France, Ministère de l'Armée de Terre, École d'État-Major, École Supérieure de Guerre et École d'État-Major, *Aide-Mémoire pour les Travaux d'État-Major*, provisional ed. [Paris: École Supérieure de Guerre et École d'État-Major, February 1952], II, ¶522 [cited hereafter as *Aide-Memoire*]).

22. *L'Arme du Train*, 26–27.

23. Motor transport assets and operations in Algeria are discussed in ibid., 26–30.

24. Ibid., 29.
25. *Expérience de la campagne d'Algérie*, 9.
26. *Aide-Mémoire*, I, ¶271.
27. *Expérience de la campagne d'Algérie*, 18–19.
28. Commandement de Région Territoriale et Corps d'Armée de Constantine, État-Major, 4e Bureau, *Fiche–Logistique du Corps d'Armée de Constantine*, No. 36/RT.CAC/4 (Constantine, 10 mars 1960), 8–9, in folder "Corps d'Armée de Constantine, État-Major, 4e Bureau—Organisation et directives logistiques, 1957–62," Dossier 1H2989 d. 1, Fonds Algérien, SHAT.
29. *Expérience de la campagne d'Algérie*, 13.
30. *L'Arme du Train*, 74–85 passim.
31. Commandement de Région Territoriale et Corps d'Armée de Constantine, État-Major, 4e Bureau, *Note de Service—Potentiel des formations du Train du Corps d'Armée de Constantine*, No. 246/RT.CAC/4/MAT (Constantine, 26 janvier 1960), Annexe ("Situation des Vehicules des Formations du Train du C. A. C. [arrêtée au 23/1/1960]"), in folder "Corps d'Armée Constantine, État-Major, 4e Bureau—Organisation et directives logistiques, 1957–62," Dossier 1H2989 d. 1, Fonds Algérien, SHAT.
32. *Expérience de la campagne d'Algérie*, 9.
33. Ibid.
34. The organization and operations of 510e GCT are described in 10e Région Militaire, Division de Constantine, Groupe de Transport de Réserve Générale du Train 510, *Rapport sur la campagne d'Algérie au GTRGT 510*, No. 221/GT510/S (Batna, 20 novembre 1956), in folder "10e Région Militaire, État-Major Interarmées, 3e Bureau—Rapports et synthèses des armes, 1956," Dossier 1H1905 d. 1, Fonds Algérien, SHAT. In 1956, the 510e GCT was formally designated as the 510th General Reserve Transport Group (*510e Groupe de Transport de Réserve Générale du Train*; 510e GTRGT).
35. Air transport organization and operations in Algeria are described in United States Air Force, Air University, Aerospace Studies Institute, Concepts Division, *Guerrilla Warfare and Airpower in Algeria, 1954–1960*, Project No. AU-411-62-ASI (Maxwell Air Force Base, AL: Concepts Division, Aerospace Studies Institute, Air University, March 1965); A. H. Peterson, G. C. Reinhardt, and E. E. Conger (eds.), *Symposium on the Role of Airpower in Counterinsurgency and Unconventional Warfare: The Algerian War* (Santa Monica, CA: Rand Corporation, July 1963); Vertol Aircraft Corporation (T. R. Pierpont, Director, European Operations), *French Army Helicopter Operations in Algeria, June 1956–September 1959*, Report No. SM-406 (Morton, PA: Vertol Aircraft Corporation, 1 November 1959); France, National Defense Committee for Scientific Action, Operations Research Group, *Report of the Operations Research Mission on H-21 Helicopter Operations in Algeria* (Morton, PA: Vertol Aircraft Corporation, April 1957); France, Ministère des Armées, *Employment of Helicopter-Borne Units in Maintaining Order in French North Africa* [*Manual T.T.A. 152: Emploi d'unites hélicoportées en maintien de l'ordre en A.F.N.*] (Translated by the Office of the Assistant Chief of Staff for Intelligence, Headquarters, Department of the Army; Washington, DC: Headquarters, Department of the Army, Office of the Assistant Chief of Staff for Intelligence, 4 May 1956); and France, Ministère des Armées, *Notice on Air Support in Operations for the Maintenance of Order in Algeria* [*Manual T.T.A. 172: Notice sur l'appui aérien dans les operations de maintien de l'ordre en Algérie*] (Translated by the Office of the Assistant Chief of Staff for Intelligence, Headquarters, Department of the Army; Washing-

ton, DC: Headquarters, Department of the Army, Office of the Assistant Chief of Staff for Intelligence, 9 December 1960).

36. *Symposium on the Role of Airpower*, 21.

37. Ibid., 20 and 23.

38. Ibid., 21.

39. Ibid., 20.

40. The organization and operation of GATAC 2 is described by its former commander, General Ezanno, in ibid., 28–31. A synopsis of his account is given here. See also *Guerrilla Warfare and Airpower in Algeria*, 44–45.

41. The organization and operations of fixed wing air transport in Algeria are described in *Guerrilla Warfare and Airpower in Algeria*, 59–60.

42. Ibid., 60. The weekly tonnage allocation on military lines in 1959 was about 153 tons.

43. Ibid., 59–60; *Symposium on the Role of Airpower*, 21. The characteristics of the various transport aircraft used by the French in Algeria are listed in Appendix B.

44. *Guerrilla Warfare and Airpower in Algeria*, 60.

45. *Symposium on the Role of Airpower*, 45.

46. Ibid., 65–70 passim.

47. *Notice on the Air Support*, 12; *L'Arme du Train*, 32, 34–35, and Charte (''Organisation des Transports Aériens en Algérie'').

48. *L'Arme du Train*, 40.

49. *Guerrilla Warfare and Airpower in Algeria*, 60–61.

50. Ibid.

51. *L'Arme du Train*, 41.

52. *Aide-Mémoire*, I, ¶278.

53. *Instructions sur le soutien logistique du corps d'armée d'Alger*, 8.

54. 10e Région Militaire, État-Major Interarmées, 3e Bureau, *Exposé du Général Allard sur la situation militaire en Algérie (Forces Terrestres) à la date du 9 février 1959 [Lecture of General Allard on the Military Situation in Algeria (Ground Forces) as of 9 February 1959]* (Alger, 9 février 1959), 10, in folder ''10e Région Militaire, État-Major Interarmées, 3e Bureau—Le problem militaire en Algérie, avril 1957,'' Dossier 1H1933 d. 2, Fonds Algérien, SHAT.

55. *L'Arme du Train*, 41.

56. Commandement de Région Territoriale et Corps d'Armée de Constantine, État-Major, 4e Bureau, *Fiche—Logistique du Corps d'Armée de Constantine (draft)*, No. 49/RT.CAC/4/ETG/F (Constantine, 16 juin 1961), 1, in folder ''Corps d'Armée de Constantine, État-Major, 4e Bureau—Organisation et directives logistiques, 1957–62,'' Dossier 1H2989 d. 1, Fonds Algérien, SHAT.

57. Luc Bourgeois, *Le Matériel pendant la guerre d'Algérie* ([Paris]: Inspection du Matériel de l'Armée de Terre, September 1987), I, Part I, 62; *French Army Helicopter Operations*, 6 (Figure 1); *Guerrilla Warfare and Airpower in Algeria*, 55. The characteristics of the various types of helicopters used in Algeria are outlined in Appendix B.

58. *French Army Helicopter Operations*, 6 (Figure 1).

59. Ibid., 5–9 passim.

60. Ibid., 10–11.

61. *Guerrilla Warfare and Airpower in Algeria, 1954–1960*, 56.

62. *Report of the Operations Research Mission on H-21 Helicopter Operations*, 12.

63. *Guerrilla Warfare and Airpower in Algeria*, 68.

64. *French Army Helicopter Operations*, 20 (Figure 5).

65. *Guerrilla Warfare and Airpower in Algeria*, 59.

66. Commandement de Région Territoriale et Corps d'Armée de Constantine, État-Major, 4e Bureau, *La Logistique dans la Région Territoriale et le Corps d'Armée de Constantine—Principes d'Organisation et de Fonctionnement*, No. 973/RT-CAC/4/ETG (Constantine, 1 février 1960), 5, in folder "Corps d'Armée de Constantine, État-Major, 4e Bureau—Organisation et directives logistiques, 1957–62," Dossier 1H2989 d. 1, Fonds Algérien, SHAT.

67. Commandement de Région Territoriale et Corps d'Armée de Constantine, État-Major, 4e Bureau, *Directive Logistique*, No. 1383/CAC-4-ETG (Constantine, 16 octobre 1957), Title II, chap. 1, in folder "Corps d'Armée de Constantine, État-Major, 4e Bureau—Organisation et directives logistiques, 1957–62," Dossier 1H2989 d. 1, Fonds Algérien, SHAT; *La Logistique*, 4–5; *Expérience de la campagne d'Algérie*, 2.

68. *Instructions sur le soutien logistique du corps d'armée d'Alger*, 4; 10e Région Militaire et Commandement des Forces Terrestres en Algérie, État-Major, 4e Bureau, *Directives Logistiques pour 1959*, No. 3127/RM.10/4/ETG (Alger, 6 avril 1959), 2, in folder "Corps d'Armée d'Oran, État-Major, 4e Bureau—Organisation logistique des forces en Algérie et au Sahara, 1959–60," Dossier 1H3215 d. 1, Fonds Algérien, SHAT.

# 5

# The Algerian National Liberation Front and Its Army

Like most rebels, the Algerian nationalists faced long odds. Their French opponent held the contested ground with long-established and fully functioning political, administrative, military, and logistical systems. The rebels, on the other hand, had all the problems inherent in welding together a heterogeneous collection of factions with often conflicting goals and of creating completely new organs of civil government and administration capable of winning the support of the Algerian people and directing their affairs. At the same time they had to establish an army capable of defeating a significant portion of one of the world's most modern and experienced armed forces and to create the means and methods for sustaining that army logistically. Thus, the path to independence was strewn with many obstacles, not all of which the leaders of the National Liberation Front (FLN) were able to overcome.

At first the FLN was quite successful in spreading the rebellion, and its first goal—gaining the support of the Algerian Moslem population—was soon achieved through propaganda and, where necessary, force and intimidation. The FLN also established credible military forces within Algeria capable of controlling the population and resources in the more remote areas, thus putting the French on the defensive. However, the French soon regained their equilibrium; they reinforced and reorganized their forces in Algeria, effectively sealed off Algeria from outside aid, and mounted an ultimately successful drive to eliminate the rebel bands inside the country one by one. Unable to reinforce its military forces inside Algeria with men and matériel, the FLN reverted to guerrilla warfare and terrorism rather than proceeding to a full-scale offensive to drive the French from Algeria by force. Thereafter, the FLN concentrated its efforts on political and diplomatic action and on building a conventional army in the safety of bases in Tunisia, Morocco, Libya, and Egypt. The sizable conventional rebel army built up outside Algeria was useful mainly as an element of prestige and as a bargaining chip in negotiations with the French; it was of little consequence militarily except insofar as it served to fix French forces on the frontiers of Algeria and thus relieve the pressure on the FLN forces of the interior.

Meanwhile, enormous effort had to be exerted to prevent the self-destruction of the nationalist movement itself. While actively combatting the French, the leaders of the FLN also had to struggle against the armed opposition of dissenting factions within the nationalist movement; several of them were not entirely suppressed until after independence. Moreover, the FLN itself faced serious internal conflicts, which eventually resulted in the rise to power of such military leaders as Houari Boumedienne, who would dominate the early years of independence after 1962.

## THE ORIGINS OF THE ALGERIAN NATIONALIST MOVEMENT

The ideological basis of the Algerian rebellion was nationalism, and the rebels' primary goal was complete independence from France. The spirit of Algerian nationalism first arose among Algerian soldiers and industrial workers in France immediately after World War I. By 1954 the nationalist movement was clearly dominated by middle-class Algerians. In the absence of a clear tradition of an independent Moslem Algeria, the mass of the Moslem population of Algeria did not share the nationalist aims of the rebel leaders and had to be convinced, often by force, of the desirability of national political and cultural sovereignty. For the most part the leaders of the Algerian rebellion were secular in their orientation, but they did not hesitate to appeal to Moslem fanaticism as well as to Arab and Berber ethnic pride to strengthen opposition to the French.[1] Although many contemporary observers, particularly in the United States and other Western countries, declared that the Algerian nationalist movement was dominated by communists and constituted part of an international communist conspiracy directed from Moscow, such was clearly not the case. In fact, the Algerian nationalists cleverly exploited the Algerian Communist Party, coopted its members and resources, and excluded it almost entirely from any role in directing the rebellion.[2]

The nationalist rebellion in Algeria from 1954 to 1962 was dominated by the National Liberation Front, which evolved from the various nationalist groups that had grown up between the first and second world wars. The three most prominent of these forerunners of the FLN were the secular nationalist Algerian People's Party (*Parti du Peuple Algérien*; PPA), the xenophobic Moslem Association of the Reformist Oulemas, and the moderate Democratic Union of the Algerian Manifesto (*Union Democratique de Manifest Algérien*; UDMA). The Algerian Communist Party (*Parti Communiste Algérien*; PCA), an offshoot of the French communist party founded in 1936, played only a peripheral role. These groups had their own goals and methods and were often in conflict with one another.

The Algerian People's Party formed the core of the hard-line Algerian nationalist movement that eventually became the FLN.[3] Established in 1923 as the North Africa Star (*Étoile Nord Africaine*; ENA), the movement was taken over in 1926 by Messali Hadj, perhaps the most prominent Algerian nationalist up to the outbreak of the rebellion in 1954. The ENA advocated independence for the "nations" of French North Africa (Morocco, Algeria, and Tunisia), the removal of all French forces from Algeria, and the formation of a national revolutionary Algerian government. The ENA was outlawed, disbanded, and re-formed several times during the 1920s and 1930s, and Messali Hadj frequently found himself in jail or exile. In March 1937 the ENA was re-formed as the PPA, and during World War II Messali Hadj sought to ally the PPA with the other two main nationalist groups, the UDMA and the Moslem Association of the Reformist Oulemas. The PPA has been accused of having planned a clash between police and an Algerian nationalist procession in Sétif in the Constantinois on VE Day, 8 May 1945, which escalated into a riot and resulted in the deaths of some one hundred Europeans.[4] Taking the riot for a nationalist uprising, the French authorities suppressed the nationalists ruthlessly. Estimates put the number of Moslems killed by the army, the police, and enraged European Algerians at about ten thousand. The PPA was again outlawed, and Messali Hadj was exiled to Brazzaville in the Congo. Messali Hadj and his party reemerged in November 1946 as the Movement for the Triumph of Democratic Liberties (*Mouvement pour le Triomphe des Libertés Democratiques*; MTLD) and achieved considerable success in elections for the Algerian Assembly, created by the French in response to Moslem demands for greater participation in the government of Algeria. Messali Hadj subsequently became even more radical, severed relations with the other nationalist parties, and turned to the Arab League for support. From 1952 on he was held in detention in France.

Perhaps the least consequential of the three original Algerian nationalist groups was the Association of the Reformist Oulemas (known generally as "the Oulema"), created in 1936 by Sheik Abdelhamid Ben Badis.[5] The Oulema focused on Moslem religious and cultural, rather than political and administrative, independence. Sheik Ben Badis died in 1940, and his movement was taken over by the even more xenophobic Brahmi-Bachir. The Oulema allied itself with the PPA and UDMA in 1944, and Brahmi-Bachir was among the nationalist leaders arrested by the French following the uprising in the Constantinois in May 1945. In April 1952 he left Algeria for the Middle East and did not return. His followers were gradually integrated into the other Algerian nationalist groups, and the virtual disappearance of the Oulema marked the clear identification of the Algerian rebels with secular nationalism. The recent emergence of a Moslem fundamentalist movement in Algeria represents a revival of the xenophobic strain of Algerian nationalism represented by Brahmi-Bachir and the Oulema.

The most moderate of the nationalist groups eventually incorporated into the FLN was the movement created in 1930 by a middle-class druggist from Sétif, Ferhat Abbas, the only prominent nationalist leader of the pre-1954 period to play an important role in the 1954–1962 war for independence.[6] Like many leaders of the Algerian rebellion, Abbas had served as a soldier in the French Army (1939–1940). Called the Congress of North African Students (*Congress des Étudiants Nord Africains*), Abbas's party was dedicated to achieving the political, economic, and social equality of the Algerian Moslem people within the framework of a French Algeria. The organization evolved under a number of different names in the 1930s and early 1940s. In March 1943 Abbas issued the so-called *Manifesto of the Algerian Moslem People*, patterned on the Atlantic Charter. Two months later the *Manifesto* was reissued with a call for the independence of Algeria at the end of World War II. Ferhat Abbas's group allied itself with the MTLD and the Oulema in 1944. He was among those nationalist leaders arrested in the wake of the uprisings in the Constantinois but he was freed in 1946, and his followers were reorganized in 1947 as the Union Democratique de Manifest Algérien (UDMA).

The reaction of the French government to the various Algerian nationalist groups was inconsistent. At times the nationalists were tolerated, but at other times they were outlawed and suppressed. Mostly they were ignored as inconsequential. Even under the pressure of wartime needs for men and political support General Charles de Gaulle brushed aside such moderate proposals as Ferhat Abbas's *Manifesto*, although some feeble moves were made to permit Moslems greater participation in the political life of Algeria.[7] The 1944 Constitution of the Fourth Republic provided for representation of Algerian Moslems in the French parliament, albeit on an unequal basis, but this provision was opposed by the Algerians of European descent (the *colons*) and was never implemented.[8] In September 1947 the French National Assembly passed a controversial Algerian Statute discarding the idea of integration and providing for representation in a new Algerian Assembly of the Moslem and European communities as separate entities, a concept accepted by neither the Moslems nor the *colons*.[9] On the whole, the halfhearted attempts by the French government to secure the continued allegiance of the Moslem population of Algeria were condescending and inadequate. The failure to make adequate provision for the rights and political participation of the majority of the inhabitants of Algeria in the post–World War II period hastened the inevitable outbreak of open rebellion and the war for independence. The limits of French resources and resolve laid bare by the success of the Viet Minh rebels in Indochina convinced the Algerian nationalists that independence was not only desirable but could be achieved by open rebellion.

## POLITICAL ORGANIZATION OF THE ALGERIAN NATIONAL LIBERATION FRONT

The attempt to unite the various Algerian nationalist groups during World War II was stillborn, and the premature uprising in the Constantinois in May 1945 resulted in French suppression of the various nationalist parties and the arrest of their leaders. Subsequently, the more radical nationalists came to believe that the only path to self-determination lay in armed rebellion. Thus, in 1947 a small group of young extremists, including Hocine Ait Ahmed and Ahmed Ben Bella, formed a secret paramilitary organization known as the Secret Organization (*Organisation Secréte*; OS) and began the recruitment and training of a small covert army.[10] The OS soon had some 1,800 men under arms, but in 1950 it was uncovered by the French; Ben Bella was arrested, and the remaining leaders of the OS escaped to Cairo or into the Algerian countryside.[11]

Following the detention of Messali Hadj in metropolitan France in 1952, serious rifts opened in the ranks of his MTLD party between those who desired to proceed to an open rebellion against the French and those who wished to continue as an underground movement. The advocates of open rebellion, many of whom were also members of the OS, proceeded to form a secret Revolutionary Committee for Unity and Action (*Comité Révolutionnaire pour l'Unité et l'Action*; CRUA) to promote their militant aims. The central governing committee of the CRUA consisted of nine members, the so-called "Club of Nine," and it met for the first time on 8 May 1954, the same day that Dien Bien Phu fell to the Viet Minh. The founding members of the Club of Nine included Mohammed Boudiaf, Ahmed Ben Bella, Belkacem Krim, Mohammed Khider, Hocine Ait Ahmed, Rabah Bitat, Larbi Ben M'Hidi, Mourad Didouche, and Mostefa Ben Boulaid.[12] The same men had created the OS and are generally considered the historical leaders of the Algerian war for independence.

The objective of the CRUA was to plan and execute an armed rebellion in Algeria that would force the withdrawal of the French and bring about the independence of a Moslem Algeria. Meeting at Clos Salembier near Algiers in mid-July 1954, the CRUA decided that the revolt would be an all-out fight to the end rather than a limited demonstration to obtain concessions from the French.[13] To that end Algeria was divided into military provinces (*wilaya*s), and commanders of those provinces were appointed. Ahmed Ben Bella, who had escaped from French custody, and two colleagues were sent to Cairo to solicit the political and logistical support of Abdel Gamal Nasser and other pan-Arabists, but the CRUA itself remained in Switzerland where, aided by two Egyptian army officers assigned for that purpose by Nasser, it pursued its program of fomenting a general insurrection in Algeria.[14]

## The Creation of the National Liberation Front

With some fourteen thousand members the MTLD was by far the strongest of the nationalist groups, and most of the leaders of both the OS and the CRUA were drawn from the more militant faction of the MTLD.[15] To clear the way for active revolutionary operations the CRUA attempted to reconcile the various factions within the MTLD, but it met with little success. The adherents of Messali Hadj and Lahouel continued to oppose overt action and subsequently formed the underground Algerian Nationalist Movement (*Mouvement National Algérien*;

MNA), which created its own rebel army to fight against both the French and the more radical nationalists.[16] The moderates of Ferhat Abbas's UDMA party also initially declined to participate in the armed revolt fomented by the CRUA.[17] The Algerian Communist Party, mistrusted by the leaders of the CRUA in any event, also refused to take part.[18] Despite the opposition of various groups, on 10 October 1954 the CRUA decided at a meeting in Algiers to launch the military revolt on 1 November 1954.[19] At the same time a new name was chosen for the movement, in order to supply an outward appearance of unity among all the nationalist groups. Henceforth, the organization that directed the rebellion would be known as the National Liberation Front (*Front de Libération Nationale*; FLN).

The armed rebellion began as scheduled on the eve of All Saints Day, 31 October 1954, with attacks on French military installations, civilians, and communications routes throughout Algeria, but especially in Kabylia and the Nementchas, where revolutionary sentiment was strongest. French reactions were confused and hesitant, allowing the FLN to spread the rebellion and gain adherents and other resources. The first phase of establishment, growth, and consolidation would last until 1957; it would see most of Messali Hadj's MNA, individual members of the PCA, Ferhat Abbas and the UDMA, many Moslem deserters from the French forces, large numbers of Algerian Moslems, and even some *colons* rally to the FLN.[20] The first twenty months of the rebellion would also see an evolution of the structure of the FLN into an organization much more complex and formal than the one that had initiated the war for independence.

The first major organizational change undertaken by the FLN occurred immediately after the revolt began. The CRUA had combined both political and military planning for the rebellion in one body, but on 1 November 1954 the political and military functions of the FLN were split in two.[21] The political direction of the rebellion and the responsibility for obtaining diplomatic, financial, and military assistance from friendly states, procuring arms and other supplies, and establishing lines of communications into Algeria were entrusted to an "External Delegation" consisting of Ahmed Ben Bella, Mohammed Khider, Mohammed Boudiaf, and Hocine Ait Ahmed, based in Cairo. The semiautonomous commanders of the military provinces (*wilayas*) within Algeria became the "Internal Delegation"; it concentrated on recruiting, arming, and training rebel fighters and support personnel and their employment in active operations against the French.

The FLN had all the characteristics of a political party (structure, discipline, and a program) except one: a coherent doctrine.[22] In fact, the FLN combined in one entity the attributes and functions of a political party, a national government, and an army. While its organizational structure changed frequently to meet new situations and requirements, the FLN's objectives remained constant: to assimilate or destroy all other Algerian political parties; to gain the support of the

Algerian Moslem population in order to obtain funds, supplies, and men for the rebellion; to drive out the French; and to achieve the independence of a Moslem Algeria free of external control.[23] FLN policy and organization evolved rapidly during the early months of the rebellion, and it was not until the summer of 1956 that a comprehensive statement of principle and policy as well as a formal political and military organizational structure were enunciated.

### The Soummam Conference

The stiffening of French resistance, heavy losses in the field, and a lack of coordination among the commanders of the *wilayas* made it necessary for the leaders of the rebellion to meet in the Soummam Valley in Kabylia on 20 August 1956 to take stock of the situation and work out solutions. The meeting lasted for twenty days and was attended by some two hundred delegates led by Belkacem Krim, Amar Ouamrane, and Ramdane Abane.[24] The attendees included the *wilaya* commanders and their representatives (that is, the Internal Delegation). Ahmed Ben Bella and other members of the External Delegation waited on the Italian Riviera for some three weeks for a signal that the way was open for them to slip into Algeria, but the organizers of the conference never gave the all-clear; the External Delegation was ultimately presented with the decisions of the Soummam meeting as a fait accompli.[25] The conference proceeded amidst vehement recriminations about military and political failures and heated debates over matters of principle and policy. It soon became clear that Ramdane Abane was the dominant personality and that the matters would be decided so as to favor the interests and opinions of the Internal Delegation over those of the External Delegation, which was represented only by Larbi Ben M'Hidi of *Wilaya* 5 (Oran).[26] Many of the key leaders greatly resented Abane's domination of the meeting, and there were even murmurings that he be eliminated before he became any more powerful.[27]

The decisions of the conference, which became known to the French in September 1956, soon after the meeting adjourned, were incorporated in a forty-page document that clarified many questions of principle, policy, objectives, and organization.[28] The three guiding principles of the FLN were codified: the primacy of the political over the military; the primacy of the Interior over the Exterior; and, as a sop to the anti-Abane elements, the concept of collective leadership at all echelons.[29] Measures to increase support among the Algerian population and to eliminate opposing groups were adopted, and terms were laid down for any future peace negotiations with the French, including the very important provision that there should be no cease-fire before the French recognized Algerian independence.[30] The negotiating points adopted at Soummam were subsequently adhered to without modification until independence was achieved in 1962.

## The National Council of the Algerian Revolution and the Committee of Coordination and Execution

The Soummam conferees also created two formal political institutions to govern the FLN and pursue the aims of the rebellion: a legislative body known as the National Council of the Algerian Revolution (*Conseil National de la Révolution Algérienne*; CNRA) and an executive body known as the Committee of Coordination and Execution (*Comité de Coordination et Exécution*; CCE). The CNRA was the supreme governing body of the FLN. In theory, the CNRA was to meet once a year to establish policies and pass other legislation, but it could also meet in extraordinary session at the call of the CCE or on demand of half the members.[31] The CNRA did meet annually except only twice, holding six sessions in all.[32] The Soummam conference is generally considered its first meeting; the second meeting was held at Cairo in August 1957, and the subsequent sessions took place in Tripoli, Libya.

The first CNRA was composed of seventeen full members and seventeen alternates, two-thirds of whom had to be chosen from members of the FLN actively fighting inside Algeria.[33] The *wilaya* commanders were members by right, and their deputies became alternates. The full members elected at Soummam included Hocine Ait Ahmed, Ramdane Abane, Ferhat Abbas, Mohammed Boudiaf, Ahmed Ben Bella, Larbi Ben M'Hidi, Rabah Bitat, Mohammed Khider, Belkacem Krim, Mohammed Lamine-Debaghine, Amar Ouamrane, Tawfik El-Madani, M'Hammed Yazid, Youssef Zirout, and two members known only by their pseudonyms, Belkacem and Mokrane.[34] Mostefa Ben Boulaid was also named a full member of the CNRA, even though he had died some five months earlier.

The other key institution created at Soummam, the Committee of Coordination and Execution, was responsible to the CNRA and was designed to carry out its decisions. The five-man CCE directed the day-to-day operations of the FLN and thus controlled its political, military, and diplomatic decision making. One of the CCE's main functions was to control and coordinate rebel activities in the *wilaya*s, which hitherto had been almost totally autonomous. The five members elected to form the original CCE were all "internals." Belkacem Krim and Larbi Ben M'Hidi had been members of the CRUA, and Ramdane Abane, Saad Dahlab, and Benyoussef Ben Khedda represented the new leaders who had emerged during the early years of the rebellion.[35] Each of the members of the CCE assumed some of the functions of a government minister: Ben M'Hidi dealt with general strategy, Abane with organization, Ben Khedda and Dahlab with propaganda and foreign relations, and Krim with liaison between the CCE and the *wilaya* commanders.[36] Although the CCE could be dissolved by a two-thirds vote of the CNRA, it in effect replaced the CRUA as the most powerful central instrument of FLN governance.

The second meeting of the CNRA, at Cairo in August 1957, reflected the loss of old leaders and the rise to prominence of new ones who had survived not only the disastrous "Battle of Algiers" and the intensification of French counter-guerrilla operations but internal intrigue as well. The CNRA itself was increased to fifty-four members, and the CCE was expanded to nine members, the functions of its individual members being redefined.[37] Only Belkacem Krim and Abane Ramdane were retained from the original five members of the CCE, and Abdelhafid Boussouf, Lakhdar Ben-Tobbal, Mohammed Lamine-Debaghine, Ferhat Abbas, Abdelaziz Mehri, Mahmoud Chérif, and Amar Ouamrane were added. Five members of the External Delegation (Ahmed Ben Bella, Hocine Ait Ahmed, Mohamed Boudiaf, Mohamed Khider, and Rabah Bitat) were also elected as honorary members of the CCE, even though they had been seized by the French en route from Morocco to Tunisia on 22 October 1956 and imprisoned in France for the duration.[38] The new nine-man CCE thus consisted of five "colonels" (Krim, Boussouf, Ben-Tobbal, Ouamrane, and Chérif) and four "politicals" (Abbas, Lamine-Debaghine, Mehri, and Abane), the proportion reflecting the rising influence of the "colonels" within the FLN.[39] The position of the military leaders of the FLN was further strengthened by the creation of an "inner circle" within the CCE consisting of the five colonels and Ramdane Abane. Although Abane's star declined as those of the "colonels" rose, he remained the most powerful political figure in the FLN and could not be totally eliminated from the high councils.[40]

## The Creation of a Provisional Algerian Government

Early efforts to form a provisional government were opposed by the FLN's Tunisian and Moroccan patrons, but on 19 September 1958 the CNRA nonetheless approved and announced the creation of a Provisional Government of the Algerian Republic (*Gouvernement Provisionale de la République Algérienne*; GPRA) to act as an Algerian government-in-exile until independence from France was achieved.[41] Ferhat Abbas was elected as the first president of the council (prime minister), the imprisoned Ahmed Ben Bella became vice president, Belkacem Krim became vice president and Minister of the Armed Forces, and Mohammed Lamine-Debaghine became Minister of Foreign Affairs.[42] The new Provisional Algerian Government was immediately recognized by Tunisia, Morocco, Libya, and Iraq; other Arab nations and developing countries soon followed suit. Communist China recognized it on 22 September 1958, but the Soviet Union and the other members of the Soviet bloc delayed formal recognition of the new Algerian shadow government until 1960.[43] The CNRA continued to be the supreme legislative body, but the CCE, all of whose members at the time became members of the GPRA, was gradually supplanted by the latter as the principal central coordinator of rebel affairs.[44] The organization of the first GPRA, which operated from Cairo, Tunis, Rabat, and Tripoli, is portrayed in Figure 5.1. The GPRA was subsequently reorganized at meetings of the CNRA at Tripoli on 18 January 1960 and again on 27 August 1961, with a view to reducing the number of its members and streamlining its structure.[45]

**Figure 5.1**
**First Provisional Government of the Algerian Republic (19 September 1958)**

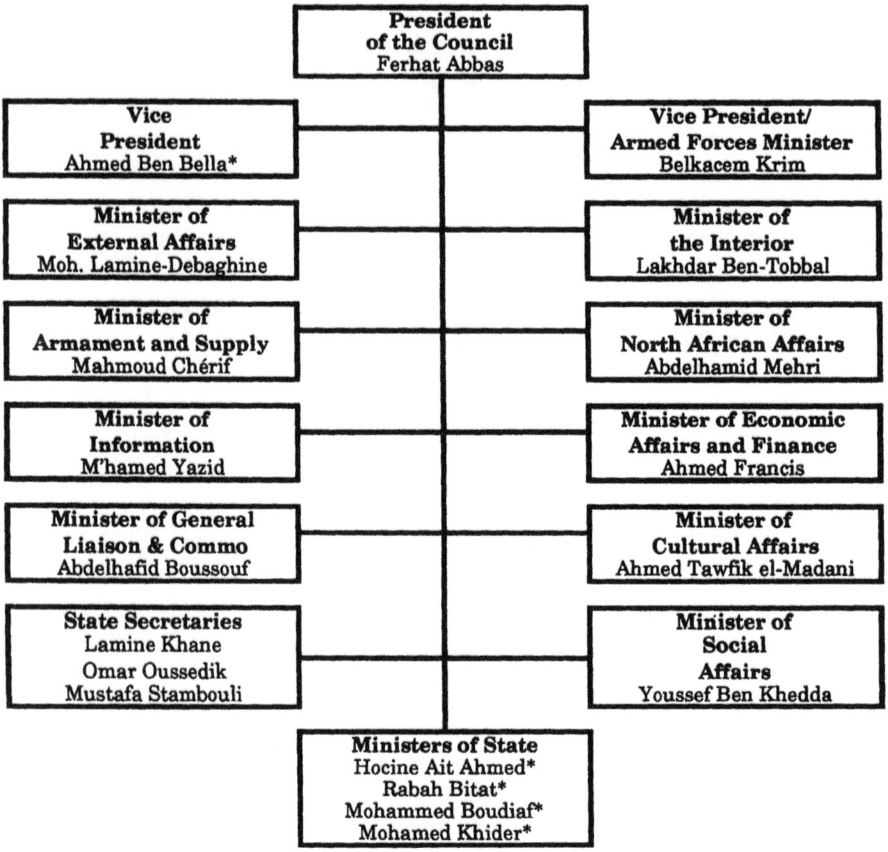


*Imprisoned by the French.

*Source*: Commandement Supérieur Interarmées et 10e Région Militaire, État-Major Interarmées, 2e Bureau, *Ordre de Bataille des bandes régulières rebelles* (Alger, 15 octobre 1957), 1–2 in folder "10e Région Militaire, État-Major Interarmées, 2e Bureau—Renseignements reçus concernant l'organisation et le potentiel des bandes rebelles, janvier–octobre 1958," Dossier 1H1692 d. 1, Fonds Algérien, SHAT.

## Territorial Organization

The GPRA exercised both political and military control of FLN affairs within Algeria through a hierarchical territorial organization with integrated political and military direction at every level. In areas under FLN control the political-military structure operated openly; in areas controlled by the French it was clandestine. Even before the outbreak of the rebellion on 1 November 1954 the

FLN had divided Algeria into six military-administrative provinces (*wilayas*), a scheme apparently borrowed from the MTLD, which had divided Algeria into six provinces (*wilayas*), thirty-three districts (*dairas*), and around a hundred sectors (*kasmas*)[46] The six *wilayas* subsequently formed the largest and most important of the FLN organizational structures within Algeria.[47] A seventh *wilaya*, sometimes called the *Wilaya* of Souk-Ahras but better known as the East Base (*Base de l'Est*), was established in the far northeastern corner of Algeria as a gateway to the supporting facilities in Tunisia.[48] By virtue of its political, economic, and cultural importance the city of Algiers was constituted as a special Autonomous Zone with three sectors, and its commander reported directly to the GPRA.[49] Due to the difficulties of communication, the *wilayas* became for all practical purposes autonomous revolutionary regions. Their sizes and internal resources were of course unequal, and while the leaders of the FLN were perfectly familiar with the boundaries of the French military and administrative divisions, the French authorities were never quite certain as to the precise dividing lines between the FLN territorial entities.[50]

Each *wilaya* was subdivided into zones (*mintakas*) regions (*nahias*), sectors (*kasmas*), communes (*douars*), *fractions*, and hamlets (*mechtas*). During the course of the war many minor changes were made in the borders and internal organization of the six *wilayas* and their various subdivisions. The usual (most representative) limits of the six *wilayas* and their subordinate *mintakas*, the East Base, and the Algiers Autonomous Zone are shown on Map 4.

The direction of military and civil affairs at each level of the hierarchical FLN territorial structure from *wilaya* to *douar* was entrusted to an integrated command committee headed by a politico-military chief (*responsable*) assisted by three deputies: for political matters, military matters, and intelligence and liaison.[51] The integrated command committees were known as Military Committees (*Comités Militaires*), or sometimes as Political Commissions (*Commissariats Politiques*). The *wilaya* commanders were normally colonels in the Army of National Liberation (the ALN, discussed below) and their assistants were majors, but at sector (*kasma*) level the chief might be a senior NCO.[52] In theory, all important decisions were taken by the chief and his three deputies collectively, but in practice the commander at each level usually acted independently.[53]

The political deputy, or commissar, played a key role in governing the territorial unit and in organizing the local civilian population to support the activities of the FLN. Although subordinate to the commander, the political deputy had primary responsibility for the functioning of what was known as the Political-Administrative Organization (*Organisation Politico-Administrative*; OPA). The OPA comprised a semicovert "parallel hierarchy" complementing the overt administrative hierarchy of *wilayas*, *mintakas*, *nahias*, *kasmas*, and *douars*. The OPA incorporated FLN-sponsored women's associations, trade unions, student groups, and similar organizations as well as members of the civilian population generally willing to work for the FLN.[54] Active participants of the OPA were organized in six-man "cells," with three cells constituting a "group" and five

**Map 4**
**Territorial Organization of the FLN in Algeria**

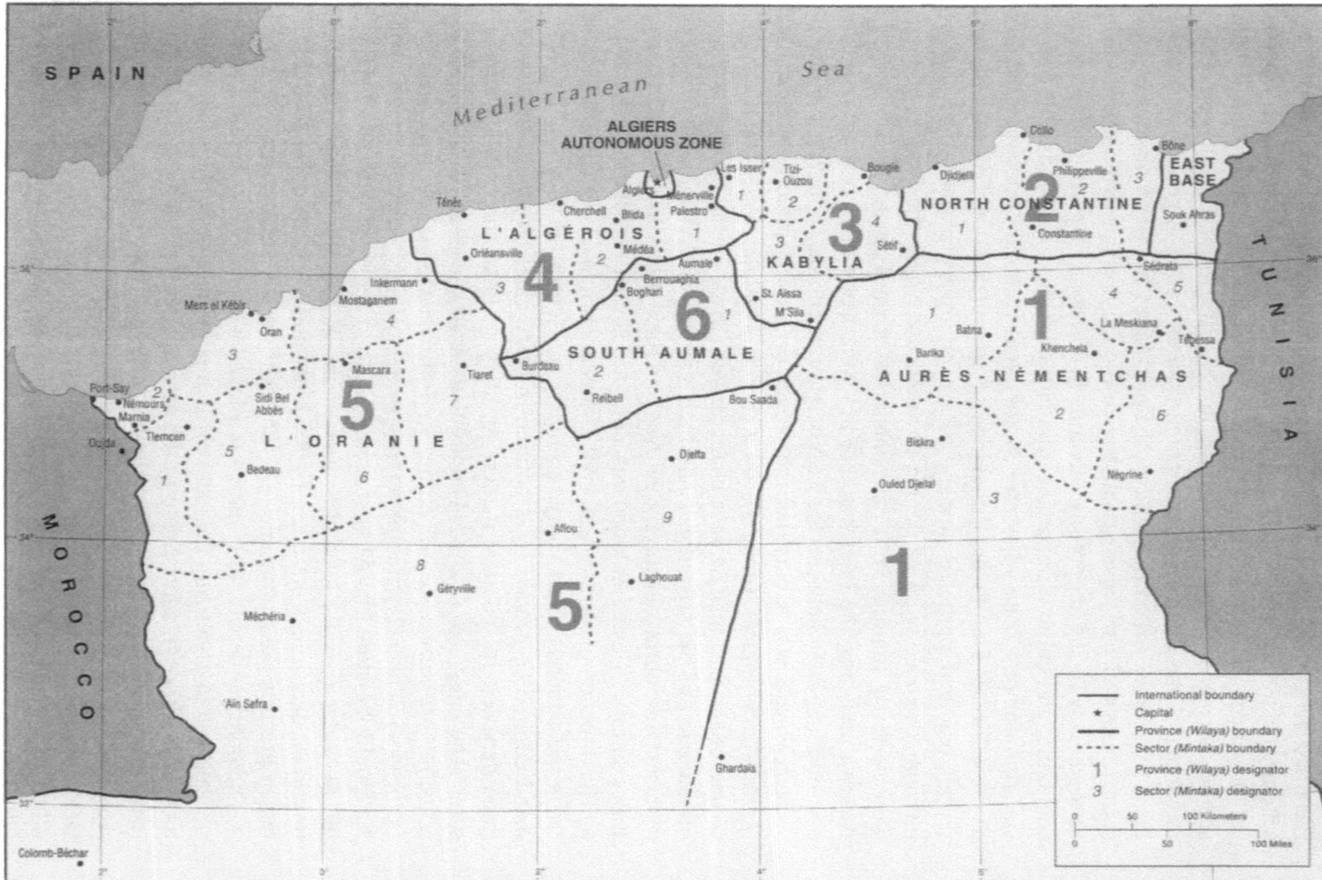

groups constituting a "section."[55] Sympathizers were organized more simply, in groups of fifteen persons. Women as well as men participated actively in the OPA, and it was through the OPA that the political deputy appointed local FLN representatives, recruited auxiliary troops and organized them for defense and sabotage actions, indoctrinated the population, rendered justice, collected funds (contributed voluntarily or through coercion), managed the financial affairs of his territorial unit, and established caches of food and munitions.[56]

The political deputy at sector (*kasma*) level was also responsible for overseeing the functioning of the two basic FLN political institutions operating at village (*douar*) level: the Committees of Three and the Popular Assemblies.[57] These organizations provided for the local administration of government in areas controlled by the FLN. The Committee of Three in each *douar* consisted of a chief (*responsable*) assisted by a political deputy and an administrative deputy; it was the executive agent of the FLN for the control and direction of the local population. The Popular Assembly (*Assemblée du Peuple* or *djemaa*) consisted of five members, each of whom was concerned with a particular administrative area (civil administration, sanitation, cultural affairs, justice, and economic matters). It gave the appearance of democratic participation but in reality was no more than a rubber stamp for the decisions of the *responsable* of the Committee of Three, who was in effect a local dictator.[58] During 1959 most of the civilians forming the Committees of Three and the Popular Assemblies were replaced by combatants of the ALN; by the end of 1960 the FLN could no longer carry out civil administration in most of Algeria.[59]

## External and Internal Dissent

From 1954 to 1962 the FLN carried out one battle against the French and another to suppress rival nationalist elements, ethnic separatists, and dissent within the FLN itself. The most important external rival faced by the FLN (other than the French, of course) was the MNA, formed by dissident elements of Messali Hadj's MTLD just before the outbreak of the rebellion. Although the influence and power of the MNA declined steadily, its armed bands operating inside Algeria constituted a serious threat to the FLN, often causing more damage to the FLN than did the French.[60] The conflict between the FLN and the MNA extended to metropolitan France, where the two groups vied for the support of the Algerian population and carried out a war of terrorism against each other. Between October 1956 and October 1957 alone some 550 Algerian Moslems were killed, and over 2,200 were wounded in terrorist incidents in France.[61] The FLN eventually triumphed over the MNA, but only after independence was achieved.

Ethnic antagonisms also plagued the FLN and created opposition both within and outside the organization. The traditional mutual antipathy of Berber and Arab in Algerian society was reflected in conflicts among personalities within the FLN, although great efforts were made to prevent such ethnic distrust from affecting the nationalist movement or becoming known to outside observers.[62] The various tribal groups in Algeria also continued their traditional feuds, in alignment with the French or one or the other of the rival nationalist groups. Such tribal conflict, especially prevalent in Kabylia and the Aurès, is reflected in the armed opposition to the FLN of such groups as the Ouled Naïl and the Beni-Illemane.[63] The Beni-Illemane tribe initially supported the MNA, but following a massacre by FLN troops of a group of Beni-Illemane at Mélouza in Kabylia on 28 May 1957, the survivors, under their chief, Bellounis, went over to the French. Armed by the French and controlled by the shadowy French 11th Shock Battalion (*11e Bataillon de Shock*), the Beni-Illemane gave the FLN considerable trouble before finally disintegrating.[64]

Perhaps the greatest threat was that posed by serious internal rifts within the FLN between the military leaders and the political leaders, rivalries that at times threatened to destroy the FLN and with it the hope of independence.[65] The Soummam conference in August 1956 did much to tamp down such conflicts, but the difficulties of communication between the *wilaya*s and the organs of FLN central government located in Cairo, Tunis, and Rabat, coupled with the necessity for quick decision making to avoid French counter-guerrilla operations, enhanced the autonomy of the *wilaya* commanders, who frequently complained of lack of support from the political leaders, safe in their Egyptian, Tunisian, and Moroccan refuges.[66] As time went on the *wilaya* commanders steadily became more independent and more dissatisfied, and some allied themselves with political factions within the FLN. The most prominent and consequential of these alliances was that between Ahmed Ben Bella and Houari Boumedienne; it was terminated only after independence was achieved, when Boumedienne deposed Ben Bella and took control of the Algerian government in his own right.

Eventually the deteriorating military situation led to open revolt. In November 1958 a group of dissidents conspired to overthrow the Provisional Government. However, the plot was discovered, and the "rebels" were tried and executed under the supervision of Colonel Houari Boumedienne.[67] Thereafter the influence and power of the hard-line Boumedienne steadily increased. Charged with restoring the morale and efficiency of the ALN, he accomplished the task expeditiously, thereby consolidating his grip on the thirty-five-thousand-man conventional army based in Tunisia and Morocco.[68] In January 1960 Boumedienne was named chief of staff of the ALN, but his ascendancy was resented by the commanders of the *wilaya*s in the interior of Algeria (*wilaya*s 2, 3, and 4). Convinced that Boumedienne was intent on increasing his personal power by building the exterior army rather than providing effective support to the hard-pressed forces fighting inside Algeria, the dissatisfied *wilaya* commanders in the summer of 1961 precipitated perhaps the greatest internal crisis of the FLN by forcing the Provisional Government to dismiss Boumedienne and dissolve the ALN General Staff.[69] Secure in his power and backed by the formidable external ALN and Ben Bella's considerable political clout, Boumedienne simply ignored the action of the Provisional Government. He subsequently came to dominate not only the FLN's military policy but its political and diplomatic policy as well, emerging as perhaps the single most powerful Algerian nationalist leader by the time independence was achieved.

## ORGANIZATION OF THE ALGERIAN ARMY OF NATIONAL LIBERATION

The Algerian Army of National Liberation (*Armée de la Libération Nationale*; ALN) began in late 1954 with a small number of uncoordinated, ill trained, and ill equipped rebel groups concentrated in Kabylia and the Aurés. Between November 1954 and mid-1957 the ALN expanded rapidly, gaining strength and operating with growing effectiveness in the back country (*bled*). However, the situation changed significantly after the disastrous "Battle of Algiers" and the completion of the Morice Line, which sealed the borders of Algeria and all but eliminated the reinforcement and resupply of the ALN elements operating inside Algeria. Forced to revert to small-scale guerrilla warfare and terrorism within Algeria, the leaders of the FLN from mid-1958 onward concentrated on creating a well armed and well trained conventional army in the safety of Tunisia and Morocco. It was charged with supporting the forces of the interior by passing men and matériel through the French barrages and fixing French military resources on the borders in order to reduce pressure on the ALN forces operating in Algeria; nevertheless, the exterior army served little purpose except as an element of prestige and a political counterweight. By 1959 the ALN was in effect two separate armies: the weak and hard-pressed guerrilla forces inside Algeria and the powerful, but largely idle, conventional forces based in Tunisia and Morocco. Neither force proved sufficient to achieve the goal of independence from France by military means; that goal was attained only later, through political negotiation.

The ALN as a coherent organization emerged only after the Soummam conference of FLN leaders in August 1956. Before August 1956 the rebel forces had evolved haphazardly, with little or no standardization as to doctrine, tactical organization, or administration. In some *wilaya*s the rebel troops were numerous, well trained, well armed, and well organized; in others they remained a hodgepodge of semi-independent rebel bands equipped in part with civilian rifles and shotguns and capable only of harassment of French outposts, sabotage, and terrorist acts.[70] At Soummam common strategy and tactical doctrine were adopted, command relationships were defined, formal organizational structures for ALN units were established, and the categories of ALN personnel, their ranks, and their administration were prescribed.[71]

### ALN Strategy and Tactical Doctrine

The strategy and tactical doctrine of the ALN as clarified at the Soummam conference were imposed, with varying degrees of success, on all ALN military forces. In general, both the strategy of the ALN and the tactics adopted to achieve the FLN's strategic goals bear a remarkable resemblance to the strategic plan and tactical doctrine of the Viet Minh in Indochina. At the same time, the military terminology employed was largely French, and French training manuals were adapted to the needs of the ALN.[72] These borrowings are not surprising given the large number of ALN combatants who had served in the French military forces, many of them in Indochina.

The political objective of the FLN, and thus the strategic objective of the ALN, was to secure the independence of Moslem Algeria by forcing the withdrawal of the French political and military administration. The strategy adopted to achieve that objective was modeled in large part on the three-phase program devised by Mao Tse-tung and the Chinese Communist theoreticians of revolutionary warfare and successfully pursued by the Viet Minh insurgents in Indochina.[73] In the strategy's classic form the first phase, essentially clandestine and defensive, involved the establishment of a viable military force and a supporting infrastructure, with overt military action limited to defensive actions and small-scale ambushes, raids, and acts of terrorism. In the second phase the rebel forces were to initiate more substantial direct action to harass and demoralize the enemy while continuing to build up their own military strength. Offensive military action during the second phase might include larger ambushes and raids, coordinated attacks on enemy facilities and lines of communications, and limited campaigns to secure resources and influence popular opinion. In the third and final phase, the rebel forces, having been organized and equipped as a conventional army, would pass over to a sustained full-scale offensive campaign to eliminate enemy military and political organs or force their withdrawal from the field.

The ALN thoroughly mixed the various phases and at first attempted to proceed too rapidly from one to the next. By 1957 the ALN inside Algeria was well into the second stage and felt sufficiently strong to proceed to the third stage. However, the completion of the Morice Line effectively cut off the ALN forces of the interior from their sources of men and matériel in Tunisia and Morocco, and the French Challe offensives in 1959–1960 mangled the ALN inside Algeria. The ALN forces in Algeria that survived the Challe offensives were forced to revert to the second, or guerrilla, stage, and the conventional forces in Tunisia and Morocco, equipped and trained for the all-out third-phase offensive, were effectively prevented from being brought into action at all. The leaders of the FLN/ALN were not unaware of their own weaknesses, and although impatient for decisive engagement, they now generally concentrated on carrying out a classic guerrilla war within Algeria, the objectives of which were to demoralize the French by keeping them always on the alert even in "safe" areas; to force the French to devote combat power to their own defenses rather than employing it in offensive action against the ALN; and in the end, to convince the French political and military authorities that the cost of continued resistance would exceed the benefits to be gained therefrom.[74]

ALN tactics were also patterned on the "hit and run" methods that had served the Viet Minh so well.[75] The main ALN tactical principle was to avoid the enemy's strength and attack his weakness, and then only when certain of victory; the main ALN tactical method was to strike rapidly and then flee. Unless trapped, the rebels seldom held their ground, preferring to avoid if at all possible being fixed and decisively engaged by the far superior French artillery and air power.[76] Heavy emphasis was placed on operational security; surprise, deception, covert movement, and variation in tactical plans were doctrinal imperatives. Night operations were considered a principal means of preserving security and achieving surprise. Inasmuch as surprise was a most important factor, the ALN emphasized rapidity of movement. Careful preparation of offensive actions was stressed, and great care was taken in the selection of objectives in order to obtain maximum effect with minimum resources. Thus ALN operations favored the use of sabotage and ambush, and French command posts, isolated detachments, officer billets and messes, and even individuals were prime targets.

In part the rebel preference for small-scale operations was due to the extremely poor communications between ALN units, a weakness that generally limited operations to the deployment of no more than one company, against a single objective.[77] Larger-scale ALN offensive actions, for example, the major battalion-size assaults on the Morice Line in the winter of 1957–1958, invariably ended in failure. Although the ALN headquarters in Tunisia and the *wilaya* headquarters inside Algeria were equipped with new German Telefunken radios by early 1959, most ALN units continued to rely on runners, and the coordination of operations did not improve substantially.[78] In any event, the effectiveness of French military countermeasures after 1957 precluded the ALN from mounting large-scale operations inside Algeria, and the ALN was never able to test the effectiveness of the large conventional army based in Tunisia and Morocco.

## Higher Command and Staff Organization

Following the Soummam conference in August 1956 the higher direction of the ALN was entrusted to the Committee of Coordination and Execution, which created two commands to carry out its plans and instructions: the Eastern Command to coordinate operations in *wilaya*s 1, 2, and 3, the Eastern Base, and the Algiers Autonomous Zone; and the Western Command to coordinate operations in *wilaya*s 4, 5, and 6.[79] With the creation of the Provisional Government in 1958 the central authority for all military operations was transferred from the CCE to the new Ministry of the Armed Forces, headed by Belkacem Krim. At the same time the Eastern and Western commands were replaced by a General Staff–East and a General Staff–West, as shown in Figure 5.2. Direction of the General Staff–East was entrusted to Colonel Mohammed Said Naceor, and Colonel Houari Boumedienne quickly rose to command the General Staff–West.

In 1959 the eastern and western general staffs were merged, and at its third

**Figure 5.2**
**Organization of the High Command of the ALN, 1958–1959**

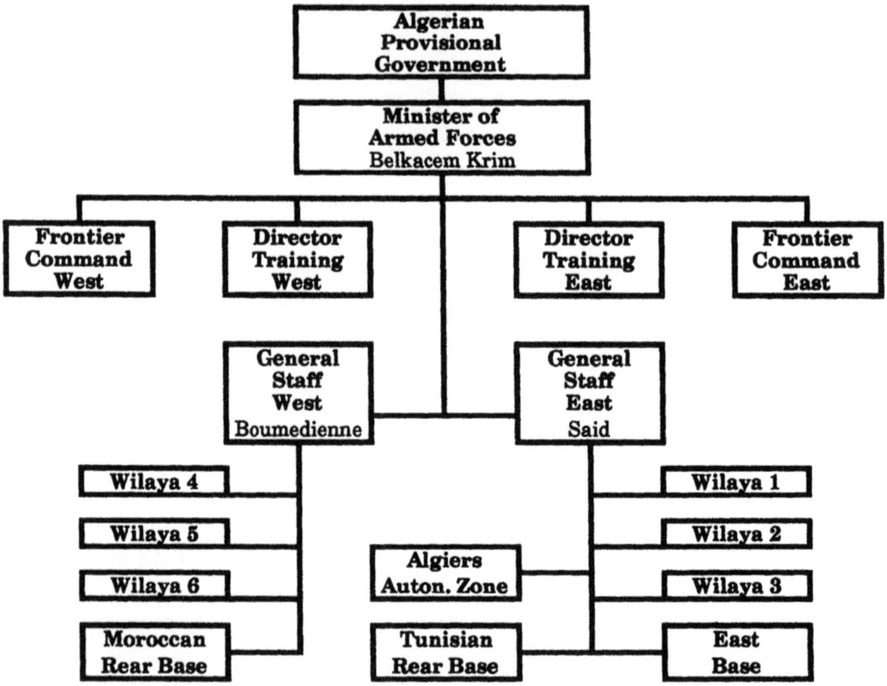


*Source*: Commandement Supérieur Interarmées et 10e Région Militaire, État-Major Interarmées, 2e Bureau, *Ordre de Bataille des bandes régulières rebelles* (Alger, 15 octobre 1957), 3, in folder "10e Région Militaire, État-Major Interarmées, 2e Bureau—Renseignements reçus concernant l'organisation et le potentiel des bandes rebelles, janvier–octobre 1958," Dossier 1H1692 d. 1, Fonds Algérien, SHAT.

congress, held in Tripoli in January 1960, the CNRA confirmed Colonel Houari Boumedienne as Chief of Staff of the ALN.[80] The powerful Colonel Boumedienne had been the de facto Chief of Staff since November 1958, and he was already busy restoring discipline and reorganizing the ALN in Tunisia. The CNRA also transferred the authority of the Minister of Armed Forces to a three-man Interministerial War Committee (IWC), consisting of Belkacem Krim (then Minister of Foreign Affairs), Abdelhafid Boussouf (then Minister of Armaments and General Communications), and Lakhdar Ben-Tobbal (then Minister of the Interior).[81] Almost immediately Boumedienne disposed of unreliable elements in the ALN high command and installed three hand-picked deputies, majors Azzedine, Slimane, and Ali Mendjli.[82] He then proceeded to reorganize the General Staff on the French pattern, placing at the head of each of the four main staff sections officers with specialist knowledge who had recently defected from

**Figure 5.3**
**Organization of the High Command of the ALN from January 1960**

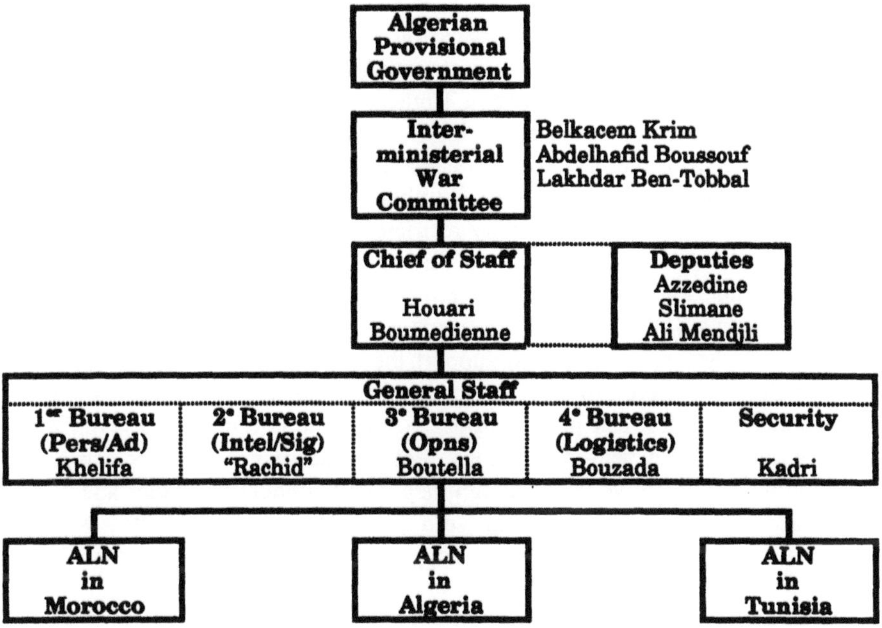


*Source*: Commandement en Chef des Forces en Algérie, État-Major Interarmées, 2e Bureau, *Organisation des Bandes Rebelles à la date du 30 septembre 1960* (Alger, 30 septembre 1960), in folder "10e Région Militaire, État-Major Interarmées, 2e Bureau—Renseignements reçus concernant l'organisation et le potentiel des bandes rebelles, janvier–octobre 1960," Dossier 1H1692 d. 3, Fonds Algérien, SHAT.

the French Army.[83] The resulting high command structure is depicted in Figure 5.3.

## Military Administration

The Soummam conferees in August 1956 also established regulations for the internal administration of the ALN. The categories of ALN soldiers were defined, a rank structure was prescribed, and provisions were made for soldier pay and the administration of military justice. The armed, active elements of the ALN were divided into three main categories: regular soldiers, part-time soldiers, and terrorists. The regulars, or *moudjahidine* (literally "those who fight for the faith"), were organized under strict military discipline into units with designated officers and NCOs. They were equipped with military weapons and uniforms, normally U.S.-type olive-drab fatigues with rubber-and-canvas boots.[84] The distinctive ALN insignia was a cap badge consisting of a red crescent and star. The *moudjahidine* normally operated in squad, platoon, company, or even battalion formations, although they might be dispersed following a major operation to avoid detection by the French. Service as a *moudjahid* was voluntary, and the usual term of service was at least two years; the FLN did not resort to the draft except on one occasion in February 1961, and then only among Algerian residents in Morocco to bring the strength of the ALN forces in Morocco up to the level of those in Tunisia.[85] The paid *moudjahidine* formed the core of armed resistance within Algeria, and they constituted the bulk of the conventional ALN forces based in Tunisia and Morocco. There were also a small number of women serving as regular members of the ALN.[86] They received the same pay as men but were generally restricted to nursing or clerical duties. In a few cases they were assigned to training or political activities.

The part-time soldiers, or *moussebiline* (literally "those who fight God's battle"), were unpaid, did not wear uniforms, and carried on their normal civilian occupations when not required to support the operations of the regular forces. The *moussebiline* were armed with an assortment of military weapons and civilian shotguns, rifles, and pistols. Loosely organized into units and given some military training, they served as guides, sentries, messengers, porters, saboteurs, and intelligence operatives.[87] Although they might conduct ambushes and small-scale raids, the *moussebiline* normally did not engage in pitched battles with the French security forces, except when trapped, or perhaps to defend their home base.

In the absence of *moudjahidine*, the *moussebiline* were the principal agents for enforcing the laws and regulations of the FLN among the indigenous population.[88] In fact a number of *moudjahidine* and *moussebiline* were assigned full-time duties outside their units to plan and direct the various command committees, intelligence networks, service organizations, and other elements of the Political-Administrative Organization.[89] In late 1959 the number of such *détachés* included some five to six thousand *moudjahadines* and twelve to fifteen thousand *moussebilines*.[90] The *détachés* formed the framework around which the numerous civilian sympathizers of the FLN in Algeria were organized. The civilian supporters assisted the *moudjahidine* and *moussebiline* by obtaining supplies, tending the wounded, gathering intelligence, spreading propaganda, and, infrequently, performing terrorist acts.

The final category of armed ALN fighter was the terrorists, or *fidayine*. The *fidayine*, many of them recruited from among the *moussebiline*, acted as saboteurs and assassins in the urban areas of Algeria, carrying out acts of terrorism to intimidate the enemies of the FLN and to demoralize French soldiers and civilians as well pro-French Moslems.[91] The *fidayine* were organized in secret commando cells and were usually well trained in their specialty. When the identities of individual *fidayine* became known, they were usually transferred to the ranks of the *moudjahidine*.

At Soummam in August 1956 formal military ranks were established in the ALN, from private (*djound*) to colonel (*sagh ethani*).[92] To minimize the possibility of the rise of a "cult of personality" contrary to the principle of collective leadership, a conscious decision was made to have no general officers until after the war for independence was concluded.[93] Officer promotions and demotions were centralized under control of the CCE, but the *wilaya* commanders were delegated the power to promote soldiers and NCOs.[94] The Soummam conferees also established a pay scale for the regular soldiers of the ALN, ranging from 1,000 FF per month for a private up to 5,000 FF per month for a colonel.[95] The families of the *moudjahidine* also received a special monthly allowance amounting to 2,000 FF for each dependent living in a rural area and 5,000 FF for each dependent in an urban area.[96] Each officer, NCO, and soldier in the ALN was also entitled to five days leave per year and a regular tobacco ration.[97]

## Strength of the Algerian Army of National Liberation

It may be impossible to make an accurate assessment of the strength of the ALN at any given point in time during the Algerian war for independence. The lack of surviving ALN records, the isolation and relative autonomy of the ALN commands within Algeria, the permeability of the barriers between the *moudjahidine* and *moussebiline* and between the *moussebiline* and civilian adherents of the FLN, and the essentially secret nature of the FLN/ALN—compounded by the fact that the FLN did not hesitate to state inflated numbers of armed and unarmed supporters—make any attempt to recapitulate the personnel strength of either the guerrilla army inside Algeria or the more conventional forces in Tunisia and Morocco very difficult indeed.[98] Table 5.1 represents an attempt to compile a useful, if incomplete, recapitulation of rebel strength at various times over the course the war. It should, of course, be used with some reserve.

The French often boasted that they had more Algerian Moslems fighting for them than did the ALN, and the best evidence seems to suggest that the total organized armed forces of the ALN never exceeded around fifty thousand men.[99] The FLN began the rebellion on 1 November 1954 with perhaps as few as four hundred armed, semitrained men.[100] Recruiting and training of volunteers for the ALN progressed rapidly in the first three years of the rebellion. By the end of 1955 the *moudjahidine* numbered perhaps six thousand, and in April 1956 French intelligence analysts put the strength of the ALN inside Algeria at about 8,050 full-time *moudjahidine* and the number of ALN "auxiliaries" (active sympathizers, not necessarily *moussebiline*) at 21,000.[101] By the beginning of 1958, just before the "Battle of the Barrages" began, the armed elements of the ALN probably numbered around 27,600 men (23,100 in Algeria, 4,000 in Tunisia, and 500 in Morocco).[102]

The ALN suffered heavy losses in the "Battle of the Barrages" in the first half of 1958 and were forced to break up the surviving units inside Algeria into company-sized and smaller packets. The French claimed to have killed or captured some three thousand rebels per month during the "Battle of the Barrages," for a total of 23,534 rebels killed or captured in the first seven months of 1958.[103] In the first seventeen months of the war the ALN had lost some 4,885 men

**Table 5.1**
**French Estimates of ALN Personnel Strength, November 1954–July 1962**

| Date | *Wilaya* 1 | *Wilaya* 2 | *Wilaya* 3 | *Wilaya* 4 | *Wilaya* 5 | *Wilaya* 6 | East Base | Inside Algeria | In Tunisia | In Maroc | Outside Algeria | Grand Total |
|---|---|---|---|---|---|---|---|---|---|---|---|---|
| Nov 1954 | | | | | | | | **400** | | | | **400** |
| Dec 1955 | | | | | | | | **6,000** | | | | **6,000** |
| Apr 1956 | | | | | | | | **8,050** | | | | **8,050** |
| **August 1956 – Soummam Conference** | | | | | | | | | | | | |
| Aug 1957 | 4,330 | 2,600 | 4,100 | 2,100 | 5,600 | 350 | 1,850 | **20,930** | | | | **20,930** |
| **Fall 1957 – Completion of the Barrages and Decision to Form an ALN "Battle Corps"** | | | | | | | | | | | | |
| Jan 1958 | 5,500 | 3,400 | 4,900 | 2,300 | 5,300 | 400 | 1,300 | **23,100** | 4,000 | 500 | **4,500** | **27,600** |
| **Summer 1958 – End of the "Battle of the Barrages" and Break Up of Large ALN Units in Algeria** | | | | | | | | | | | | |
| Aug 1958 | 5,130 | 3,000 | 4,000 | 2,400 | 4,000 | 700 | 1,300 | **20,530** | | | | **?** |
| Oct 1958 | 5,120 | 3,170 | 3,760 | 3,700 | 5,420 | 1,110 | 1,700 | **23,980** | | | | **?** |
| **February 1959 – Start of the Challe Offensives** | | | | | | | | | | | | |
| Apr 1959 | 5,482 | 6,150 | 9,990 | 5,300 | 5,800 | 3,070 | 2,450 | **38,342** | 4,360 | 2,100 | **6,460** | **44,702** |
| Mar 1959 | 3,127 | 2,967 | 4,119 | 1,940 | 2,788 | 1,415 | ? | **16,356** | 8,875 | 4,310 | **13,185** | **29,541** |
| Jul 1959 | 3,295 | 3,603 | 8,640 | 4,125 | 4,160 | 2,480 | 45 | **26,368** | 9,900 | 5,100 | **15,000** | **41,368** |
| Nov 1959 | 3,310 | 7,500 | 6,550 | 3,800 | 3,500 | 2,800 | 60 | **27,520** | 9,900 | 5,200 | **15,150** | **42,620** |
| **April 1960 – End of Challe Offensives** | | | | | | | | | | | | |
| May 1960 | | | | | | | | **9,000** | 10,000 | 5,000 | **15,000** | **24,000** |
| Sep 1960 | 3,320 | 4,700 | 5,075 | 2,890 | 2,324 | 1,350 | ? | **19,659** | 14,500 | 8,880 | **23,380** | **43,039** |
| Jan 1961 | 2,925 | 4,650 | 4,600 | 2,600 | 1,950 | 1,350 | 35 | **18,110** | 17,925 | 8,200 | **26,125** | **44,235** |
| Apr 1961 | 2,575 | 4,225 | 4,325 | 2,375 | 1,975 | 1,200 | 25 | **16,700** | 19,540 | 9,600 | **29,140** | **45,840** |
| Jul 1961 | 2,325 | 3,925 | 3,950 | 2,300 | 1,800 | 1,160 | 30 | **15,490** | 20,000 | 9,500 | **29,500** | **44,990** |
| Oct 1961 | 2,460 | 4,010 | 4,100 | 2,370 | 1,770 | 1,150 | in *W.*2 | **15,860** | 20,900 | 9,700 | **30,600** | **46,460** |
| Jan 1962 | 2,370 | 4,290 | 3,850 | 2,310 | 1,660 | 1,040 | in *W.*2 | **15,520** | 21,000 | 9,800 | **30,800** | **46,320** |
| **March 1962 – Cease Fire in Algeria** | | | | | | | | | | | | |
| Mar 1962 | 2,430 | 4,220 | 3,140 | 2,055 | 1,525 | 1,025 | in *W.*2 | **14,395** | 22,000 | 9,850 | **31,850** | **46,245** |
| **July 1962 – Algerian Independence** | | | | | | | | | | | | |

*Source*: Compiled by the author from various sources, primarily French intelligence documents on ALN order of battle contained in Dossiers 1H1682 d. 2, 1H1689 d. 1, and 1H1692 d. 1, Fonds Algérien, SHAT. N.B.: Includes *moudjahidine, moussebiline, fidayine*, and *détachés*.

killed in action, and in the first three years of the war ALN casualties had totaled some thirty thousand killed and thirteen thousand captured.[104] Nevertheless, overall ALN strength appears to have peaked in the summer of 1959 at 46,900 men (31,600 in Algeria, 10,100 in Tunisia, and 5,200 in Morocco).[105] At the same time there were in Algeria perhaps as many as ninety thousand active unarmed FLN sympathizers, who provided important support services.[106]

The so-called "Challe offensives," which began in February 1959 and ended in April 1960, came close to wiping out the rebel forces in the *wilayas*. The toll of ALN leaders was particularly high. Between April 1959 and April 1960 five *wilaya* commanders were killed by the French and another by his own men. Some 55 percent of the rebel leaders in *Wilaya* 5 were rendered *hors de combat* in 1959, and in the first seven months of 1960 some 1,120 leaders, from *wilaya* commanders to sector chiefs, were eliminated.[107] In March 1960 the French counted twenty-five thousand rebel fighters inside Algeria, some of them unarmed, plus perhaps 10,500 men in Tunisia and another 5,500 in Morocco.[108] By mid-1960 the ALN in Algeria was reduced to only about ten to twelve thousand fighting men.[109] The ALN recovered quickly in terms of raw numbers but not in terms of men trained, armed, and organized into coherent fighting units. In July 1961 the French counted 15,490 rebels inside Algeria, 20,000 in Tunisia, and another 9,500 in Morocco, for a total of 44,990.[110] By the time of the cease-fire in March 1962 the overall total had risen slightly to 46,245 men, of whom 15,520 were in Algeria, 22,000 were in Tunisia, and 9,850 were in Morocco.[111]

Both the ALN and the French security forces had to contend with a small number of dissident elements that took the field in Algeria. The most significant were the armed bands of the MNA adherents of Messali Hadj. The number of MNA combatants was never large; it declined from around five hundred men with fifteen crew-served weapons in November 1959 to only 270 men and ten or twelve such weapons in March 1962.[112] The Beni-Illemane followers of Bellounis numbered perhaps a thousand men at their peak, and the strength of the other dissident tribal groups (the Ouled Naïl and those in the Aurès) was less than five hundred.[113]

## Tactical Organization

The tactical organization of the ALN forces evolved over the course of the war for independence in accordance with political decisions taken by the FLN leadership and the military situation. Until the Soummam conference in August 1956 there was no standardization among the fighting groups in the various *wilayas*, and the size and composition of units and the types and amounts of weapons and other equipment authorized varied widely. At Soummam, manning and equipment standards were adopted for tactical units ranging in size from a fire team (*demi-groupe*) of one corporal and four men to a battalion (*faïlek*) of three companies and a total of twenty officers and 330 men, as shown in Figure 5.4.[114] The infantry company (*katiba*) was to consist of about five officers and 105 men equipped with a few automatic weapons and light mortars; it was

**Figure 5.4**
**ALN Tactical Unit Hierarchy, 1956**

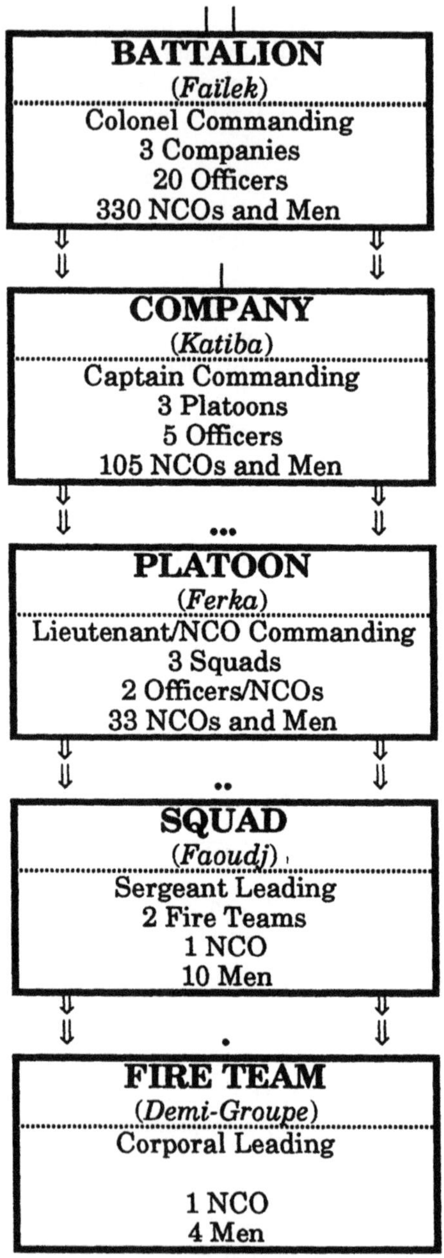


*Sources*: Rémi Brocart, "L'Organisation politico-administrative et militaire du F.L.N.: vue à travers les archives du 5e bureau de l'E.M.I.," in *Introduction à l'étude des archives de l'Algérie*, ed. Service Historique de l'Armée de Terre (Château de Vincennes: Service Historique de l'Armée de Terre, 1992), 154; Arslan Humbaraci, *Algeria: A Revolution That Failed* (New York: Frederick A. Praeger, 1966), 38–39.

anticipated that as recruitment and the receipt of heavy weapons proceeded, the *katiba* would be organized into battalions (*faïleks*), which would form the basic tactical unit of the ALN.[115] Each *faïlek* would be composed of three *katibas* and a headquarters element and would include mortars, machine guns, and antitank rocket launchers. Some consideration was also given to the eventual consolidation of *faïleks* into brigades. Although the leaders of the ALN in 1956 anticipated the progressive escalation of operations inside Algeria, they had not yet adopted the policy of creating a separate conventional army based outside the country. Thus the tactical organizations prescribed at Soummam were applicable to all ALN forces then extant or planned.

In the months following the Soummam conference the ALN leaders made considerable progress in reorganizing the hodgepodge of forces in the *wilayas*. Despite the vulnerability to French air power and artillery of larger and thus less mobile and less flexible units, the leaders of the ALN, eager to escalate the war to its third and final stage, made plans in the fall of 1957 for increasing the size of the battalion to six hundred men.[116] However, the continuing lack of competent officers made a six-hundred-man battalion impractical, and in the winter of 1957 the ALN began to reorganize its battalions with 380 officers and men, in preparation for a planned offensive against the French in the spring of 1958.[117] Although the ALN created a number of *faïleks* on paper, and each *wilaya* supposedly boasted several such units, they played little or no operational role, most ALN operations being conducted by units of company size or smaller.[118] One of the few occasions in which ALN battalions were actually used was when in 1957 a few of them were deployed near the Tunisian and Moroccan borders.[119]

In September 1957 the French completed the construction of the Morice Line, and in February 1959 the commander in chief in Algeria, General Maurice Challe, initiated a series of strong offensives to root out the ALN units inside Algeria. Effectively cut off from reinforcements and resupply from their bases in Tunisia and Morocco and battered by the Challe offensives, the ALN units inside Algeria were forced to abandon any plans for the formation and operation of battalion-size units. By June 1959 the ALN battalions (*faïleks*) in Algeria had been broken up; thenceforth the basic ALN tactical unit inside Algeria would be the company (*katiba*).[120] In some cases even the reduced-strength *katiba* proved too large and unwieldy to avoid decisive contact with the French security forces, and the ALN organized a new formation, the independent *commando* of from ten to fifteen men, which proved very difficult for the French to find and destroy.[121] At the end of 1959 *Wilaya* 4 dissolved all of its remaining *katibas* and conducted operations solely in formations of platoon size or smaller.[122] In the summer of 1958 there were some 121 companies (*katibas*) and fifty-three independent platoons (*ferkas*) inside Algeria; by the summer of 1960 there were only thirty-five *katibas* (of which twenty-one were at less than 50 percent strength), about one hundred understrength *ferkas*, and about 140 *commandos* or groups of similar size.[123] In the later stages of the war there was some effort to re-form the forces within Algeria into *katibas* of a new, smaller (thirty–forty men), more mobile type better able to outrace the highly mobile French *commandos de chasse*.[124] However, independence was achieved before substantial reorganization of the small, scattered ALN units within Algeria took place.

With the isolation, disintegration, and reversion to guerrilla action of the ALN forces in the *wilayas*, the leaders of the ALN—most particularly Houari Boumedienne, the Chief of Staff after January 1960—turned their attention to the creation in the safety of Tunisia and Morocco of a conventional army equipped with heavy weapons. The formation of such a battle corps (*corps de bataille*) had been initiated by the Minister of the Armed Forces, Belkacem Krim, in early 1959 for the purpose not of entering the battle against the French inside Algeria but of constituting an element of prestige and a political bargaining chip that might also be useful for controlling events in Algeria once independence was achieved.[125] The ALN *corps de bataille* also served to fix large numbers of French troops on the Tunisian and Moroccan barrages against the possibility of its employment. On only two occasions was the *corps de bataille* actually engaged in sustained operations against the French: in Operation DIDOUCHE in September–October 1959 and again in Operation AMIROUCHE in November 1959–January 1960. Both were designed to relieve the pressure of the French Operation JUMELLES on the battered ALN forces in *wilayas* 2 and 3, and both operations ended in failure, with heavy losses for the ALN.[126]

The ALN battle corps was organized along far more conventional lines than were the ALN companies (*katibas*) and platoons (*ferkas*) operating inside the *wilayas*. In general, its units were larger and much better supplied with automatic weapons, antitank rockets, and mortars. As support from abroad increased, the battle corps received heavy mortars (120 mm and larger), antiaircraft artillery (12.7 mm antiaircraft machine guns and 20/40 mm automatic antiaircraft guns), and even some light (75 and 105 mm) artillery. The battle corps also received radio equipment and some motor transport, although apparently no armored vehicles. Under direction of Colonel Boumedienne a new "type" of infantry battalion was designed with organic heavy weapons support, as shown in Figure 5.5.

The development of the ALN conventional army was completed after the achievement of independence by political and diplomatic means in July 1962. At the time of the Algerian cease-fire in March 1962 the ALN conventional army in Tunisia and Morocco numbered some 14,875 combat troops, plus headquarters and support personnel, in thirty-two infantry battalions (320–575 men each), eight heavy infantry companies (100–350 men each), four artillery battalions (100 men each), two 120 mm mortar batteries (100 men each), one section of heavy mortars (100 men), and a land mine section, plus supporting transport and other logistical elements.[127] The ALN *corps de bataille* was quickly moved into Algeria after independence was declared and immediately transformed into the Algerian national army. The irregular forces that had borne the brunt of the battle inside Algeria were either absorbed into the conventional

**Figure 5.5**
**ALN Infantry Battalion—Type "Boumedienne"**

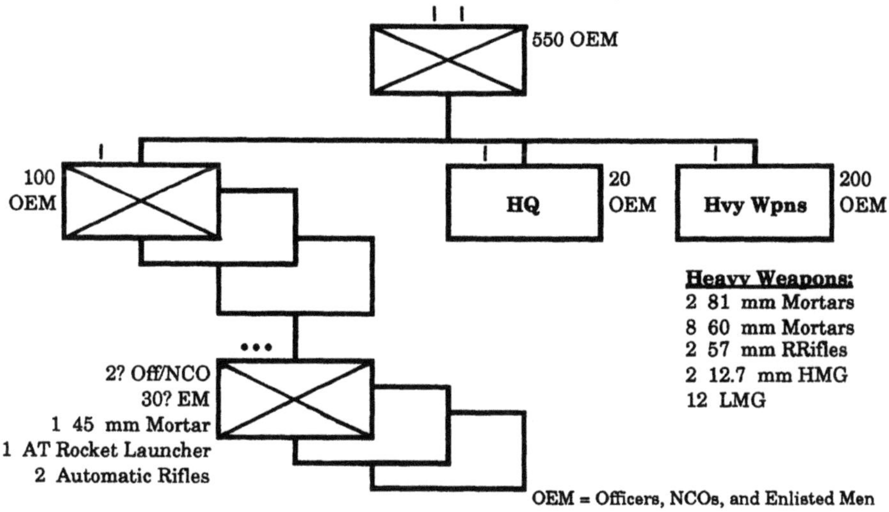


*Source*: Commandant en Chef des Forces en Algérie, État-Major Interarmées, 2e Bureau, Section "Opérations," *Evolution des structures et des méthodes de la rébellion de 1957 à 1960* No. 4515/EMI/2/OPE, (Alger, 18 août 1960), 17, in folder "10e Région Militaire, État-Major Interarmées, 3e Bureau—Enseignements de la lutte contre les rebelles d'AFN-ESG, 10 mars 1958," Dossier 1H1942 d. 3, Fonds Algérien, SHAT.

forces or disbanded. As the only remaining organized military force of any consequence, the ALN conventional army became the arbiter of post-independence internal political conflicts, just as its founder, Belkacem Krim, had intended.

## CONCLUSION

Although rival nationalist groups, ethnic dissidents, internal conflicts, and above all, effective French military action to seal off Algeria's borders and hunt down the rebels denied the FLN the achievement of its goals by military means, such factors had little impact on the rebel effort to win at the conference table what they had been unable to secure on the battlefield. In the end the FLN succeeded in achieving its primary goal of independence from France through political and diplomatic means. That achievement was made possible, however, by the amazing feat of creating *ex nihilo* an effective political structure that was able to supress internal dissent, overcome external competitors, and gain the support of outside forces. That same newly created political entity also proved itself able to raise, organize, equip, train, and direct military forces capable of posing a significant threat to French control inside Algeria and at the same time of creating a conventional army of imposing presence, which, although never fully committed in battle, played an important role in achieving the overall victory.

## NOTES

1. France, Ministère de la Défense Nationale et des Forces Armées, *Counter Guerrilla Operations for Maintaining Order in French North Africa [Manual T.T.A. 123: Operations de contre guerilla dans le cadre du maintien de l'ordre en A.F.N.]* (Edited and translated by the Office of the Assistant Chief of Staff for Intelligence, Headquarters, Department of the Army; Washington, DC: Headquarters, Department of the Army, Office of the Assistant Chief of Staff for Intelligence, 24 August 1956), 11.

2. Alistair Horne, *A Savage War of Peace* (New York: Viking Press, 1977), 138; Edgar O'Ballance, *The Algerian Insurrection, 1954–62* (London: Faber and Faber, 1967), 54 and 56.

3. The development of the PPA is described in *Counter Guerrilla Operations*, 7–8.

4. O'Ballance, 32. The historian Alistair Horne (*A Savage War of Peace*, 28) has called the shots fired at Sétif "the first volley of the Algerian War."

5. The development of the Oulema is described in *Counter Guerrilla Operations*, 8.

6. The development of the UDMA is described in ibid., 9.

7. O'Ballance, 32.

8. Concepts Division, Aerospace Studies Institute, Air University, *Guerrilla Warfare and Airpower in Algeria, 1954–1960*, Project No. AU-411-62-ASI (Maxwell Air Force Rose, AL: Concepts Division, Aerospace Studies Institute, Air University, March 1965), 6–7.

9. O'Ballance, 34.

10. Norman C. Walpole et al., *U.S. Army Area Handbook for Algeria*, Department of the Army Pamphlet No. 550–44 (Washington, DC: Headquarters, Department of the Army, January 1965), 491.

11. Alf Andrew Heggoy, *Insurgency and Counterinsurgency in Algeria* (Bloomington, IN: Indiana University Press, 1972), 39, Chart 2.

12. Arslan Humbaraci, *Algeria: A Revolution That Failed* (New York: Frederick A. Praeger, 1966), 272.

13. Horne, 79.

14. Ibid., O'Ballance, 38.

15. O'Ballance, 38.

16. *Counter Guerrilla Operations*, 8.

17. Ibid., 9. It was not until April 1956 that Ferhat Abbas and the UDMA rallied to the FLN and took an active part in the armed rebellion.

18. O'Ballance, 54.

19. O'Ballance (*The Algerian Insurrection*, 38) states the meeting took place in Switzerland, but it appears that the decision was actually taken in Algiers by a committee of the *wilaya* commanders chaired by Mohammed Boudiaf, who subsequently flew to Geneva and Cairo to inform the other members of the CRUA (see Heggoy, 65). The other nationalist groups were not informed of the decision.

20. Horne, 128.
21. O'Ballance, 41.
22. Rémi Brocart, "L'Organisation politico-administrative et militaire du F.L.N.: vue à travers les archives du 5e bureau de l'E. M.I.," in *Introduction à l'étude des archives de l'Algérie*, ed. Service Historique de l'Armée de Terre (Château de Vincennes: Service Historique de l'Armée de Terre, 1992), 161.
23. Victor J. Croizat, "The Algerian War," *The Marine Corps Gazette* 41, no. 12 (December 1957), 47.
24. O'Ballance, 71.
25. Horne, 143–144.
26. Ibid., 144.
27. O'Ballance, 71.
28. Brocart, 155.
29. Commandant en Chef des Forces en Algérie, État-Major Interarmées, 2e Bureau, Section "Opérations," *Evolution des structures et des méthodes de la rébellion de 1957 à 1960*, No. 4515/EMI/2/OPE (Alger, 18 août 1960), 1, in folder "10e Région Militaire, État-Major Interarmées, 3e Bureau—Enseignements de la lutte contre les rebelles d'AFN-ESG, 10 mars 1958," Dossier 1H1942 d. 3, Fonds Algérien, SHAT.
30. Horne, 145.
31. Brocart, 162.
32. *U.S. Army Area Handbook for Algeria*, 234.
33. Ibid.
34. Humbaraci, 274.
35. *U.S. Army Area Handbook for Algeria*, 234; Horne, 145.
36. Humbaraci, 50.
37. Ibid., *U.S. Army Area Handbook for Algeria*, 234.
38. *U.S. Army Area Handbook for Algeria*, 234.
39. Horne, 226–227.
40. Ibid., 227. Abane was killed in May 1958.
41. O'Ballance, 122–123; *U.S. Army Area Handbook for Algeria*, 234.
42. O'Ballance, 123.
43. Humbaraci, 51.
44. *U.S. Army Area Handbook for Algeria*, 234–235.
45. Ibid., 235; Humbaraci, 274–275.
46. Heggoy, 292.
47. *Wilaya* 6 was the smallest and least active of the six and at various times almost disappeared. Several writers, notably Humbaraci (*Algeria: A Revolution That Failed*, 40), have indicated that *Wilaya* 6 incorporated all of the Sahara. It is evident from examination of the detailed evidence that responsibility for the Sahara was divided between the commanders of *Wilaya* 1 and *Wilaya* 5, while the commander of *Wilaya* 6 was responsible for only the area to the immediate south of *wilaya*s 3 and 4.
48. George Armstrong Kelly, *Lost Soldiers: The French Army and Empire in Crisis, 1947–1962* (Cambridge, MA: MIT Press, 1965), 169. Metropolitan France was sometimes referred to as "*Wilaya* 7," due to the political and terrorist activities carried out there by the FLN.
49. Brocart, 155–156.
50. Heggoy, 109. This situation was, of course, an advantage for the FLN.
51. *U.S. Army Area Handbook for Algeria*, 235; Brocart, 162. The intelligence and

liaison deputy at each territorial level was responsible for the development of an intelligence network in his assigned area, complete with agents, secret lines of communications, safe houses, and letter drops. His work also included the subversion, compromise, or elimination of Moslem administrators appointed by the French (see *Evolution des structures et des méthodes*, 9).

52. Heggoy, 109 and 111.

53. Brocart, 160.

54. Kelly, 171–172.

55. Brocart, 151.

56. *Counter Guerrilla Operations*, 15.

57. Brocart, 151–152; Heggoy, 119.

58. Brocart, 152. The president of the Popular Assembly was normally also a member of the local Committee of Three (see Heggoy, 125).

59. Brocart, 153.

60. *U.S. Army Area Handbook for Algeria*, 38–39; O'Ballance, 89–90.

61. O'Ballance, 93.

62. Horne, 5 and 143. The conflict within the FLN was between Arabs, such as Ben Bella, and Kabylies (Berbers), such as Abane, Krim, and Ouamrane.

63. Horne, 88.

64. Horne, 222.

65. The historian Alistair Horne (*A Savage War of Peace*, 128) has noted that "throughout the war internal dissent and personal animosities were the F.L.N.'s single greatest enemy."

66. O'Ballance, 68.

67. Horne, 325–326.

68. Ibid., 476.

69. Ibid., 536.

70. Brocart, 153; Horne, 111–112.

71. O'Ballance, 72–73 and 207–208; Horne, 144–145.

72. Heggoy, 112. The use of French military terminology by the ALN can cause confusion. For example, an ALN *bataillon* (*faïlek*) bore little or no resemblance to its French namesake.

73. *Counter Guerrilla Operations*, 17; O'Ballance, 10.

74. *Counter Guerrilla Operations*, 13.

75. ALN tactical doctrine is described in ibid., 12–15.

76. Peter Braestrup, "Partisan Tactics—Algerian Style," *Army* (August 1960), 40–41.

77. Braestrup, 39.

78. Ibid.

79. Humbaraci, 40.

80. Horne, 328.

81. Humbaraci, 40–41 and 274. Krim, Boussouf, and Ben-Tobbal were known as "the trio."

82. Horne, 412.

83. Ibid.

84. *Counter Guerrilla Operations*, 15–16; Humbaraci, 38 and 41. The French generally referred to all types of armed rebels as *fellagah* ("bandits").

85. Humbaraci, 38.

86. O'Ballance, 75.
87. Ibid., 74–75; Heggoy, 114.
88. *Counter Guerrilla Operations*, 16.
89. *Evolution des structures et des méthodes*, 8–9.
90. Lieutenant Colonel Nougues, *Caracteristiques générales des opérations en Algérie* (Algiers: SDRR/EMI, [1960D]), 13, in folder "10e Région Militaire, État-Major Interarmées, 3e Bureau—Enseignements de la lutte contre les rebelles d'AFN-ESG, 10 mars 1958," Dossier 1H1942 d. 3, Fonds Algérien, SHAT.
91. Heggoy, 114–115 and 293–294; Brocart, 150–151. O'Ballance (*The Algerian Insurrection*, 75) mistakes the *fidayine* for the unarmed civilian sympathizers of the FLN.
92. Humbaraci, 38, note. The prescribed ranks and their corresponding insignia were: *djound* (private); *djound el-aouel* (corporal; one reversed red chevron); *aarif* (sergeant; two reversed red chevrons); *aarif el-aouel* (sergeant-major; three reversed red chevrons); *moussaad* (adjutant; one red chevron underlined in white); *moulazem* (officer-cadet; one white star); *moulazem ethani* (second lieutenant; one red star); *dhabet ethani* (first lieutenant; one red and one white star); *dhabet ethani* (captain; two red stars); *sagh el-aouel* (major; two red and one white star); *sagh ethani* (colonel; three red stars).
93. O'Ballance, 72; Horne, 144–145.
94. O'Ballance, 73.
95. Ibid., 74. Humbaraci (*Algeria: A Revolution That Failed*, 38) gives the rate of pay as twenty FF per month for both officers and men.
96. O'Ballance, 74.
97. Humbaraci, 38. Colonel Boumedienne, a chain-smoker, was allotted a double cigarette ration.
98. Brocart, 153; O'Ballance, 49.
99. O'Ballance, 165.
100. Heggoy, 79 and 86. Other estimates range as high as three thousand men.
101. Luc Bourgeois, *Le Matériel pendant la guerre d'Algérie* (Paris: Inspection du Matériel de l'Armée de Terre, September 1987), I, Part 1, 22; Heggoy, 79–80.
102. Commandement Supérieur Interarmées et 10e Région Militaire, État-Major Interarmées, 2e Bureau, *Organisation des Bandes Rebelles mis à jour à la date du 15 janvier 1958* (Alger, 15 janvier 1958), in folder "10e Région Militaire, État-Major Interarmées, 2e Bureau—Renseignements reçus concernant l'organisation et le potentiel des bandes rebelles, janvier–octobre 1958," Dossier 1H1692 d. 1, Fonds Algérien, SHAT.
103. Horne, 321; O'Ballance, 120 and 209–210.
104. O'Ballance, 209–210.
105. 10e Région Militaire, État-Major Interarmées, 2e Bureau, *"F.L.N.—Annexe: Potentiel Rebelle F.L.N. au 1er août 1959"* (Alger, septembre 1959), in folder "10e Région Militaire, État-Major Interarmées, 2e Bureau-Étude générale sur l'évolution du potentiel de l'A.L.N. (Armements et Effectifs), 1957–62," Dossier 1H1682 d. 2, Fonds Algérien, SHAT.
106. Horne, 321–322; O'Ballance, 88. The overall ALN strength would thus have been about 140,000, a level that was maintained for only a short time.
107. *Evolution des structures et des méthodes*, 9.
108. Ministère de la Défense Nationale et des Forces Armées, Secretariat d'État au Forces Armées "Terre," État-Major de l'Armée, 2e Bureau, *La Rébellion Algérienne—Étude de la Menace (Mars 1960)* (Paris, 23 mai 1960), in folder "10e Région Militaire,

État-Major Interarmées, 3e Bureau—Enseignements de la lutte contre les rebelles d'AFN-ESG, 10 mars 1958 [*sic*]," Dossier 1H1942 d. 3, Fonds Algérien, SHAT.

109. *Guerrilla Warfare and Airpower in Algeria*, v–vi. As early as May 1960 the French intelligence services were reporting only nine thousand armed *moudjahidine* inside Algeria (see *La Rébellion Algérienne*, 5).

110. Commandement Supérieur des Forces en Algérie, État-Major Interarmées, 2e Bureau, *Estimation du potentiel rebelle à la date du 1er juillet 1961*, No. 2,910/CSFA/EMI/2/OPE/PLIT (Alger, 1 août 1961), in folder "10e Région Militaire, État-Major Interarmées, 2e Bureau—Étude générale sur l'évolution du potentiel de l'A.L.N. (Armements et Effectifs), 1957–62," Dossier 1 H 1682 d. 2, Fonds Algérien, SHAT.

111. Commandement Supérieur des Forces en Algérie, État-Major Interarmées, 2e Bureau, *Estimation du potentiel d'A.L.N. à la date du 1er mars 1962*, No. 990/CSFA/EMI/2/OPE (Alger, 3 avril 1962), in folder "10e Région Militaire, État-Major Interarmées, 2e Bureau—Étude générale sur l'évolution du potentiel de l'A. L. N. (Armements et Effectifs), 1957–62," Dossier 1H1682 d. 2, Fonds Algérien, SHAT.

112. Commandement en Chef des Forces en Algérie, État-Major Interarmées, 2e Bureau, *Estimation du potentiel rebelle à la date du 1er novembre 1959*, No. 6,847/EMI/2/OPE/PLIT (Alger, 19 novembre 1959), in folder "10e Région Militaire, État-Major Interarmées, 2e Bureau—Étude générale sur l'évolution du potentiel de l'A.L.N. (Armements et Effectifs), 1957–62," Dossier 1H1682 d. 2, Fonds Algérien, SHAT; *Estimation du potentiel d'A.L.N. à la date du 1er mars* 1962.

113. Commandement des Forces Terrestres en Algérie et de la 10e Région Militaire, État-Major, 2e Bureau, *Estimation du potentiel rebelle à la date du 1er mars 1959*, No. 1,760; RM.10/2/OPE/PLIT (Alger, 24 mars 1959), in folder "10e Région Militaire, État-Major Interarmées, 2e Bureau—Étude générale sur l'évolution du potentiel de l'A.L.N. (Armements et Effectifs), 1957–62," Dossier 1H1682 d. 2, Fonds Algérien, SHAT; *"F.L.N.—Annexe: Potentiel Rebelle F.L.N. au 1er août 1959."*

114. Brocart, 154.

115. O'Ballance, 73–74.

116. Ibid., 88–89. O'Ballance notes that the desire of the leaders of the FLN/ALN to demonstrate their strength to their international supporters also played a role in the decision to move to larger formations.

117. Ibid., 89.

118. *Evolution des structures et des méthodes*, 7.

119. Heggoy, 111–112.

120. O'Ballance, 121.

121. A. H. Peterson, G. C. Reinhardt, and E. E. Conger (eds.), *Symposium on the Role of Airpower in Counterinsurgency and Unconventional Warfare: The Algerian War*, Rand Memorandum 3653-PR (Santa Monica, CA: Rand Corporation, July 1963), 14.

122. *Evolution des structures et des méthodes*, 10.

123. Ibid., 8.

124. O'Ballance, 165.

125. *Evolution des structures et des méthodes*, 16; O'Ballance, 192.

126. *Evolution des structures et des méthodes*, 16.

127. *Estimation du potentiel d'A.L.N. à la date du 1er mars 1962.*

# 6

# The Logistics of the Algerian Army of National Liberation

The Algerian Army of National Liberation (ALN) can be compared to a lobster or crab, in that its supporting structures were almost entirely external. Indeed, the most striking characteristic of the rebel logistical system was the disproportion between the means accumulated outside Algeria and those available to the fighting forces of the Interior. The first—and in many ways the most difficult—challenge faced by the political and military leaders of the Algerian National Liberation Front (FLN) was the creation of a logistical system capable of sustaining the nationalist fighters and political operatives within Algeria along with, after the summer of 1958, the parallel conventional forces based in Tunisia and Morocco.

On the whole the FLN was successful in its efforts to create *ex nihilo* a viable system for the procurement, reception, storage, maintenance, distribution, and issue of food, clothing, weapons, ammunition, vehicles and animals, fuel, and other supplies. Soon after the rebellion began in November 1954 the leaders of the FLN obtained the active financial and material support of Egypt, Tunisia, Morocco, and the members of the Arab League. The communist regimes in Yugoslavia and Eastern Europe also became eager supporters of the rebels, and the backing of Communist China was soon obtained as well. Given the necessary financial credits and access to the arms markets of the world, the FLN had little trouble in procuring what was needed and moving it to the logistical bases on Algeria's borders, bases they were able to establish with the aid of friendly regimes. However, although largely successful in obtaining the necessary war matériel and in creating an extensive exoskeleton of safe bases, depots, and training centers in nearby Egypt, Libya, Tunisia, and Morocco, the Algerian rebel leaders were in the end unable to resupply their forces inside Algeria adequately in the face of French sea, air, and ground interdiction efforts.

## ORGANIZATION OF ALN LOGISTICS

When the war for independence began in November 1954 the ragtag rebel bands inside Algeria were armed with a hodgepodge of civilian rifles, shotguns, and knives and a few old weapons gleaned from the World War II battlefields of North Africa or purchased at high cost in nearby Tunisia and Libya.[1] Food, medicine, and other supplies had to be foraged from the countryside or purchased secretly in the cities. Most of the early operations of the ALN had as a principal objective the seizure of arms and ammunition from the French; additional stocks were obtained from Algerian Moslems who deserted from the French forces.

Until the Soummam conference in August 1956 the *wilaya*s were largely autonomous operationally and logistically, and in some respects they remained so right up to the end of the war for independence. One result of this autarky was that the *wilaya*s in the interior of the country were chronically undersupplied as compared to those on the borders. There was little or no cross-leveling of equipment and supplies among the *wilaya*s, and at any given time there was likely to be significant differences in the supply status of any two. The External Delegation, led by Ahmed Ben Bella operating from Cairo, took the lead in developing sources of supply among countries friendly to the FLN and in establishing lines of communications and a system of camps and depots in Tunisia and Morocco for training and the staging of men and matériel for onward movement to the *wilayas*.[2] Despite the difficulties posed by French secret agents and the overt action of French military forces to intercept shipments to the rebels, by 1955 Ben Bella and his associates in the External Delegation had erected a workable system of bases and routes for providing the ALN forces in Algeria with weapons, ammunition, and other supplies.[3] In the early years of the rebellion the ALN logistical system was focused in Egypt and Libya. It was not until the French occupation of Tunisia and Morocco ended in 1956 that the elaborate system of bases in those countries was firmly established.

Ben Bella and his associates in the External Delegation were excluded from the Soummam conference in the summer of 1956 but continued to facilitate the flow of supplies from friendly countries in the Middle East and Eastern Europe. However, the Soummam conferees placed the primary responsibility for the logistical support of the ALN in the hands of the newly created National Council of the Algerian Revolution (CNRA) and its executive body, the Committee of Coordination and Execution (CCE). With the creation of a Provisional Government of the Algerian Republic (GPRA) on 19 September 1958, responsibility for most logistical support of the ALN was passed to the new Minister of Armaments and General Supply (MARG), Mahmoud Chérif.[4] The MARG, later renamed the Ministry of Armaments and General Liaison (MALG), was also responsible for the logistical support of civilian elements of the FLN, including refugees. Curiously, responsibility for the supply and maintenance of radio equipment was controlled by the Ministry of General Liaison and Communi-

**Figure 6.1**
**ALN Logistical Organization, October 1958**

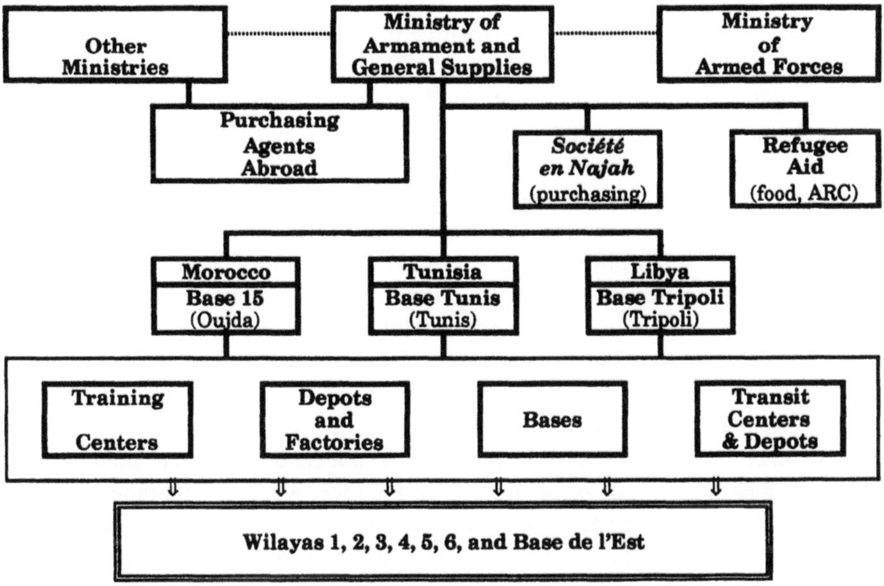


*Source*: Commandement Supérieur Interarmées pour la 10e Région Militaire, État-Major Interarmées, 2e Bureau, *Logistique rebelle du ler juillet au 30 septembre 1958*, No. 6969/RM.10/2/PLIT (Alger, 10 octobre 1958), 3 and charte ("Subordination des Differents Organismes Logistiques F.L.N."), in folder "Fiches sur la logistique rebelle, decembre 1957–decembre 1958," Dossier 1H1689 d. 1, Fonds Algérien, SHAT.

cations, under Abdelhafid Boussouf, one of the most powerful members of the GPRA after 1960 and a key player in the logistical development of the ALN. The resulting logistical organization of the ALN is outlined in Figure 6.1.

The Ministry of Armament and General Supply, as well as the other ministries of the Algerian Provisional Government, maintained a network of agents in the Arab countries, Eastern Europe, and elsewhere to facilitate the purchase and transportation of arms, ammunition, radios, food, clothing, medical supplies, and other war matériel for the ALN. Equipment and supplies were purchased through official government agencies of friendly countries in the Middle East and Eastern Europe, on the open market, and from clandestine arms dealers. Payments were made with credits advanced by friendly governments and with cash obtained from collections in Algeria and France. The equipment and supplies thus obtained were shipped to ports in Egypt, Syria, Libya, Tunisia, and Morocco and subsequently moved forward to ALN depots in Tunisia and Morocco under the supervision of representatives of the Ministry of Armament and General Supplies.

The exact point at which responsibility for logistical support of the ALN passed from political (Ministry of Armament and General Supply) to military (Ministry of the Armed Forces/General Staff) hands is unclear, in part because the political and military agencies of the FLN/ALN were so thoroughly integrated at every level and the personalities involved did not restrict themselves to solely "political" or solely "military" functions. It is certain, however, that once supplies were staged in depots in Tunisia and Morocco, the responsibility for their movement across or around the French barrages to the *wilaya*s in the Interior fell principally to the military authorities.

Similarly, little is known about the logistical organization and procedures within the *wilaya*s and within the rebel units of various sizes. The available French documents pertaining to rebel organization and logistics do not address in any detail logistical organization at lower levels. The omission is probably an indicator of defects in the French intelligence system rather than evidence that such matters were left unattended to by the rebel commanders. Certainly the principal of command responsibility was applied, and a small number of rebel personnel concerned themselves exclusively with matters of supply and maintenance. The commanders of the *wilaya*s independently determined the distribution within their area of operations of whatever weapons and other supplies reached them from outside or were obtained internally, and at least one officer on the staff of each *wilaya* commander was designated to handle logistical matters. For example, in September 1960 the staff of *Wilaya* 1 (Aurès-Nementchas), commanded by Lakdar Hadj, included Major Tahar Zebiri in charge of logistics and four officer candidates—Ahmed Abidi, Ali Belbar, Mohamed Mellouni, and Hocine Soltan—in the Intendance section.[5] At the same time the staff of the commander of the Eastern Frontier at Le Kef in Tunisia included a chief of general services (Babaye Ali), a chief of finance (Mohamed Chérif Salhi), a chief of medical service (Dr. Nakkache), a chief of transport and automotive maintenance (Abdelouahab), a chief of armaments and munitions (Captain Abid Said), and a chief of Intendance (Tahar Oussedik) assisted by a deputy (Chérif Djoghri), a supply chief (Ali Ainouz), and a clothing chief (Mohamed Taieb Bourezzaz).[6] The new heavy infantry battalion of the external battle corps designed by Houari Boumedienne had only twenty men in its headquarters company, but it is probably safe to assume that at company (*katiba*) level ALN units had at least a supply sergeant and an armorer who specialized in the repair of weapons and oversaw the supply of ammunition.

## SOURCES OF SUPPLY

The first major task of the ALN logisticians was to obtain the food, clothing, weapons, ammunition, and other supplies necessary to support the rebel forces in the field inside Algeria. The FLN was extraordinarily successful in exploiting the financial and other resources of the Moslem population in both Algeria and France and in attracting support for the nationalist movement from the countries of the Arab League, the Eastern bloc, Communist China, and even some Western nations. From these sources they received substantial amounts of money, weapons, equipment, and expertise as well as political and diplomatic support, despite intense French efforts, both overt and covert, to inhibit such assistance.

Financial support was essential, and the level of external financial aid alone was about five billion FF in 1957, ten billion FF in 1958, and seventeen billion FF in 1959, most of it provided by the Arab countries.[7] By the end of 1960 the annual budget of the FLN was around 1.36 billion FF, of which the greatest part was provided by Communist China and the Arab countries.[8] To help pay for arms and other equipment, significant sums—perhaps as much as 510 million FF over the course of the war—were also extracted from Algerian workers in France and from the Moslem population in Algeria itself.[9] Metropolitan France, sometimes known as *Wilaya* 7, was well organized by the FLN, and it was a particularly rich source of cash for the nationalist movement. The Algerian population in France, which included students and professionals as well as industrial workers, was concentrated in Paris with significant numbers in the east, north, and Rhone Valley and on the Mediterranean coast. Aided and abetted by leftist intellectuals and in competition—often directly and violently—with the agents of Messali Hadj's MNA, the FLN collectors in France made regular monthly forced collections from even the humblest Algerian Moslem factory worker.[10] The aggregate sums were considerable: 600 million FF in 1957, eventually reaching 2.5 billion FF annually.[11] The cash collected in France was smuggled out of the country and deposited in Swiss banks, from which funds were withdrawn for the purchase of arms and supplies. The most important of the currency smuggling operations was run by a French Marxist professor, François Jeauson. In one year alone, the Jeanson network smuggled more than ten billion FF out of the country.[12] Lesser ''contributions'' were obtained from the Moslem population in Algeria, and the FLN also concocted various complex and often illegal schemes, including the counterfeiting of various currencies, to obtain funds.

### Internal Sources

Although the rebel forces inside Algeria were dependent for nearly all of their logistical support on bases located outside the country, the ALN exploited local resources of funds, shelter, food, medicine, and other supplies to the maximum extent possible. Unlike the Viet Minh in Indochina, however, the ALN apparently manufactured little of its own military equipment and supplies locally, perhaps because obtaining the raw materials was as difficult as smuggling the finished products through the French defensive lines. From the beginning of the war the ALN obtained a good deal of its arms, ammunition, and other supplies from the French forces, by theft, capture on the battlefield, and Moslem deserters from the French forces. Between November 1954 and August 1960 the French lost some 10,579 weapons either in combat or stolen by deserters.[13] Of course, the weapons and other equipment so acquired by the ALN were more than offset by French captures from the rebels, particularly in the later years of the war.

For example, in 1955 the French lost 644 weapons and the ALN lost 688, but in 1959 the French lost 1,892 and the ALN 8,356.[14]

Carelessness and cupidity on the part of some French soldiers also played a part in arming the rebels. One young French parachute sergeant, Pierre Leulliette, recalled that in October 1955

> we learn that large-scale trafficking in weapons and ammunition has started in our new regiment. During the reorganization, thousands of cartridges and a certain number of weapons were more or less carelessly left lying around. The cartridges are everywhere: in the pockets of uniforms, in the bottoms of packs, and even in the mud, where they sink out of sight. After we leave the camp, Arab children pick up quantities of them. Meanwhile, the traffickers are right among us. The customers are not only the Europeans of the city, most of whom are armed, but also the Arabs. Cartridges bring their highest price in the brothels of Constantine. They have become a medium of exchange.[15]

## External Sources

The earliest, most enthusiastic, and most consistent supporters of the Algerian nationalist movement were the Arab countries of North Africa and the Middle East, which provided funds, arms and equipment, bases, technical expertise, and diplomatic support. Motivated by the spirit of pan-Arabism and anticolonialism, the new nations of Egypt, Libya, Tunisia, and Morocco were the most important backers of the FLN. Egypt under Gamal Abdel Nasser took the lead in promoting the cause of the Algerian nationalists with propaganda and diplomatic support, but the actual provision of material support from Nasser's Egypt, not to mention armed intervention, was somewhat disappointing.[16] Nasser's bark was worse than his bite, especially in the critical early years of the rebellion, but Egyptian propaganda and diplomatic maneuvering did align the Egyptian-dominated Arab League squarely behind the FLN and greatly facilitated the subsequent conclusion of aid arrangements. By 1952, well before the outbreak of the Algerian war for independence, the Arab League had created an Algerian section in its "Maghreb Office," and Algerian nationalists were receiving training in Middle Eastern countries.[17] In 1958 Arab League contributions to the FLN totaled around twelve billion FF, 75 percent of which came from the United Arab Republic (Egypt).[18] Paradoxically, the first non-Arab Moslem country to provide aid to the Algerian rebels was Turkey, a NATO ally of France. Turkish military aid to the FLN was entirely covert. It began with a shipload of arms and ammunition for the ALN that arrived at Tripoli in November 1957 and that was accompanied by a Turkish military mission headed by General Nadji Sezen.[19] A second shipment of arms was dispatched through Tunis in November 1959. The Turkish people strongly supported the FLN, but the Turkish government continued to support outwardly its NATO ally, France, by voting in its favor, or abstaining, in United Nations resolutions on the Algerian situation.

The most substantial support to the Algerian rebels was provided by the neighboring Arab countries of Tunisia and Morocco, which sent important amounts of financial and material aid and facilitated the acquisition of weapons and other supplies and equipment for the ALN. Once Tunisia and Morocco were granted their independence from France in March 1956, the Algerian rebels had secure bases on the borders of Algeria in which to train, stockpile supplies, and mount both combat operations and movements of supplies to the ALN forces of the interior. Although the support of Tunisia and Morocco was not entirely smooth or altruistic, an armed Algerian insurrection would not have been possible without it.[20]

It was a common misperception in the West, and particularly in the United States, that the Algerian rebels were supplied and controlled by the Soviet Union as part of a larger communist program of international subversion and "wars of liberation."[21] The high proportion of Soviet-type weapons captured by the French from the ALN probably contributed to the idea that the rebellion was communist directed.[22] In point of fact, the Algerian rebels would have preferred Western arms and equipment, but communist-bloc matériel was more easily obtained.[23] The fact of the matter is that the FLN was not at all a communist-led conspiracy. The rebels distrusted and exploited the Algerian communist party, and the French communist party provided little or no support of any kind to the nationalist rebels; on the other hand, it was in the interest of the French to perpetuate the myth of a communist-inspired and supported rebellion in Algeria in order to garner solid support from the United States and NATO.

Although the Soviet Union itself was slow either to recognize the GPRA or provide military aid, the Algerian rebels did receive substantial material support from the bloc countries of Eastern Europe. Yugoslavia, for example, was one of the FLN's earliest and most faithful supporters, and it was the source of a good deal of the arms and ammunition purchased for the ALN.[24] An FLN delegation to Moscow, however, in December 1958 produced little in the way of material support from the Soviet government.[25] Subsequent missions to Moscow in September 1959 and again in September 1960 produced much hope but few real returns, although the visit of Ferhat Abbas to Moscow in September–October 1960 did result in recognition of the GPRA on 7 October 1960 and "a promise to supply arms once the G.P.R.A. was in control of a piece of 'liberated' Algerian territory—a highly unlikely prospect in the prevailing military state of the war."[26] It was not until 1960 that the Soviet Union finally began supplying military equipment to the FLN (initially through Bulgaria and Czechoslovakia), but once committed the Soviet Union became a major supplier of arms and equipment, including heavy weapons.[27]

The People's Republic of China (PRC) was the first communist government to recognize the newly formed Algerian Provisional Government in 1958, and in December of that year the first FLN delegation, headed by Ben Khedda and Mahmoud Chérif, arrived in Peking to discuss PRC support.[28] The first FLN mission to China laid the groundwork for subsequent FLN delegations that negotiated financial and technical aid agreements and the delivery of matériel. In

1959 the Communist Chinese agreed to train Algerian officers and arranged for the purchase of arms in the Middle East for the FLN.[29] Aid for the FLN from the PRC began to arrive as early as April 1959, and it soon reached substantial proportions, eventually amounting to several million dollars' worth of arms and other equipment.[30] A Chinese consulate was established in 1959 in Oujda, a key ALN staging area on the Moroccan-Algerian border, and in April 1959 the steamship *Ottobane* at Casablanca discharged 192 tons of green tea, 37,500 pairs of canvas combat shoes (*pataugas*), and forty-one tons of kakhi cloth for the ALN.[31]

Early in the rebellion the FLN made significant efforts to obtain support for the Algerian nationalist cause among the friendly nations of the Third World. From 1955 the FLN was represented unofficially at the United Nations, and FLN leaders participated in the summit meeting of Maghreb nations in 1956, the Afro-Asian Conference in Cairo in 1957, and the Accra Conference of Independent African States in April 1958. Although the Arab League was an early and enthusiastic supporter of the FLN, it was not until February 1960 that the FLN was represented regularly at Arab League meetings, and only in August 1962 did independent Algeria become the thirteenth member of the Arab League and a signatory of its defense pact.[32] With the creation of the GPRA in September 1958 fourteen nations extended immediate recognition, and by 1960 the GPRA had diplomatic representation in twelve countries.[33] By the time independence was achieved in July 1962 the GPRA had "been recognized by some 30 nations; it had participated in numerous international conferences, and had, in effect, been carrying on relations with other sovereign states for almost eight years."[34] Diplomatic recognition was not always translated into financial or material aid, but the FLN gained significant strength from the psychological and diplomatic support of friendly countries.

FLN agents also arranged for purchases of arms and equipment in a large number of other countries in Europe, Asia, and the Middle East, including some ostensibly allied to France. Most often such acquisitions were made through arms dealers of less than sterling reputation, but many purchases were made from reputable firms, often in the name of and for consignment to legitimate firms in Egypt, Libya, Tunisia, or Morocco. For example, the firm of Mittemeyer in the Federal Republic of Germany supplied radios, weapons, and other equipment (600 million FF worth in April 1959 alone), and a Dutch firm headed by a Mr. Roy of Haarlem extended to the FLN a credit of twenty million dollars for the purchase of foodstuffs and explosives.[35] Purchases were also made in Finland and other Western countries, but there is no indication that the FLN purchased any significant quantity of matériel directly from the United States, although much of what found its way into ALN hands was manufactured in the United States.[36]

## LOGISTICAL INFRASTRUCTURE

Important ALN training camps and some logistical facilities were located in Egypt and Libya, but the bulk of the ALN logistical infrastructure was concentrated in Tunisia and Morocco. As soon as Tunisia and Morocco were granted their independence from France in March 1956, the External Delegation of the FLN opened offices in Rabat and Tunis and received the overt support of the Tunisian and Moroccan governments, which had been restrained up to that time by the presence of French troops.[37] The achievement of national sovereignty by Tunisia and Morocco thus "radically altered the course of the Algerian War," by providing the ALN with safe areas in which to assemble matériel; rest, train, and outfit forces; and prepare supplies for onward movement to the rebel forces in Algeria.[38] The existence of secure bases outside Algeria was critical to the FLN cause; without them the French would have been able to suppress the military activity of the nationalist rebels completely and in a much shorter time.[39]

In September 1956 the intelligence staff of the 10th Military Region issued a study on the potential outside aid to the Algerian rebels that might be provided in the next four months.[40] The authors of the study concluded (correctly) that the rebels were dependent on support from Tunisia, Morocco, and to a lesser extent from Libya and the other Arab countries, and that the most likely course of events would be a progressive movement toward more direct and open support for the Algerian nationalists but without direct intervention by the armed forces of any of the Arab countries. The study estimated that as of September 1956 the ALN in Algeria was receiving some 250 to three hundred weapons per month from Morocco and four to five hundred per month from Tunisia. It noted further that the aid provided by Morocco consisted of training camps at Nador and around Oujda, rest camps around Bou Beker, and hospitals in Oujda and Figuig; that although western Tunisia did not yet constitute a rear zone for the ALN comparable to that in eastern Morocco, Tunisia had potentially the better facilities for the transit of arms and other supplies to the rebels; and that the principal termini of the LOCs from Egypt and Libya were at Souk el-Arba and Ghardimaou for North and East Constantinois and Thelepte and Redeyef for the Nementchas and Aurès.

The French estimated that the number of weapons in rebel hands as of 31 August 1956 was around twenty light mortars; twelve machine guns, 160–170 automatic rifles, two-thousand submachine guns, eight or nine thousand military rifles; and twenty-five thousand hunting weapons. Rebel personnel strength was estimated to be about 150,000 men (20,000 in regular units, 25,000 auxiliaries armed with hunting weapons, and around 100,000 in the OPA) plus perhaps another eight to nine hundred thousand sympathizers. The French authorities also estimated that only 50–60 percent of the personnel in the regular ALN units were properly armed—except in the Oranais where the percentage was near one hundred—and that the crucial problem for the rebels was not the recruitment of men but their arming. Given the potential support available to the rebels, French intelligence officers expected that by the end of 1956 the rebel forces would include 36,000–37,000 men in regular units, twenty-five thousand auxiliaries armed with hunting weapons, five to six hundred automatic rifles, four or five thousand submachine guns, twenty thousand military rifles, twenty-five thousand hunting weapons, and a number of mortars, heavy machine guns, and antitank weapons. Such figures would represent an increase between 1 September 1956 and 1 January 1957 of 100–120 percent in weapons and 80 percent in personnel. Such augmentation would give the rebels the potential to spread the rebellion throughout Algeria, begin to establish a battle corps in its Moroccan refuges and perhaps in western Tunisia as well, and—through the installation of radio equipment at its Tunisia and Moroccan bases—change substantially the coordination of its military operations inside Algeria.

French intelligence estimates of the value to the Algerian rebels of the support bases in Tunisia and Morocco were subsequently proven to be well founded. By 1958, ALN logistical facilities in Tunisia and Morocco had reached the extent show on Map 5. Of course, the number of weapons and the amounts of ammunition stocked in the ALN depots in Egypt, Libya, Tunisia, and Morocco varied from time to time, but French intelligence sources were able to provide "snapshots" at various points. For example, among the military weapons believed to be in ALN depots in Egypt, Libya, Tunisia, and Morocco in September 1959 were at least six 20 mm antiaircraft guns, 410 Breda machine guns, twenty 12.7 mm heavy machine guns, 1,280 automatic rifles, and 61,000 military rifles and submachine guns, plus an unknown number of recoilless rifles, mortars, 7.62 mm machine guns, and antitank rocket launchers.[41] Between January 1957 and September 1959 the ALN had also obtained some 200–250 AN/GRC-9 radios, of which thirty-six had been captured by the French.

On 1 November 1959 French intelligence officers estimated that the ALN had some 74,000 military weapons and 120 million rounds of various types of ammunition outside Algeria.[42] Of the seventy-four thousand weapons, there were sixty thousand individual weapons and about three thousand crew-served weapons in depot stocks, and about ten thousand individual and one thousand crew-served weapons in the hands of ALN troops stationed outside Algeria. The totals on 1 November 1959 included some 21,000 rifles and carbines, 1,400 automatic rifles and machine guns, and seventy-five million rounds of ammunition in the ALN depot at Marsa Matrouh in Egypt; eighteen thousand rifles and carbines, six hundred automatic rifles, six to fourteen million rounds of ammunition, and an unknown number of mortars and machine guns in depots in Libya (at Pissida, Zaoura, and Oum el-Ksob); at least 9,500 rifles and automatic rifles, several hundred machine guns, fifty 57 mm recoilless rifles, fifteen 20 mm antiaircraft guns, forty million cartridges, and an unknown quantity of mortar ammunition in depots in Tunisia (at Kairouan, Le Kef, Ghardimaou, Pichon, and Souk el-Arba); and some five thousand weapons and several hundreds of thousands of rounds of ammunition in Morocco (at Oujda and in hands of the troops).[43]

The large quantities of arms and ammunition in ALN depots outside Algeria

**Map 5**
**ALN Logistical Facilities outside Algeria, 1958**

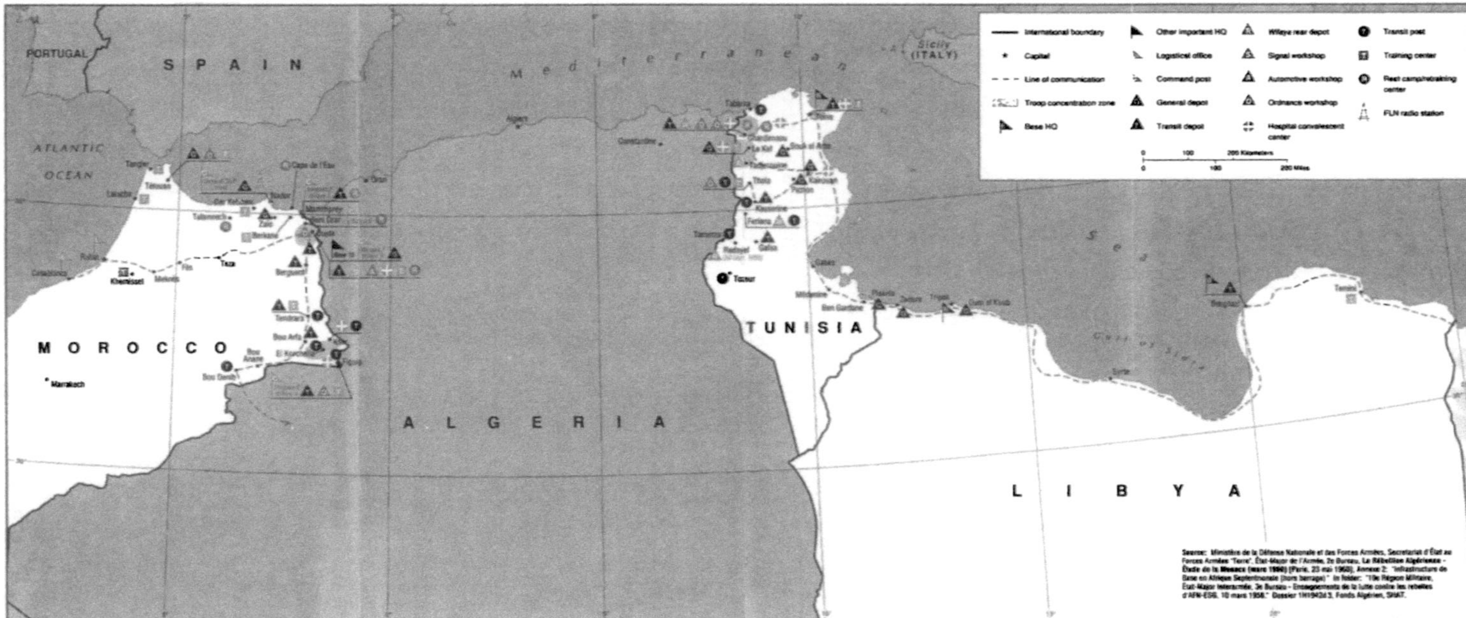


Source: Ministère de la Défense Nationale et des Forces Armées, Secretariat d'État au Forces Armées "Terre", État-Major de l'Armée, 2e Bureau, La Rébellion Algérienne - Étude de la Menace (mars 1960) (Paris, 23 mai 1960), Annexe 2: "Infrastructure de Base en Afrique Septentrionale (hors barrage)." In folder: "10e Région Militaire, État-Major Interarmée, 3e Bureau - Enseignements de la lutte contre les rebelles d'AFN-ESB, 10 mars 1958." Dossier 1H1942d 3, Fonds Algérien, SHAT.

caused some speculation among French intelligence officers as to the intent of the ALN. One theory was that the ALN was preparing to arm volunteers from other Arab and African states, a thoroughly alarming prospect.[44] Other theories were that the ALN was selling arms for profit to other rebel groups (in Guinea, for example); that the ALN was stockpiling arms as a tool for use in negotiations with the French or to provide for an Algerian national army once independence was achieved (both probably correct); or that the ALN intended to support a rebellion in Tunisia and Morocco.

### Bases in Egypt and Libya

Both Egypt and Libya provided training bases and logistical facilities for the ALN. There was an important depot of arms and ammunition at Marsa Matrouh, but Egypt was most important for providing specialized training facilities. In January 1958 French intelligence officers identified a number of ALN training facilities in Egypt, including a naval school (ten student officers), a parachute school (ten to twenty students), a frogman school (about ten students), an aviation school (about forty student mechanics and Air Force signal personnel), a signals school, a school for combat medics (twenty to thirty students), and an infantry heavy weapons (mortars) school (forty student officers).[45]

Libya contained some of the most important ALN logistical facilities outside Algeria. ALN logistical activities in Libya were controlled by the Tripoli Base, under the command of Mohammed El-Hadi, who reported to the Ministry of Armament and General Supplies (MARG) rather than through the military chain of command.[46] The Tripoli Base, which had its own budget for personnel and the purchase of vehicles and spare parts, consisted of depots at Pissida, Zaoura, and Oum el-Ksob, a transport element (about thirty trucks), and maintenance facilities in several farms in the vicinity of Tripoli.[47] The Benghazi Base, actually no more than a transit center, was subordinate to the Tripoli Base.

### The Tunisian Base

The achievement of Tunisian national sovereignty in 1956 and the subsequent withdrawal of French troops in the course of 1957 opened the way for overt support of the Algerian rebels by the government of President Habib Bourguiba. Almost immediately the FLN External Delegation transferred its headquarters from Cairo to Tunis and began the process of building a system of FLN training and logistical bases, many of which were handed over by the Tunisian army.[48] Eventually, Tunisia provided a haven for some thirty thousand ALN troops and numerous logistical and training facilities.

Although Bourguiba and the Tunisian government provided diplomatic and financial support to the Algerian rebels as well as bases on Tunisian territory, the relationship between Tunisia and the FLN was not without strain. Desirous of retaining French aid and fearful of a return of French military forces and the loss of Tunisia's own newly gained freedom, Bourguiba maintained a certain distance from the FLN leaders and consistently refused to join his own forces to those of the ALN in active operations against the French (although the Tunisian National Guard transported supplies for the ALN to the western areas of Tunisia occupied by the ALN forces).[49] Bourguiba's fears were proven well founded when formal support agreements between Tunisia and the FLN were concluded in April 1957: two months later France suspended its economic aid program for Tunisia. [50] From time to time the French forces operating in Algeria also found it expedient to violate the territorial sovereignty of Tunisia in hot pursuit of ALN bands. The most striking incident was the French bombing of the Tunisian village of Sakiet-Sidi-Youssef in February 1958, after which Tunisia broke off diplomatic relations with France.[51]

Although an effort was made to isolate the uniformed troops of the ALN from the Tunisian towns and villages, the presence of so many armed foreign troops on Tunisian soil produced the predictable incidents. Moreover, the numerous Algerian refugees who flooded into Tunisia—over sixty thousand by the end of 1957—were an additional burden for the new Tunisian state.[52] Several bloody clashes between the minuscule Tunisian armed forces and the ALN in 1959 hardened Bourguiba's attitude toward the Algerian rebels. He adopted a policy that included a grant of near-total liberty of action for the ALN in the frontier zone; prohibition of ALN facilities south of Tamerza, to avoid interference with the construction of an oil pipeline in that area; formal rejection of the formation of "International Brigades" on Tunisian territory; and opposition, for several months in 1959–1960, to the introduction into Tunisia by the FLN of heavy weapons (75 mm recoilless rifles and 120 mm mortars).[53] The restriction on heavy weapons was lifted after the FLN entered into negotiations with the French as they had been advised to do by Bourguiba.

By mid-1958 the FLN infrastructure in Tunisia had become by far the largest and most important element of the FLN/ALN logistical support system. Commanded by *Commandant* (Major) Kaci and directly subordinate to the Ministry of Armament and General Supplies (MARG), the Tunis Base was responsible not only for the supply of ALN forces in Tunisia and eastern Algeria but for various FLN civilian organizations as well (the Algerian Red Crescent, refugee relief, etc.).[54] The Tunis Base controlled a wide variety of administrative and logistical facilities, including general and transit depots; the rear command posts and rear bases of *wilaya*s 1, 2, 3, and 4 (in the vicinity of Tunis at Mathildeville, Hammanlif, and Le Bardo); personnel and matériel management agencies; an intendance service; recruitment, medical, financial, and social services; a purchasing service (the *Société en Najah*); and an organization linking the liaison officers of the various *wilayas*.[55] For administrative purposes Tunisia was divided into a rear zone in eastern Tunisia and two operational zones on the Algerian border, the Northern Operational Zone (*Zone Opérationnelle Nord*; ZON) being separated from the Southern Operational Zone (*Zone Opérationnelle Sud*; ZOS) by a line between Le Kef and Tadjerouine. [56] French intelli-

**Table 6.1**
**FLN/ALN Support Facilities in Tunisia, 15 May 1959**

| | |
|---|---|
| **Eastern Tunisia** | |
| Hospital/Convalescent Centers (3) | Tunis; between Tunis and Ghardimaou; E. of Kairouan |
| Training Center | Tunis |
| General Depots (2) | Pichon; Kairouan |
| Transit Depot | Tunis |
| **Northern Operational Zone** | |
| Hospital/Convalescent Centers (2) | Ghardimaou; Le Kef |
| Training Centers (2) | Le Kef; Thala |
| General Depots (2) | Le Kef; Souk el-Arba |
| Transit Depot | Ghardimaou |
| Rear Depots of *Wilayas* 2, 3, 4, and 6 | Ghardimaou, Tadjerouine, Feriana, and Redeyef |
| Mine/Grenade/Bangalore Workshops (3) | Ghardimaou (2); near Thala |
| Rest Camp/Retraining Centers (2) | near Ghardimaou |
| Transit Posts (2) | near Tabarka; near Thala |
| **Southern Operational Zone** | |
| Transit Depots (2) | Kasserine; Gafsa |
| Transit Posts (4) | W. of Kasserine; near Feriana; Tamerza; Tozeur |

*Source*: Commandement en Chef des Forces en Algérie, État-Major Interarmées, 2e Bureau, *Logistique Rebelle au cours du mois de mai 1959*, No. 3445/EMI/2/OPE/PLIT (Alger, 11 juin 1959), "Situation F.L.N. en Tunisie au 15-5-59," in folder "10e Région Militaire, État-Major Interarmées, 2e Bureau—Fiches sur la logistique rebelle, decembre 1958–decembre 1959," Dossier 1H1689 d. 2, Fonds Algérien, SHAT.

gence officers estimated that as of 15 May 1959 there were some 10,730 ALN troops in Tunisia, including 4,500 combat troops on the frontier, 4,730 transits, and 1,500 infrastructure personnel.[57] The support facilities located in Tunisia at that time are shown in Table 6.1.

The four general depots (at Kairouan, Pichon, Le Kef, and Souk el-Arba) and the four transit depots (at Tunis, Ghardimaou, Kasserine, and Gafsa) were operated by the Tunis Base under direction of the MARG. The rear depots (at Ghardimaou, Tadjerouine, Feriana, and Redeyef) were the responsibility of the *wilaya* commanders. The decision in July 1958 to centralize control of the distribution of arms and ammunition for the *wilaya*s under the Ministry of Armament and General Supplies had the practical effect of giving the MARG responsibility for the *wilaya* depots as well, since thereafter the general and transit depots could refuse requests for arms and ammunition that had not been approved by the MARG.[58] The stockage levels in the depots varied from time to time according to the receipt of arms and equipment from FLN supporters and the ability of the ALN to move matériel on into Algeria. Between 1 December 1957 and 31 January 1958 the ALN in Tunisia received some 16,805 military rifles, 329 submachine guns, 296 automatic rifles, 380 machine guns, eighteen 81 mm mortars, twelve 120 mm mortars, 150 bazookas, and 300 grenade launchers, as well as over 6.5 million cartridges of various calibers, 6,610 mortar rounds, 2,000 grenades, and 2,500 antitank rockets.[59] In the first half of 1958 the ALN accumulated some seventeen thousand weapons in Tunisia and passed into Algeria about six thousand of them.[60]

### The Moroccan Base

Until the end of 1956 Morocco was the principal supporter of the Algerian rebellion, providing funds, arms, and other equipment and permitting limited use of Moroccan territory by the ALN. Following its independence in 1956 and the subsequent withdrawal of French troops, Morocco openly supported the FLN, and the FLN established a major base area in eastern Morocco to support *Wilaya* 5. The principal mission of the so-called Western Base was to support ALN forces in Algeria; thus the FLN activities inside Morocco remained much more restricted than in Tunisia, where the ALN battle corps was assembled. ALN forces in Morocco probably never exceeded nine thousand men, including recruits and logistical personnel.[61] Moreover, FLN activities in Morocco were restricted by chronic shortages of arms and equipment, caused in part by Moroccan government sequestration of matériel shipped to Morocco for the FLN and in part by effective French naval interdiction.[62]

Possessing a substantial army, the Moroccan government of King Mohammed V was more successful than the Tunisian government of Habib Bourguiba in controlling the FLN. FLN relations with Morocco were affected by the same considerations as were FLN relations with Tunisia—the Moroccan desire to maintain good relations with France and the burdensome presence of Algerian refugees (some ninety thousand) in eastern Morocco.[63] FLN operations in Morocco were also affected by tribal conflicts, particularly in the border areas on the edge of the Sahara.[64] For example, FLN activities in the area around Figuig were complicated during 1957–1958 by open conflict with the Moroccan Liberation Army over the allegiance of the Doui M'ni and Ouled Djerir tribes; at the end of 1959 and beginning of 1960 FLN operations in the area of Oujda were similarly affected. The rivalry of the FLN and the Moroccans over the Sahara led the government of Prime Minister Ahmed Balafrej to oppose all importation of arms for the FLN, but relations improved during the government of Prime Minister Abdallah Ibrahim and the personal government of King Mohammed V.[65] Eventually, many of the conflicts of the FLN with the Moroccan authorities were resolved at a meeting in Tangiers of representatives of the FLN, Morocco, and Tunisia, at which the common interests of the three were defined.[66]

Like the Eastern Base in Tunisia, the Western Base in Morocco was organized with a rear area (in western Morocco) and two operational zones on the frontier with Algeria—a northern zone (*Zone Opérationnelle Nord*; ZON) focused on Oujda and a southern zone (*Zone Opérationnelle Sud*; ZOS) focused on Figuig.[67] Important logistical facilities were located in all three zones, but the combat forces were concentrated in the two operational zones on the border with Algeria. In mid-1960 the ALN in Morocco had eleven companies (katibas) in the ZON between the Mediterranean and Tenbrara and five small (250–300 men) battalions around Figuig.[68]

The FLN/ALN logistical facilities in Morocco were controlled by Base 15 at Oujda, commanded by Colonel Larbi Ben M'Hidi.[69] Base 15 was colocated with the command post and base of *Wilaya* 5, which was responsible for providing personnel for the operation of depots and other facilities in Morocco.[70] On 15 May 1959 French intelligence officers estimated that there were only some 4,720 ALN troops in Morocco, including 900 combat troops on the frontier, 1,780 invalids, 800 recruits, and some 3,820 transits and support personnel.[71] The number of ALN personnel assigned to logistical duties in Morocco was quite small. In October 1958 there were only five hundred men assigned to the FLN infrastructure in Morocco.[72] The FLN headquarters and support facilities located in Morocco at that time are as shown in Table 6.2. By October 1959 additional MARG general depots were operating at Kenitra and Casablanca in western Morocco, the headquarters of *Wilaya* 5 had been moved to Bou Arfa (on 10 October 1959), and the FLN was seeking authorization from the Moroccan government to relocate the caravan departure base at Bou Denib to Erfoud or Aoufouss in order to accommodate the shift of caravan routes from Morocco into Algeria farther to the south.[73]

As was the case with the FLN depots in Tunisia, the stockage levels of the depots in Morocco fluctuated according to the arrival of shipments from abroad and the success of the ALN in moving matériel onward to the forces in Algeria. At the end of October 1959 the MARG depots in Morocco had in stock at least 5,000 weapons, 1,068 tons of wheat, 822 tons of rice, 129 tons of tea, 8,000 meters of cloth, and 9,850 pairs of shoes.[74] By 1 December 1959 the level of rations and quartermaster items had risen by virtue of shipments received from Communist China to 2,000 tons of wheat; 1,000 tons of rice; 155 tons of tea; 45,800 pairs of shoes; 366,000 meters of cloth; and 2,500 blankets.[75] The capacity of FLN manufacturing workshops in Morocco was quite limited, but the grenade factory at Tetouan was equipped with German machinery, and in the last half of 1957 it had a production capacity of several thousand grenades per month.[76]

## In the *Wilayas*

Comparatively little is known about the logistical facilities located inside Algeria and controlled by the various *wilaya* commanders. There do not appear to have been any substantial logistical bases inside Algeria, but sizable caches of weapons and other supplies were found frequently by French units, and there

**Table 6.2**
**FLN/ALN Support Facilities in Morocco, 15 May 1959**

| | |
|---|---|
| **Western Morocco** | |
| Logistical Office | Rabat |
| General Depot | Tetouan |
| Mine/Grenade/Bangalore Workshop | Tetouan |
| Training Centers (4) | Khemisset; Larache; Tangier; Tetouan |
| **Northern Operational Zone** | |
| Command Post-General Staff-West | Nador |
| Command Post-*Wilaya* 5 | Beni Drar-Oujda |
| Command Post-*Mintaka* 2, *Wilaya* 5 | Martimprey |
| Command Post-*Mintaka* 1, *Wilaya* 5 | Oujda |
| Headquarters, Base 15 | Oujda |
| General Depots (3) | Nador; Zaio; Oujda |
| Transit Depots (4) | Martimprey; Oujda; S. of Oujda; Berguent |
| Signal Workshop | Oujda |
| Automotive Workshop | Oujda |
| Mine/Bangalore Workshop | Beni Drar-Oujda |
| Hospital/Convalescent Center | Oujda |
| Rest Camp/Retraining Centers (5) | Talamrech; Martimprey; Beni Drar; Oujda; Cape de l'Eau |
| Training Centers (3) | Dar Kebdani; Berkane; Oujda |
| Transit Post | El-Korchefia |
| **Southern Operational Zone** | |
| Command Post-Mintaka 8, Wilaya 5 | Figuig |
| Transit Depots (4) | Figuig; Bou Arfa; Tendrara; Ain souf el-Kser |
| Mine/Bangalore Workshop | Figuig |
| Hospital/Convalescent Centers (2) | Kser; near Figuig |
| Training Centers (2) | Tendrara; Figuig |
| Transit Posts (4) | E. of Tendrara; Bou Denib; N. of Figuig (2) |

*Source*: Commandement en Chef des Forces en Algérie, État-Major Interarmées, 2e Bureau, *Logistique Rebelle au cours du mois de mai 1959*, No. 3445/EMI/2/OPE/PLIT (Alger, 11 juin 1959), ''Situation du F.L.N. au Maroc au 15-5-59,'' in folder ''10e Région Militaire, État-Major Interarmées, 2e Bureau—Fiches sur la logistique rebelle, decembre 1958–decembre 1959,'' Dossier 1H1689 d. 2, Fonds Algérien, SHAT.

almost certainly were small workshops engaged in the repair of weapons and the manufacture of munitions (bombs and grenades) and other items. For example, the bombs used by rebel terrorists in Algiers and other Algerian cities are known to have been manufactured in local workshops from materials brought in from Tunisia and Morocco or obtained locally. However, the ALN, unlike the Viet Minh in Indochina, did not create an extensive network of clandestine factories and depots but relied instead on the receipt of matériel from the exterior bases and temporary dumps. Nevertheless, some of the ALN base camps and storage areas were quite extensive and well appointed. Ex-''para'' sergeant Pierre Leulliette described one such base that he and his comrades found in a cave in the Constantinois in April 1956.

The floor is covered with thick straw, with sheepskins, sometimes even with rugs. Hundred-pound bags of couscous—one wonders if they were dropped by planes—water in goatskin bottles, sugar, and even live goats and sheep, have been stockpiled here on a very large scale. There are storerooms cut in the rock where, besides equipment and piles of new uniforms of Egyptian make, we find quantities of leaflets and documents (most of which we burn), portable typewriters, Red Cross boxes with syringes, packages of bandages, etc., and blankets much like ours, together with army cots.[77]

Some rebel storage areas found by the French forces were more bizarre. Leulliette described one such cache found by the "paras" during an operation in the Constantinois in early 1956.

One day one of us climbs for fun into one of the olive trees that surround the village, and we soon find that most of these innocent symbols of peace are hollow, and absolutely full of weapons. Just put your arm in and you can pull out magnificent guns, all carefully wrapped and most of them brand-new. A great deal of matériel is coming in by sea in spite of our patrols, and this region is a veritable supply base. Seventeen weapons are found this way. Since we are promised a reward of 500 francs a gun, we go at this work with a will. All the trees are taken apart. We also find cartridges, ammunition belts, clothing, and sacks of beans and wheat. It's the enchanted forest.[78]

## QUARTERMASTER AND ORDNANCE SUPPLIES

Very little definitive information is available on the actual operation of the ALN logistical system at the lowest levels or on consumption rates for rations or munitions. In part this gap in information results from the lack of ALN records (or of western access thereto), the usual secretiveness of the FLN, and the autonomous operations of the *wilaya*s, which inhibited the establishment of standard tables of organization and equipment and standard logistical procedures. The lack of information is also due to the inadequacies of French intelligence collection during the Algerian war. The focus of French collection, and the yardstick by which French success in degrading the rebel logistical system was measured, was the number of weapons and amount of ammunition in the hands of the rebels. In that area more detailed data is available, even if scattered and somewhat contradictory.

Some basic facts are clear. The Algerian rebel forces inside Algeria received the greater portion of their clothing, arms, ammunition, and other equipment as well as a substantial amount of their rations from outside Algeria. Supplies of any kind remained scarce throughout the war, and as a result the ALN never progressed to the "general offense" phase and remained throughout the war in most respects a mobile guerrilla army. ALN consumption rates for most items were relatively low, and the rebels were adept at exploiting the civilian population and living off the countryside, thereby reducing requirements for some items. Because the rebels were lightly armed, did not possess heavy weapons such as artillery pieces or heavy antiaircraft guns, and relied on animals and human porters for the movement of supplies and equipment, requirements for arms, ammunition, and petroleum products remained quite low in terms of the tonnages required.

### Food and Clothing

The ALN obtained rations and clothing from a number of sources, including local exactions and purchases in Algeria and shipments from friendly nations like the People's Republic of China. In general, the diet of all ALN soldiers was austere, as much from natural inclination as from the prevailing scarcity of supplies. The tough soldiers of the ALN were inured to hardship, and several days in the field without regular meals were normal. The rebel soldiers operating in Algeria were usually obliged to content themselves with unleavened bread, *couscous*, peppers, and coffee, with an occasional issue of mutton, rice, or goat's milk.[79] However, the ALN units in Tunisia, particularly those formed as part of the ALN battle corps, appear to have been somewhat better provisioned. On 19 June 1959 French forces operating five kilometers northeast of Lamy intercepted an ALN convoy enroute from Tunisia. Among the items captured were packaged ALN field rations, several components of which were of American manufacture. Packed in a 15 by 15 by 18 cm cardboard carton marked "*G.P.R.A.—MINISTÈRE DU RAVITAILLEMENT*," each ration contained three cans, with cocoa, jam, cookies, crackers, and one packet of powdered milk, manufactured by McKay-Davis Chemical Corporation, Toledo, Ohio; one six-gram packet of granulated sugar, packed by Van Brode Milling Company, Clinton, Massachusetts; and one packet of soluble coffee, manufactured by the George Washington Division, American Home Foods, Morris Plains, New Jersey, and packaged by Maison Keller Corporation, Roseland, New Jersey. Additional items included in the ration were one 1⅞ ounce packet of chewy chocolate caramels (eight pieces), manufactured by the York Caramel Division, Curtiss Candy Company, York, Pennsylvania; one 125-gram can of sardines, marked *SARDINELLES Ä L'HUILE MIRA* (Tunisian); one can of cooked peaches; one five-hundred-gram packet of dried figs, marked *FIGUES SÈCHES D'ALGÉRIE*; one packet of *biscuits de guerre* (hardtack); four 1.25 ounce packets of dehydrated pea soup, manufactured by Pillsbury Mills, Minneapolis, Minnesota, and packed by Sanna Dairies, Menemonie, Wisconsin; and an accessory packet containing one packet of "Chicle Treets" spearmint chewing gum; one packet of "Lucky Strike" cigarettes; two boxes of matches; one packet of toilet paper; three plastic spoons; and one vial of iodine water-purification tablets, manufactured by Harrisson Products, San Francisco, California.[80]

Regular ALN troops wore kakhi and olive-drab fatigue uniforms similar to those worn by U.S. and French troops, and they carried canteens, messkits, blankets, and other equipment manufactured in France, the United States, Eastern Europe, Communist China, or other locations. A good deal of the shoes, clothing, uniform cloth, and camp equipment used by the ALN in the later stages of the war appears to have come from the People's Republic of China.

### Arms and Ammunition

The availability of arms and ammunition was the critical restraining factor on the growth of the ALN and on ALN operations inside Algeria throughout the war. The Algerian nationalists began the rebellion with only a few old firearms, and for the first year of the war the bulk of rebel weapons were obtained by theft or capture from the French.[81] Weapons and ammunition did not begin to flow to the ALN from outside until the fall of 1955, following an arms agreement concluded with Czechoslovakia by Egyptian president Nasser on behalf of the FLN.[82] From the fall of 1955 until mid-1958, when the French barrages began to take hold, the ALN imported significant amounts of arms and ammunition through Morocco and Tunisia. After the summer of 1958, however, the outside sources were all but shut off, and the number of weapons available to the ALN inside Algeria declined drastically, as did the amount of ammunition percolating through the barrages. By the beginning of 1959 the munitions situation inside Algeria had become very difficult; deliveries from Tunisia and Morocco were low, especially for units in the center of Algeria (*Wilaya* 4), and the ALN forces in Algeria were forced to stockpile automatic weapons, reduce the number and size of units, and restrict combat operations.[83]

Most of the weapons and ammunition obtained by the ALN came from Czechoslovakia, Yugoslavia, and West Germany.[84] A study of the arms taken from the rebels in November–December 1957 revealed that about 36 percent were weapons lost by the French and 64 percent came from other sources.[85] As for the weapons themselves, 29 percent were of British manufacture, 34 percent of German manufacture, 13.5 percent each of Italian and Spanish manufacture, 3 percent of Belgian manufacture, and 9 percent of Czech manufacture. On the whole these weapons were of rather recent manufacture and in good condition. French intelligence officers estimated that the number of weapons available to the ALN in Algeria in January 1958 at the beginning of the "Battle of the Barrages" was about 232 crew-served and 14,870 individual weapons, as shown in Table 6.3.

There does not appear to have been any widely applied standard authorization of weapons or basic loads of ammunition for the ALN units operating inside Algeria. However, ALN documents captured by the French forces inside Algeria in early 1958 indicated that the usual basic load for ALN units was seven rounds per rocket launcher; a thousand rounds per machine gun; five hundred rounds per automatic rifle; and one hundred rounds per submachine gun or rifle.[86] Other documents captured during the first quarter of 1959 reflected the declining amounts of ammunition reaching the rebels inside Algeria, in that the "normal" authorization for a company (*katiba*) was now only four hundred rounds per

**Table 6.3**
**Armament of the Rebel Commands in Algeria, 15 January 1958**

| *Wilaya* | Commander | Troops | RL | Mort | MG | AR | SMG | Rifle |
|---|---|---|---|---|---|---|---|---|
| 1 | M. Amouri | 5,500 | 4 | 21 | 22 | 79 | 360 | 2,820 |
| 2 | B. Lakdar | 3,400 | 8 | 13 | 11 | 101 | 363 | 2,770 |
| 3 | *Amirouche* | 4,900 | 0 | 13 | 6 | 55 | 343 | 1,340 |
| 4 | D. Slimane | 2,300 | 0 | 0 | 13 | 73 | 340 | 730 |
| 5 | H. Boumedienne | 5,300 | 0 | 7 | 24 | 124 | 412 | 3,616 |
| 6 | T. Djoghlali | 400 | 0 | 0 | 2 | 3 | 20 | 170 |
| E. Base | L. Amara | 1,700 | 10 | 31 | 12 | 27 | 220 | 1,300 |
| | **Total** | **23,000** | **22** | **100** | **110** | **470** | **2,000** | **12,400** |

*Source*: Commandement Supérieur Interarmées et 10e Région Militaire, État-Major Interarmées, 2e Bureau, *Organisation des Bandes Rebelles mis à jour à la date du 15 janvier 1958* (Alger, 15 janvier 1958), in folder ''Renseignements reçus concernant l'organisation et le potentiel des bandes rebelles, janvier–octobre 1958,'' Dossier 1H1692 d. 1, Fonds Algérien, SHAT.

machine gun, two hundred rounds per automatic rifle, and twenty–fifty rounds per rifle.[87]

In March 1960 French intelligence officers completed a study of the amounts of arms and ammunition that the ALN would need to pass into Algeria from their exterior bases in order to (1) avoid asphyxiation (that is, maintain the status quo), (2) fully arm all of their combat personnel in Algeria with military weapons, or (3) sustain their forces inside Algeria with all combat personnel fully armed.[88] For the purposes of the study it was assumed that the ALN had twenty-eight thousand combat troops inside Algeria, of which nine thousand were fully armed with military weapons, eleven thousand were partially armed with a mix of military and hunting weapons, and eight thousand were unarmed; that ALN weapons losses would be four hundred weapons per month (based on the fact that the decline in rebel armament had averaged 350–400 weapons per month during 1959); that 1 percent of the ALN's weapons would require replacement each month due to damage, wear, or other factors; and that ALN monthly ammunition consumption would be two hundred rounds per machine gun or automatic rifle, fifty rounds per submachine gun, and ten rounds per rifle. The study concluded that in order to maintain the status quo the ALN would need to provide their forces inside Algeria 350–400 weapons (three to four tons) and 190,000 cartridges (four tons), for a total of seven–eight tons per month. To arm fully all twenty-eight thousand ALN combat troops in Algeria would require delivery of 140 tons of arms (19,000 weapons) and 163 tons of ammunition, for a total of 303 tons. To sustain a force of twenty-eight thousand fully armed men the ALN would be required to pass through or around the barrages four to five tons of arms (680 weapons) and twenty-two tons of ammunition, or a total of twenty-seven tons per month. In fact, because of the constantly increasing effectiveness of the barrages, heavy French pressure on ALN forces inside Algeria, and the steadily declining numbers of weapons in the hands of the ALN forces inside Algeria, even the sustainment rate of seven to eight tons per month was to prove impossible. For example, in 1959 the ALN managed to infiltrate into Algeria about fifty thousand cartridges per month, only about 25 percent of what was required.

As far as French intelligence agencies were concerned, the real measurement of Algerian rebel potential was the number of weapons of various types available to the ALN at any given time, either in the hands of the troops in Algeria or in depots in Morocco, Tunisia, Libya, Egypt, or elsewhere. Such measurements depended on three variables: the number of weapons lost by the French forces to the rebels; the number of weapons lost by the rebels to the French forces; and the number of weapons delivered to the ALN inside Algeria from Tunisia and Morocco. The first two factors are summarized for the main period of the Algerian war in Table 6.4 and depicted graphically in Figure 6.2.

In addition to tracking the steady diminution of rebel armaments, the French intelligence agencies also assiduously tracked the number of weapons available to the ALN inside Algeria in absolute terms. Table 6.5 summarizes the French

**Table 6.4**
**Balance of French and ALN Weapon Losses, November 1954–July 1960**

| Year | Lost by French | Lost by Rebels | Difference (for Rebels) |
|---|---|---|---|
| November–December 1954 | 9 | 300 | –291 |
| 1955 | 644 | 688 | –44 |
| 1956 | 3,326 | 3,349 | –23 |
| 1957 | 1,687 | 6,792 | –5,105 |
| 1958 | 1,851 | 9,095 | –7,244 |
| 1959 | 1,892 | 8,356 | –6,464 |
| January–August 1960 | 1,170 | 4,385 | –3,215 |
| **Cumulative Total** | **10,579** | **32,965** | **–22,386** |

*Source*: Commandant en Chef des Forces en Algérie, État-Major Interarmées, 2e Bureau, Section "Opérations," *Evolution des structures et des méthodes de la rébellion de 1957 à 1960*, No. 4515/EMI/2/OPE (Alger, 18 août 1960), Tableaux ("Armes de guerre perdues" et "Armes de guerre prisés"), in folder "10e Région Militaire, État-Major Interarmées, 3e Bureau—Enseignements de la lutte contre les rebelles d'AFN-ESG, 10 mars 1958," Dossier 1H1942 d. 3, Fonds Algérien, SHAT.

estimates of rebel armament inside Algeria on selected dates between August 1957 and March 1962. The dramatic decline in the number of weapons available to the ALN in Algeria brought about by the "Battle of the Barrages" and the subsequent Challe offensives is displayed graphically in Figure 6.3.

## TRANSPORTATION

The weakest link in the ALN logistical chain was without doubt transportation. The rebels were completely outclassed by the French with respect to logistical and operational mobility. The French forces were well provided with ships, boats, transport aircraft, helicopters, rail transport, and motor vehicles; the ALN had only a few trucks, camels, horses and mules, and the backs of its soldiers and civilian supporters. The Algerian rebels were poorly equipped for the transport battle, even when compared with the Viet Minh, who by the end of the Indochina war had over eight hundred trucks and innumerable small watercraft and were even using sections of the railroads in northern Tonkin. Not only were the ALN transport resources slim and relatively inefficient, but they were also subject to very intense and effective French interdiction by air, sea, and land over terrain that offered little cover or concealment from aerial observation and attack. That the ALN was able to move men and supplies at all is a tribute to its ingenuity and determination.

### Ocean and Coastal Water Transport

Matériel obtained for the FLN in Europe, Asia, and the Middle East was shipped by sea aboard commercial vessels to Egypt, Libya, Tunisia, or Morocco

**Figure 6.2**
**French and ALN Weapon Losses, November 1954–July 1960**

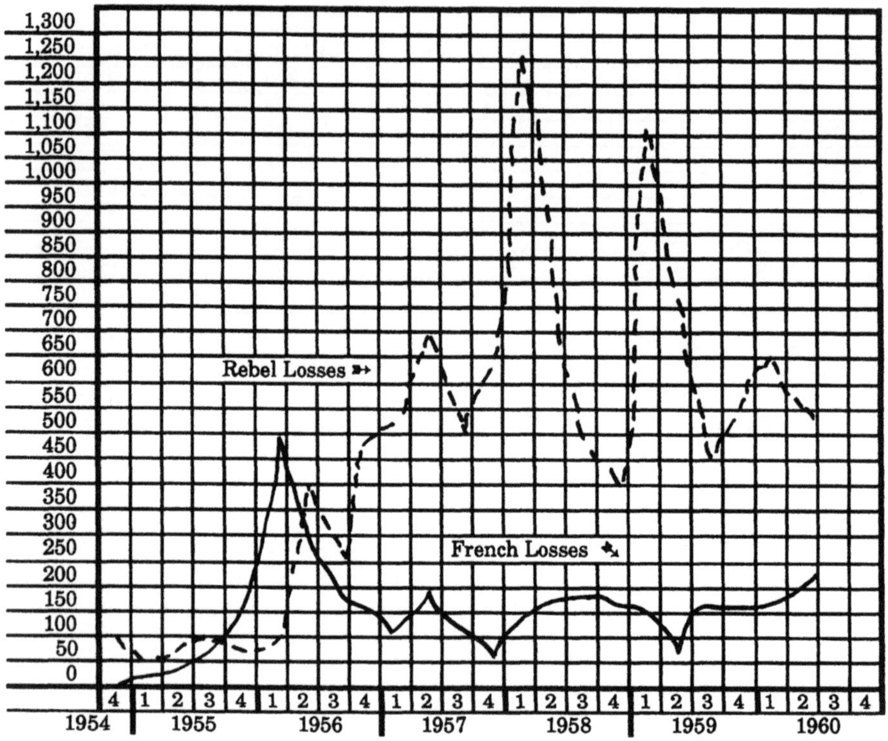


*Source*: Commandant en Chef des Forces en Algérie, État-Major Interarmées, 2e Bureau, Section "Opérations," *Evolution des structures et des méthodes de la rébellion de 1957 à 1960*, No. 4515/EMI/2/OPE (Alger, 18 août 1960), Table, in folder "10e Région Militaire, État-Major Interarmées, 3e Bureau—Enseignements de la lutte contre les rebelles d'AFN-ESG, 10 mars 1958," Dossier 1H1942 d. 3, Fonds Algérien, SHAT. Data points are end of each quarter.

for land transport to ALN bases on the borders of Algeria and subsequent infiltration into the *wilayas*. The vessels employed in this traffic flew a number of different flags and usually moved ALN cargoes under false documentation. For example, ALN cargoes of arms and ammunition were frequently consigned to the Tunisian or Moroccan armed forces, and weapons such as flamethrowers were manifested as "crop sprayers." A considerable quantity of arms, ammunition, and other military equipment reached the ALN by this route. A few examples of shipments that in fact reached the ALN are shown in Table 6.6.

In February 1955 the FLN landed a cargo of arms on the coast of Spanish Morocco from the yacht *Dina*, which had been borrowed from the Queen of Jordan; in general, however, coastal infiltrations were apparently rare despite Algeria's extensive and rugged coastline.[89] Effective French coastal surveillance seems to have inhibited the ALN from attempting to land cargoes of arms,

**Table 6.5**
**Summary of French Estimates of Rebel Weapons in Algeria on Selected Dates**

| Date | Mortar | RL | MG | AR | Total Crew Wpns | SMG | Rifle | Total Wpns |
|---|---|---|---|---|---|---|---|---|
| Aug 1957 | 71 | ? | 97 | 388 | 556 | 2,085 | 10,175 | **12,816** |
| Oct 1957 | 74 | ? | 74 | 315 | 463 | 1,975 | 10,485 | **12,923** |
| Jan 1958 | 85 | 22 | 90 | 462 | 659 | 1,718 | 12,746 | **15,123** |
| Apr 1958 | 83 | 45 | 197 | 522 | 847 | 2,237 | 12,495 | **15,579** |
| Aug 1958 | 85 | 35 | 250 | 550 | 920 | 2,300 | 12,490 | **15,710** |
| Mar 1959 | 66 | 26 | 213 | 540 | 845 | 2,198 | 10,743 | **13,786** |
| Apr 1959 | 77 | 60 | 336 | 599 | 1,072 | 2,732 | 12,647 | **16,451** |
| Jul 1959 | 27 | 15 | 191 | 399 | 632 | 1,626 | 8,168 | **10,426** |
| Aug 1959 | 25 | 16 | 186 | 398 | 625 | 1,570 | 8,040 | **10,235** |
| Nov 1959 | 21 | 18 | 176 | 379 | 594 | 1,480 | 7,430 | **9,504** |
| Sep 1960 | 27 | 16 | 91 | 294 | 428 | 1,472 | 6,100 | **8,000** |
| Jan 1961 | 25 | 15 | 123 | 271 | 434 | 1,335 | 5,240 | **7,009** |
| Apr 1961 | 24 | 14 | 117 | 267 | 422 | 1,275 | 6,400 | **8,097** |
| Jul 1961 | 26 | 15 | 123 | 258 | 422 | 1,200 | 4,300 | **5,922** |
| Oct 1961 | 24 | 15 | 130 | 262 | 431 | 1,295 | 4,360 | **6,086** |
| Jan 1962 | 18 | 15 | 131 | 254 | 418 | 1,305 | 4,232 | **5,955** |
| Mar 1962 | 27 | 12 | 128 | 258 | 425 | 1,452 | 4,255 | **6,132** |

*Source*: Compiled by the author from various sources, principally the monthly reports on rebel organization and logistics prepared by the 2e Bureau, Joint General Staff. Civilian weapons are excluded. Data points are the first of each month.

**Figure 6.3**
**Decline in Potential Armament of the ALN in Algeria, May 1958–July 1960**

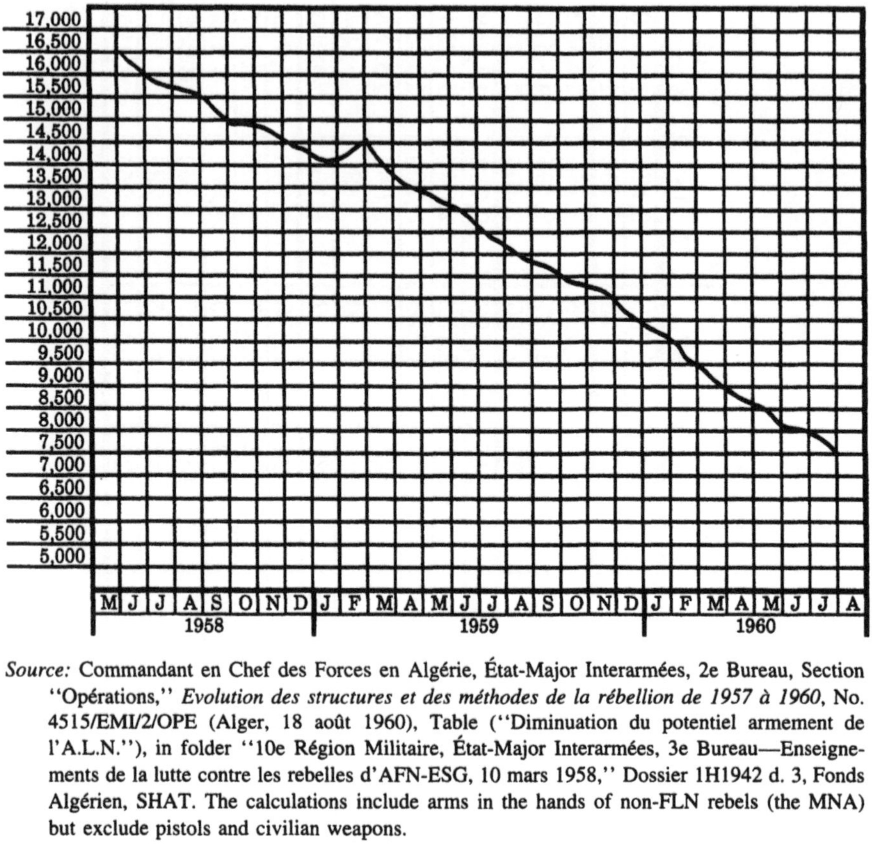


*Source:* Commandant en Chef des Forces en Algérie, État-Major Interarmées, 2e Bureau, Section "Opérations," *Evolution des structures et des méthodes de la rébellion de 1957 à 1960*, No. 4515/EMI/2/OPE (Alger, 18 août 1960), Table ("Diminuation du potentiel armement de l'A.L.N."), in folder "10e Région Militaire, État-Major Interarmées, 3e Bureau—Enseignements de la lutte contre les rebelles d'AFN-ESG, 10 mars 1958," Dossier 1H1942 d. 3, Fonds Algérien, SHAT. The calculations include arms in the hands of non-FLN rebels (the MNA) but exclude pistols and civilian weapons.

ammunition, and other matériel directly on the Algerian coast to any substantial degree.

## Air Transport

Despite frequent reports by French soldiers of nighttime overflights and landings of unidentified aircraft and a constant preoccupation on the part of French intelligence officers with the possible air-landing or parachuting of supplies to the rebels, there is no reliable evidence that the ALN in Algeria received any of its arms, ammunition, or other supplies by air. The ALN had some pilots in training in Egypt and may have been able to call on Arab countries for limited airdrops, but Constantin Melnik, who as a member of the staff of Prime Minister Michel Debré would have been in a position to know, stated flatly that "the FLN never had any aircraft at its disposal, even at its rear bases in Morocco,

**Table 6.6**
**Representative Successful Shipments for the ALN**

| Date | Vessel | Registry | Loading Port | Discharge Port | Cargo/Remarks |
|---|---|---|---|---|---|
| Nov 1957 | Ardahan | Turkish | Turkey | Tripoli | arms and ammunition |
| Apr 1959 | Ottobane | unknown | PRC | Tunis and Casablanca | 192 tons green tea; 37,500 prs shoes; 41 tons kakhi cloth |
| May 1959 | Vigrafjord | Norwegian | PRC | Casablanca | 1,000 tons rice |
| May 1959 | Rikskou | unknown | PRC | Casablanca | 2,000 tons wheat; 55 tons blankets; 10 cases wine |
| May 1959 | Omalgora | unknown | Saudi Arabia | Tripoli | 30 tons explosives; 220 tons inf munitions; 17 vehicles |
| Sep 1959 | Tacito | Italian | Trieste | Tunis | 6,651 raincoats |
| Sep 1959 | Vittoria | Italian | Trieste | Tunis | 8,246 raincoats |
| Sep 1959 | Minima | Italian | Trieste | Tunis | 7,000 wirecutters; 1,200 belts; 2,300 prs pants; 510 meters cloth; 6,000 razors; 5,040 prs shorts; 1,200 plastic glasses; 5,000 messkits |

*Source:* Compiled by the author from various sources, principally the monthly reports on rebel logistics prepared by the 2e Bureau, Joint General Staff.

Tunisia, and Libya.''[90] However, one French Air Force commander in Algeria, General Ezanno, later admitted that there were ''some very hazy instances of air intrusion from Tunisia and Morocco. They could have brought, at a maximum, radio sets, radio spares, literature, and money. We had, of course, a little smuggling activity by air. They were just small intrusions, nothing much.''[91] French intelligence officers were constantly on the lookout for ALN infiltration of Algeria by air, even though they concluded that the technical demands of such a resupply system were beyond the capabilities of the rebels.[92] However, some cargoes were moved from abroad to ALN bases outside Algeria by air. For example, on 16 October 1959 a shipment of arms, ammunition, and radios was delivered by air from Iraq to Benghazi via Cairo, and a second such delivery was made on 29 October 1959.[93]

## Ground Transport

From time to time the ALN forces in Algeria were able to commandeer and utilize civilian motor vehicles, but most tactical and logistical movements inside Algeria were conducted on foot or with pack animals, such as camels, horses, and mules. Both man-pack and animal transport were limited in scale and very inefficient.[94] As a rule, such movements took place by the infiltration of small groups rather than in large convoys, which could be observed and attacked by the ever-vigilant French aircraft.[95] The practical effect of the lack of any heavy transport was to reduce the tactical mobility of rebel forces and restrict their armament to light and easily transportable weapons that did not consume tremendous amounts of ammunition. Consequently the ALN inside Algeria consistently found itself outmaneuvered and outgunned by the French forces.

Highway movements outside Algeria itself, though they might be observed by French aerial and ground surveillance, were immune from attack, and the FLN had a substantial fleet of motor vehicles supporting its bases in Egypt, Libya, Morocco, and Tunisia. The FLN depots in Tunisia, for example, received a good portion of the arms, ammunition, and other supplies provided by FLN supporters by highway from Alexandria and the depot at Marsa Matrouh, via Tripoli and Tunis. This well established LOC is indicated on Map 6. The average travel time was about five hours from Marsa Matrouh to Tripoli, one hour from Tripoli to Bengardine, and two or three hours from Bengardine to Tunis.[96] The GPRA Ministry of Armament and General Supplies controlled three sections of transport vehicles assigned to the Tripoli Base and responsible for motor transport operations on the main LOC.[97] Two of these transport sections were equipped with a command vehicle (Land Rover) and six Mercedes twelve-ton trucks, and the third section had a jeep, one Bedford five-ton truck, two Mercedes seven-ton trucks, two Fiat fifteen-ton tractor-trailers, and two Lancia twenty-five-ton tractor-trailers.[98] Between 27 January and 16 March 1959, for example, the two truck sections moved some three hundred tons of cargo (thirty-five to forty truckloads) in shuttle operations between Marsa Matrouh in Egypt and FLN depots in Tunisia.[99] During the month of October 1959 the FLN transports moved some forty tons of munitions, camp equipment, and clothing from Libya into Tunisia.[100]

In addition to the trucks used to maintain the main supply line from Egypt through Libya into Tunisia, the FLN also possessed a number of light vehicles (automobiles and light trucks) for support of the Provisional Government; transport and liaison vehicles belonging to the MARG, the Ministry of Liaison and General Communications, and Tunis Base; and, in Tunisia and Morocco, a number of vehicles belonging to the ALN and used to satisfy the daily requirements of the ALN units in those areas.[101] In all, the FLN had some 269 vehicles of various types in November 1959, a number that had increased to about 350 vehicles in February 1960, as shown in Table 6.7.[102] Maintenance for FLN motor vehicles was accomplished by automotive workshops in Tripoli, Tunis, and Oujda.[103]

## CONCLUSION

There is much we still do not know about the logistical system of the FLN/ALN, and given the lack of surviving records it is unlikely that we shall ever be able to reconstruct Algerian rebel logistical operations at the unit level or determine with any degree of accuracy the actual amounts of arms, ammunition, and other supplies the rebels succeeded in passing through the French defensive barrages. But it is clear that on the whole the FLN/ALN was amazingly successful in accomplishing the first two of its logistical tasks. The necessary out-

## Map 6
## ALN Infiltration Routes into Algeria, January–March 1959

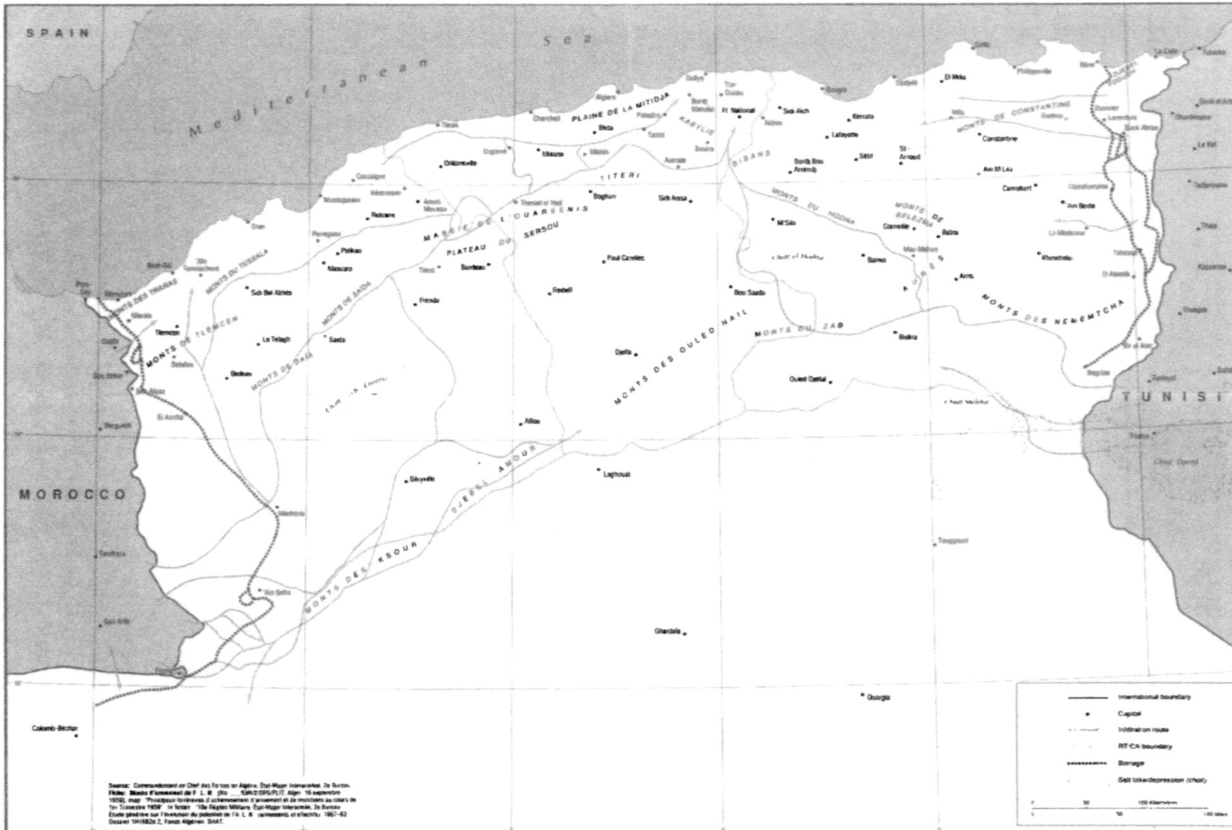

**Table 6.7**
**Distribution of FLN Motor Vehicles by Type, February 1960**

| Location | Jeeps & Land Rovers | Light Trucks | Trucks (1.5T–25T) | Misc Vehicles | Total Vehicles |
|---|---|---|---|---|---|
| Morocco | 6 | 24 | 29 | 3 | **62** |
| Tunisia | 30 | 100 | 117 | 16 | **263** |
| Libya | 2 | 0 | 21 | 0 | **23** |
| **Total** | **38** | **124** | **167** | **19** | **348** |

*Source*: Ministére de la Défense Nationale et des Forces Armées, Secretariat d'État au Forces Armées "Terre," État-Major de l'Armée, 2e Bureau, *La Rébellion Algérienne—Étude de la Menace (mars 1960)* (Paris, 23 mai 1960), Annexe 6, in folder "10e Région Militaire, État-Major Interarmées, 3e Bureau—Enseignements de la lutte contre les rebelles d'AFN-ESG, 10 mars 1958," Dossier 1H1942 d. 3, Fonds Algérien, SHAT. The category of "Miscellaneous Vehicles" includes ambulances, buses, cars, and tankers.

side financial support and sources of weapons, ammunition, and other war material were obtained quickly in the early months of the rebellion and increased as time went on. Also, the establishment of bases in Tunisia and Morocco and the staging of supplies for the Interior at those bases did not prove very difficult. However, the rebel logistical system had a striking weakness—its distribution system. The lack of efficient transport within Algeria and effective French sea, air, and ground interdiction made it extremely difficult for the rebels to move men and supplies from their safe bases in Tunisia and Morocco to the points where they were needed inside Algeria, or between the *wilayas*, or even within each *wilaya*. This lack of logistical and operational mobility came close to bringing about the total collapse of the rebellion as a serious military threat to the French.

## NOTES

1. Alf Andrew Heggoy, *Insurgency and Counterinsurgency in Algeria* (Bloomington: Indiana University Press, 1972), 115.
2. Ibid., 116–117.
3. Ibid., 117.
4. Ministère de la Défense Nationale et des Forces Armées, Secretariat d'État au Forces Armées "Terre," État-Major de l'Armée, 2e Bureau, *La Rébellion Algérienne—Étude de la Menace (mars 1960)* (Paris, 23 mai 1960), 13–14, in folder "10e Région Militaire, État-Major Interarmées, 3e Bureau—Enseignements de la lutte contre les rebelles d'AFN-ESG, 10 mars 1958," Dossier 1H1942 d. 3, Fonds Algérien, SHAT.
5. Commandement en Chef des Forces en Algérie, État-Major Interarmées, 2e Bureau, *Organisation des Bandes Rebelles à la date du 30 septembre 1960* (Alger, 30 septembre 1960), page W.1–1, in folder "10e Région Militaire, État-Major Interarmées, 2e Bureau—Renseignements reçus concernant l'organisation et le potentiel des bandes rebelles, janvier–octobre 1960," Dossier 1H1692 d. 3, Fonds Algérien, SHAT.
6. Ibid., page C.D.F2.

7. 10e Région Militaire, État-Major Interarmées, 2e Bureau, *Fiche sur la Rébellion Algérienne* (Paris, 16 mars 1959), 1, in folder "10e Région Militaire, État-Major Interarmées, 2e Bureau—Étude générale sur l'évolution du potentiel de l'A.L.N. (armements et effectifs), 1957–62," Dossier 1H1682 d. 2, Fonds Algérien, SHAT.

8. Edgar O'Ballance, *The Algerian Insurrection, 1954–62* (Hamden, CT: Archon Books, 1967), 160.

9. Ibid.

10. Arslan Humbaraci, *Algeria: A Revolution That Failed* (New York: Frederick A. Praeger, 1966), 52–53; Alistair Horne, *A Savage War of Peace* (New York: Viking Press, 1977), 236–237. Students were required to contribute five hundred FF, workers three thousand FF, and shopkeepers up to fifty thousand FF or more per month. In 1959 French authorities arrested some 848 FLN fund collectors (see Humbaraci, 54).

11. Horne, 237. Around 300 million FF per month were collected in the Paris region alone in 1960 (see p. 409).

12. Ibid., 237–238.

13. Commandant en Chef des Forces en Algérie, État-Major Interarmées, 2e Bureau, Section "Opérations," *Evolution des structures et des méthodes de la rébellion de 1957 à 1960*, No. 4515/EMI/2/OPE (Alger, 18 août 1960), Table ("Armes du guerre perdues"), in folder "10e Région Militaire, État-Major Interarmées, 3e Bureau—Enseignements de la lutte contre les rebelles d'AFN-ESG, 10 mars 1958," Dossier 1H1942 d. 3, Fonds Algérien, SHAT.

14. Ibid., Tableaux "Armes du guerre perdues" and "Armes du guerre prisés."

15. Pierre Leulliette, *St. Michael and the Dragon: Memoirs of a Paratrooper* (Boston: Houghton Mifflin, 1964), 161.

16. Horne, 85.

17. O'Ballance, 36. The Maghreb, "the land of the setting sun," included Tunisia, Algeria, and Morocco.

18. Ibid., 123.

19. Humbaraci, 20–21.

20. Horne, 130.

21. George A. Kelly, *Lost Soldiers: The French Army and Empire in Crisis, 1947–1962* (Cambridge, MA: MIT Press, 1965), 162.

22. Horne, 547.

23. Ibid., 246, quoting FLN spokesman M'hammed Yazid.

24. Humbaraci, 44.

25. Horne, 404–405.

26. Ibid., 405.

27. Humbaraci, 44.

28. *Fiche sur la Rébellion Algérienne*, 7; Horne, 317.

29. Norman C. Walpole et al., *U.S. Army Area Handbook for Algeria*, Department of the Army Pamphlet No. 550–44 (Washington, DC: Headquarters, Department of the Army, January 1965), 318.

30. Humbaraci, 44; *U.S. Army Area Handbook for Algeria*, 319.

31. Commandement en Chef des Forces en Algérie, État-Major Interarmées, 2e Bureau, *Logistique Rebelle au cours du mois d'avril 1959*, No. 2880/EMI/2/OPE/PLIT (Alger, 16 mai 1959), 3, in folder "10e Région Militaire, État-Major Interarmées, 2e Bureau—Fiches sur la logistique rebelle, decembre 1958–decembre 1959," Dossier 1H1689 d. 2, Fonds Algérien, SHAT.

32. *U.S. Army Area Handbook for Algeria*, 309.

33. Ibid., 305.

34. Ibid.

35. *Logistique Rebelle au cours du mois d'avril 1959*, 1.

36. Commandement en Chef des Forces en Algérie, État-Major Interarmées, 2e Bureau, *Logistique Rebelle au cours du mois de mai 1959*, No. 3445/EMI/2/OPE/PLIT (Alger, 11 juin 1959), 1, in folder "10e Région Militaire, État-Major Interarmées, 2e Bureau—Fiches sur la logistique rebelle, decembre 1958–decembre 1959," Dossier 1H1689 d. 2, Fonds Algérien, SHAT.

37. O'Ballance, 67.

38. Kelly, 146. The Tunisian and Moroccan bases were manned primarily by personnel recruited by the FLN from among the Algerian refugees who had fled from the border areas when in 1957 the French began to clear the border regions and erect the barrages (see Heggoy, 228–229).

39. *Evolution des structures et des méthodes*, 21: "Sans l'hospitalité accordée aux Rebelles par la TUNISIE et le MAROC, il y a longtemps que la Paix et la Liberté règneraient en ALGERIE."

40. The details that follow are drawn from this study: 10e Région Militaire, État-Major Interarmées, 2e Bureau, *Étude sur l'aide exterieure susceptible d'être apportée à la rébellion algérienne (periode du 1.9.56–31.12–56)* (Alger, 14 septembre 1956), in folder "10e Région Militaire, État-Major Interarmées, 3e Bureau—Étude sur l'aide exterieure susceptible d'être apportée à la rébellion algérienne, 1956," Dossier 1H1874 d. 7, Fonds Algérien, SHAT.

41. Commandement en Chef des Forces en Algérie, État-Major Interarmées, 2e Bureau, *Fiche: Stocks d'armement du F.L.N.* (Alger, 16 septembre 1959), 1, in folder "10e Région Militaire, État-Major Interarmées, 2e Bureau—Étude générale sur l'évolution du potentiel de l'A.L.N. (armements et effectifs), 1957–62," Dossier 1H1682 d. 2, Fonds Algérien, SHAT.

42. 10e Région Militaire, État-Major Interarmées, 2e Bureau, "*Charte: Armement de Guerre Détenu par le F.L.N. en dehors de l'Algérie* (au 1er novembre 1959 et au 1er octobre 1958)" (Alger, [1959]), in folder "10e Région Militaire, État-Major Interarmées, 2e Bureau—Étude générale sur l'évolution du potentiel de l'A.L.N. (armements et effectifs), 1957–62," Dossier 1H1682 d. 2, Fonds Algérien, SHAT.

43. Ibid. See also: Commandement en Chef des Forces en Algérie, État-Major Interarmées, 2e Bureau, *Logistique Rebelle au cours du mois de septembre 1959*, No. 6025/EMI/2/OPE/PLIT (Alger, 5 octobre 1959), Annexe 2, in folder "10e Région Militaire, État-Major Interarmées, 2e Bureau—Fiches sur la logistique rebelle, decembre 1958–decembre 1959," Dossier 1H1689 d. 2, Fonds Algérien, SHAT.

44. *Fiche: Stocks d'armement du F.L.N.*, 1–2.

45. Commandement Supérieur Interarmées, 10e Région Militaire, État-Major Interarmées, 2e Bureau, *Étude générale sur l'evolution du potentiel de l'A.L.N. (armement-personnel) au cours du 2° semestre 1957*, No. 1,655/EM.10/2/OPE/EXPL (Alger, 30 janvier 1958), 13–14, in folder "10e Région Militaire, État-Major Interarmées, 2e Bureau—Étude générale sur l'évolution du potentiel de l'A.L.N. (armements et effectifs), 1957–62," Dossier 1H1682 d. 2, Fonds Algérien, SHAT.

46. Commandement Supérieur Interarmées pour la 10e Région Militaire, État-Major Interarmées, 2e Bureau, *Logistique rebelle du 1er juillet au 30 septembre 1958*, No. 6969/RM.10/2/PLIT (Alger, 10 octobre 1958), Annexe 1, 1, in folder "Fiches sur la

logistique rebelle, decembre 1957–decembre 1958,'' Dossier 1H1689 d. 1, Fonds Algérien, SHAT.

47. Ibid., Annexe 1, 1–2.
48. O'Ballance, 85.
49. Horne, 130; O'Ballance, 85.
50. *U.S. Army Area Handbook for Algeria*, 308–309.
51. Ibid., 309.
52. O'Ballance, 85–86.
53. *Evolution des structures et des méthodes*, 18.
54. *Logistique rebelle du 1er juillet au 30 septembre 1958*, Annexe 1, 2.
55. Ibid.
56. *Evolution des structures et des méthodes*, 17.
57. *Logistique Rebelle au cours du mois de mai 1959*, ''Situation F.L.N. en Tunisie au 15-5-59.'' By August 1960 there were in the ZON about 6,200 men in ten battalions plus another 1,000 in convoy units, and in the ZOS around 2,600 men in four battalions and another 300 in convoy units. The General Staff and service units accounted for another five hundred men, and recruits in training numbered between four and five thousand (see *Evolution des structures et des méthodes*, 17).
58. *Logistique rebelle du 1er juillet au 30 septembre 1958*, Annexe 1, 2.
59. *Étude générale sur l'evolution du potentiel de l'A.L.N.*, 10.
60. Délégation Générale du Gouvernement et Commandement en Chef des Forces en Algérie, 10e Région Militaire, Cabinet Militaire, *Memoire sur la situation militaire* (Alger, 13 juin 1958), 4, in folder ''10e Région Militaire, État-Major Interarmées, 3e Bureau—Le problem militaire en Algérie, avril 1957,'' Dossier 1H1933 d. 2, Fonds Algérien, SHAT.
61. *Organisation des Bandes Rebelles à la date du 30 septembre 1960*, 3.
62. *Evolution des structures et des méthodes*, 13.
63. Ibid., 14; *U.S. Army Area Handbook for Algeria*, 308.
64. *Evolution des structures et des méthodes*, 14.
65. Ibid., 13.
66. O'Ballance, 86.
67. *Evolution des structures et des méthodes*, 14.
68. Ibid.
69. *Organisation des Bandes Rebelles à la date du 30 septembre 1960*, page ALNM.2.
70. *Logistique rebelle du 1er juillet au 30 septembre 1958*, Annexe 1, 3.
71. *Logistique rebelle au cours du mois de mai, 1959*, ''Situation du F.L.N. au Maroc au 15-5-59.''
72. *Logistique rebelle du 1er juillet au 30 septembre 1958*, 4. The number included headquarters personnel and trainers as well as logistical support personnel.
73. Commandement en Chef des Forces en Algérie, État-Major Interarmées, 2e Bureau, *Logistique Rebelle au cours du mois d'octobre 1959*, No. 6550/EMI/2/OPE/PLIT (Alger, 5 novembre 1959), 8–9, in folder ''10e Région Militaire, État-Major Interarmées, 2e Bureau—Fiches sur la logistique rebelle, decembre 1958–decembre 1959,'' Dossier 1H1689 d. 2, Fonds Algérien, SHAT.
74. Ibid., 10.
75. Commandement en Chef des Forces en Algérie, État-Major Interarmées, 2e Bureau, *Logistique Rebelle au cours du mois de decembre 1959*, No. 0101/EMI/2/OPE/

PLIT (Alger, 8 janvier 1960), 12, in folder "10e Région Militaire, État-Major Interarmées, 2e Bureau—Fiches sur la logistique rebelle, decembre 1958–decembre 1959," Dossier 1H1689 d. 2, Fonds Algérien, SHAT.

76. *Étude générale sur l'evolution du potentiel de l'A.L.N.*, 9.

77. Leulliette, 197.

78. Ibid., 177.

79. Peter Braestrup, "Partisan Tactics-Algerian Style," *Army*, August 1960, 38.

80. Commandement en Chef des Forces en Algérie, État-Major Interarmées, 2e Bureau, *Logistique Rebelle au cours du mois de juin 1959*, No. 4036/EMI/2/OPE/PLIT (Alger, 3 juillet 1959), Annexe 9, in folder "10e Région Militaire, État-Major Interarmées, 2e Bureau—Fiches sur la logistique rebelle, decembre 1957–decembre 1959," Dossier 1H1689 d. 2, Fonds Algérien, SHAT.

81. O'Ballance, 49. Alistair Horne (*A Savage War of Peace*, 84) notes that Ahmed Ben Bella claimed that the FLN began the rebellion with only 350–400 weapons, the heaviest a machine gun.

82. O'Ballance, 50.

83. *Fiche sur la Rébellion Algérienne*, 10.

84. O'Ballance, 50.

85. *Étude générale sur l'evolution du potentiel de l'A.L.N.*, 7.

86. Commandement Supérieur Interarmées pour la 10e Région Militaire, État-Major Interarmées, 2e Bureau, *Logistique rebelle du 1er janvier au 30 mars 1958*, No. 3,641/EM/10/2/OPE/PLIT (Alger, 19 mai 1958), 4, in folder "Fiches sur la logistique rebelle, decembre 1957–decembre 1958," Dossier 1H1689 d. 1, Fonds Algérien, SHAT.

87. *Fiche sur la Rébellion Algérienne*, 10.

88. Commandement en Chef des Forces en Algérie, État-Major Interarmées, 2e Bureau, *Fiche au sujet d'un renforcement éventuel de l'A.L.N. de l'Interieur*, No. 1562/EMI/2/OPE/PLIT (Alger, 18 mars 1960), in folder "10e Région Militaire, État-Major Interarmées, 2e Bureau—Fiches sur la logistique rebelle, decembre 1958–decembre 1959," Dossier 1H1689 d. 2, Fonds Algérien, SHAT.

89. Horne, 129–130. Horne notes that Moroccan shepherds covered up the traces of the operation by driving their sheep back and forth along the beach.

90. Constatin Melnik, *The French Campaign against the FLN*, RAND Memorandum RM-5449-ISA (Santa Monica, CA: Rand Corporation for the Office of the Assistant Secretary of Defense–International Security Affairs, September 1967), 44.

91. A. H. Peterson, G. C. Reinhardt, and E. E. Conger (eds.), *Symposium on the Role of Airpower in Counterinsurgency and Unconventional Warfare: The Algerian War* (Santa Monica, CA: Rand Corporation, July 1963), 11.

92. *La Rébellion Algérienne*, 16: "La nécessité de ravitailler l'interieur peut prendre pour lui, dans un avenir assez proche, une urgence telle qu'il pourra être contraint de confier cette mission à des moyens aériens. . . . La possibilité technique d'un tel system de ravitaillement n'est pas viable."

93. *Logistique Rebelle au cours du mois d'octobre 1959*, 7.

94. Horses and mules can carry about two hundred pounds, and camels can carry about three hundred pounds for long distances. However, a portion of their load must be reserved for the animal's own forage: about twenty-four pounds of grain and fodder and eight gallons of water per day per horse or mule; some thirty-five pounds of grain and straw and ten gallons of water per day per camel (see Donald W. Engels, *Alexander the Great and the Logistics of the Macedonian Army* [Berkeley, CA: University of California

Press, 1978], 14 and 126–129). A seasoned North Korean porter was able to carry a load of perhaps sixty pounds for a distance of fifteen miles per day, while Viet Minh porters normally carried a load of about forty-five pounds for 15.5 miles per twenty-four-hour day (see E. L. Atkins, H. P. Griggs, and Roy T. Sessums, *North Korean Logistics and Methods of Accomplishment*, ORO Technical Memorandum ORO-T-8 [EUSAK] [Chevy Chase, MD: Operations Research Office, Johns Hopkins University, 1951], 7; D.M.O. Miller, " 'A Handful of Rice?': Logistics in the Viet Minh Campaign," *The Army Quarterly and Defence Journal* 100, no. 1 [April 1970], 110).

95. Vertol Aircraft Corporation (T. R. Pierpont, Director, European Operations), *French Army Helicopter Operations in Algeria, June 1956–September 1959*, Report No. SM–406 (Morton, PA: Vertol Aircraft Corporation, 1 November 1959), 30.

96. *Étude générale sur l'evolution du potentiel de l'A.L.N.*, 12.

97. *Logistique Rebelle au cours du mois de juin 1959*, Annexe ("Situation du Parc Auto du F.L.N.").

98. Ibid. In October 1959 the two Mercedes seven-ton trucks were transferred from the base at Marsa Matrouh to Tripoli, and the two twenty-five-ton heavy transporters were transferred from Tripoli to Tunis (see *Logistique Rebelle au cours du mois d'octobre 1959*, 8).

99. *Fiche sur la Rébellion Algérienne*, 7.

100. *Logistique Rebelle au cours du mois d'octobre 1959*, 7–8.

101. *Logistique Rebelle au cours du mois de juin 1959*, Annexe ("Situation du Parc Auto du F.L.N.").

102. 10e Région Militaire, État-Major Interarmées, 2e Bureau, "*Vehicles—Situation au 1 novembre 1959*" (handwritten notes, Alger, novembre 1959), in folder "10e Région Militaire, État-Major Interarmées, 2e Bureau—Étude générale sur l'évolution du potentiel de l'A.L.N. (armements et effectifs), 1957–62," Dossier 1H1682 d. 2, Fonds Algérien, SHAT; *La Rébellion Algérienne*, Annexe 6.

103. *Logistique Rebelle au cours du mois de juin 1959*, Annexe ("Situation du Parc Auto du F.L.N.").

# 7

# Mobility, Counter-Mobility, and Interdiction

Despite aggressive French counter-action, the Algerian National Liberation Front (FLN) was successful in completing its first two logistical tasks: obtaining the necessary supplies and equipment and staging them at safe bases on the borders of Algeria. However, the FLN and the Algerian Army of National Liberation (ALN) were far less successful in forwarding the arms and supplies collected at the bases in Tunisia and Morocco to the rebel forces inside Algeria. Until the completion of the French barrages in September 1957 the onward movement of war matériel to its ultimate users was difficult but not impossible. However, in the so-called "Battle of the Barrages," which occurred in the first half of 1958, the French effectively sealed off the *wilayas* of the interior from their external bases of support. Consequently, much of the matériel collected for the use of the rebels fighting in Algeria was diverted to the buildup of large conventional forces in Tunisia and Morocco, forces that were never effectively committed to the military struggle against the French. Cut off from their sources of supply, the hard-pressed rebel forces inside Algeria were no longer able to sustain large formations but were forced to split up into platoon and even squad-size units, hide many of their crew-served weapons, and reduce the scale and tempo of their operations. What little remained of the ALN fighting forces and logistical infrastructure in Algeria was nearly destroyed in the subsequent Challe offensives between February 1959 and April 1960. The inability of the ALN leaders to evade either the French interdiction of logistical movements across the barrages or the mopping-up operations inside Algeria led directly to what can only be considered a French military victory over the ALN—a victory that proved hollow, however, when FLN political and diplomatic initiatives obtained the independence their military efforts could not secure.

## "STRATEGIC" INTERDICTION

The French government resorted to a number of overt and covert measures to reduce the diplomatic, financial, and matériel support being provided to the Algerian rebels. All the persuasiveness of French diplomacy was brought to bear to inhibit the granting of credits and the sale of arms to the rebels. Where possible, French financial sanctions were also employed, as in the termination in June 1957 of the French economic aid to Tunisia. In June 1959 the French government threatened to publish a list detailing the arms supplied to the Algerian rebels by legitimate companies in such countries as Italy and West Germany.[1] The threat was sufficient to dry up most of the rebel supply sources in Europe, and the FLN was reduced to reliance on the small quantities of arms and equipment that could still be obtained from Arab countries or from illegal arms dealers in Europe and smuggled out. Subsequently, FLN agents in Europe frequently found it necessary to intimidate the arms dealers and smugglers to get the shipments to Algeria.[2]

The French also carried out a program of intimidation against the illegal arms dealers who trafficked with the FLN.[3] French intelligence and covert action agencies carried on a virtual secret war of their own, with no holds barred. The lead in such activities was taken by *Bureau* 24 of the French security service (*Service d'Espionage et Countre-Espionage*; SDECE), led by "Colonel Lamy" and connected to the infamous 11th Shock Battalion (*11e Bataillon de Choc*) of French covert commandos. *Bureau* 24 employed a full range of covert actions to disrupt the illegal arms trade with the FLN, including the use of paid killers and a variety of less drastic, but still quite effective, "dirty tricks." For example, *Bureau* 24 appears to have controlled the mysterious operative known as "The Killer," who "took care of" a number of "problems" for the French security services, and its bag of tricks included surreptitious substitution of cargoes and the operation of arms factories in Spain and Switzerland to manufacture defective arms for the FLN.[4] Another shadowy organization very active in preventing the shipment of arms to the Algerian rebels was known as the "Red Hand" (*Main Rouge*). The "Red Hand" apparently was responsible for assassinating FLN and MNA agents, arms dealers, and smugglers and for blowing up ships transporting or about to load cargoes for the rebels.[5] Little is known about the "Red Hand," other than that it was a French underground terrorist group. It may well have been a "black" program run by *Bureau* 24.

Many of the illegal arms dealers supplying the FLN met untimely deaths by car bomb, poison dart, and silenced pistol. Inadequate arms export controls made West Germany the focal point of the lucrative illegal arms trade with the Algerian rebels, and the port of Hamburg was the origin of many such shipments. Consequently, much of the "counter-action" took place in Germany. For example, the FLN chief in West Germany, Ait Achène, was shot down in Bonn in November 1957, and one of the principal German dealers providing arms to the FLN, George Puchert, who operated out of Tangiers, was the victim of a bomb, full of ball bearings, that exploded under the driver's seat of his automobile during a visit to Frankfurt in early 1959.[6] Six months earlier, in September 1958, one of Puchert's partners, a Swiss explosives expert named Marcel Léopold, had met his end by means of a poisoned dart in the hallway of a Geneva hotel.[7] Some dealers exhibited uncommon good sense and severed their connections with the FLN after surviving assassination attempts. For example, Otto Schlüter, a respectable Hamburg arms maker, ceased supplying the FLN in 1958 after the fourth car-bomb attack in two years wounded him and killed his mother.[8] Former Nazis in exile in Cairo were also prominent in the arms trade, and one of them, an ex–SS member named Ernst-Wilhelm Springer, was particularly active until a mysterious attempt on his life apparently convinced him that the game was perhaps not worth the candle.[9] In some cases the "target" simply disappeared, as was the case with the Italian oil magnate Enrico Mattei, whose private plane crashed mysteriously in October 1962. Mattei appears to have been involved in negotiations to supply the FLN with money and arms in return for special considerations regarding the development of the newly opened Saharan oil fields.[10] The leading candidates for having eliminated Mattei from the scene include the Secret Army Organization (*Organisation Armée Secrète*; OAS) and the SDECE.

## MARITIME INTERCEPT OPERATIONS

French maritime surveillance and intercept operations were very effective, and a significant number of important shipments intended for the ALN never reached their destination. The 4th Maritime Prefecture (*PRÉMAR IV*), responsible for coastal surveillance and intercept operations, all but eliminated deliveries to the ALN by coastal shipping, using a combination of surface vessels and long-range maritime surveillance aircraft. The zone controlled by PRÉMAR IV extended seaward for ninety kilometers from the North African coast, between the meridians of Oujda, Morocco, and Cap Bon.[11] Other elements of the French Navy were responsible for surveillance and intercept operations in the Atlantic and in the Mediterranean Sea. Coastal surveillance was facilitated by the establishment of exclusion zones and by coastal ground patrols by French marines.

Cooperation among the French intelligence services, the Navy, and the Customs Service was excellent, and sufficient assets were available, even at the beginning of the war, to carry out the surveillance and intercept mission successfully. In response to an inquiry by the French commander in chief in Algeria, the commander of PRÉMAR IV reported in January 1956 that the resources then available to him were sufficient to fulfill the tasks with which he was charged.[12] As of January 1956, the assets available included three patrol craft, three subchasers, and five avisos supplemented by four destroyer escorts and five patrol craft on loan from metropolitan France (four of them always in Algerian waters), four vessels from the *Inscription Maritime*, and the twenty-knot customs cutter *Capitaine Rose*. PRÉMAR IV's aerial surveillance assets included four old and worn-out Catalinas based at Lartigue, a squadron of eight TBM-3E single-engine antisubmarine aircraft based at Algiers and Bône, and a monthly allocation of 150 flight hours of long-range Neptune maritime surveillance aircraft based at Lartigue. Additional assets were provided as required later in the war.

French zeal in intercepting suspected cargoes for the ALN in the Mediterranean frequently provoked diplomatic protests from countries whose vessels were hove to and boarded by French naval personnel. Among the countries that registered diplomatic protests with France were Great Britain, Yugoslavia, Bulgaria, Rumania, Poland, and Czechoslovakia; a diplomatic crisis with West Germany arose in December 1960 after France had stopped some seventeen West German ships on the high seas.[13] The number of such intercepts was quite high. About six hundred ships were "controlled" each month, and of that number inquiries were run on about three hundred, and thirty to sixty were actually boarded.[14] In 1959 alone the French Navy in the western Mediterranean tracked some 41,300 vessels, stopped and investigated 2,565, and rerouted eighty-four under escort to French ports for closer examination.[15]

The results of such high-handed methods were excellent, and the French maritime surveillance program did a great deal to prevent arms, ammunition, and other military supplies from reaching the rebels in Algeria. As shown in Table 7.1, the two most significant maritime intercepts were the *Athos* in October 1956—the first major shipment of arms to the rebels from Egypt—and the *Lydice* in April 1959—the largest known shipment and the occasion of a major diplomatic row. The French were aided in their maritime interception program by other countries, such as Spain and even Tunisia and Morocco, who seized or detained cargoes intended for the ALN for their own reasons.

## CONTROL OF ALGERIAN AIR SPACE

The commander of the 5th Air Region was responsible for control of the air space over Algeria, and it acted aggressively to keep out rebel or unidentified aircraft. In response to an inquiry by the French commander in chief in Algeria, the commander of the 5th Air Region, General R. Frandon, replied in January 1956 that his means were not sufficient to preclude clandestine rebel flights over Algeria's borders with Tunisia and Morocco.[16] Notwithstanding, General Frandon was successful in his assigned mission. In some respects the French aerial blockade of Algeria may have been a bit too efficient. The Soviet foreign minister, Andrei Gromyko, accused the French of "international banditry" following an incident on 9 February 1960 in which a French fighter over international waters 130 kilometers north of Algiers buzzed and fired warning shots at a Soviet Il-18 carrying the General Secretary, Leonid Brezhnev, to Morocco for an official visit.[17] The most notorious example of aggressive French air control—perhaps more correctly classified as an act of air piracy—was the interception and forced landing in Algerian territory on 22 October 1956 of an Air

**Table 7.1**
**Representative Shipments for the ALN Intercepted or Seized**

| Date | Ship | Registry | Loading Port | Intended Port | Cargo/Remarks |
|---|---|---|---|---|---|
| Oct 1956 | **Athos** | Sudanese | Alexandria | Tunis? | Intercepted by the French: over 70 tons of arms and ammo (72 mortars; 40 MG; 74 AR; 240 SMG; 2,300 rifles; 2,000 mortar rounds; 600,000 rounds of SAA) |
| Jun 1957 | **Andros** | Greek? | unknown | Casablanca | Seized by Morocco: 7 cases of rifles |
| Jun 1957 | **Juan Illuenca** | Spanish | Alexandria | Casablanca? | Seized by Spain: 3,000 rifles; 550 AR; 1,595 SMG; 750 mortars; 1,350 mortar rounds; 60,660 handgrenades; 6.6 million cartridges; 5 cases of frogman equipment |
| Jun 1957 | **Korsoe** | Finnish | Finland | Casablanca | Seized by Morocco: 20 tons of arms (including 20 81 mm mortars) |
| Jan 1958 | **Slovenija** | Yugoslavian | Yugoslavia | Casablanca | Intercepted by the French: 55 tons of arms (including 4,000 rifles; 200 MG; 1,000 SMG; 1,500 pistols; 15 81 mm mortars; 40 rocket launchers); 95 tons of ammo and explosives (including 350 bangalore torpedoes) |
| Apr 1959 | **Lydice** | Czech | Gdynia | Casablanca | Intercepted by the French: 3,563 cases (12,000) of rifles and munitions; 433 cases (2,000) of MG and parts; total of 581 tons |
| Dec 1959 | **Biesbach** | W. German | Hamburg? | Tripoli | Intercepted by the French: explosives |

*Source*: Compiled by the author from various sources, principally the monthly reports on rebel logistics prepared by the 2e Bureau, Joint General Staff of the commander in chief in Algeria.

Maroc DC-3 en route from Rabat to Tunis and carrying Ahmed Ben Bella and other important members of the FLN External Delegation, who were subsequently imprisoned in France for the duration of the war.[18]

## CROSSING THE BARRAGES

Given the meager transport assets of the ALN and the highly mobile, ever-improving French surveillance and blocking forces, the failure of the ALN logistical system to pass its most important test—to get men and matériel through or around the French border defenses and distribute them to the rebel forces fighting in Algeria—is not surprising. Until the completion of the system of French defensive barriers on the eastern and western frontiers of Algeria in

September 1957 the ALN was in fact able to infiltrate significant numbers of men and quantities of matériel through the screen of French troops covering the borders. However, once the barrages were completed the ALN had only three choices: to infiltrate small groups of men and supplies through the barrage, hoping to evade French detection and pursuit; to skirt the barrages to the south through the open desert, where even small convoys of men and camels were extremely vulnerable to French aerial observation and attack by aircraft, mechanized ground elements, or heliborne units; or to blast through the barriers of wire, mines, radar, and mobile French firepower in major combat operations involving large forces. The ALN attempted all three methods, without appreciable success. By mid-1958 the ALN units inside Algeria began to suffer serious shortages of newly trained recruits, replacement arms, ammunition, and other supplies, and they were forced to scale back the size of formations and the tempo of operations.

### Movements into Algeria before September 1957

Before the completion of the French barrages (barriers) on the eastern and western borders of Algeria in September 1957, the ALN was able to arm new recruits and carry out an aggressive program of attacks on vital French installations, lines of communications, and personnel. Although the French commanders recognized the importance of the logistical support being received by the rebels from outside Algeria, the sealing of the borders was just one of many operational problems and did not receive the highest priority. Once it recovered from the initial shock of the rebellion and significantly increased its combat and support forces in Algeria, the French high command in Algeria concentrated on protecting persons and property and hunting down the rebel bands operating in the interior. The enormous resources required for putting in place and maintaining the defensive *quadrillage* system and protecting some 7,500 ''key points,'' 4,500 kilometers of railroads, and 30,000 kilometers of highways left few troops and little equipment to seal off the 3,500 kilometers of border that Algeria shared with its neighbors.[19]

Despite the high cost the French authorities undertook a number of active and passive measures designed to reduce the amount of matériel reaching the ALN forces of the interior from Tunisia and Morocco. Intensive intelligence and air-ground surveillance activities were carried out, the population was removed from certain key areas, control zones were established, known ALN way stations and depots on the lines of communication from Tunisia and Morocco were attacked, a line of frontier posts was established, and work was started on a double barrage of mines and electrified barbed wire on the northern end of the frontier with Morocco. In December 1956 the French forces in Algeria were granted authority by the Secretary of State for Foreign Affairs to pursue rebel bands over the borders of Algeria into Tunisia and Morocco under certain circumstances (for example, when in ''hot pursuit'').[20] Guarding the frontiers consumed a consid-

**Table 7.2**
**Frontier Deployment of French Forces, April 1957**

| Location | Kilometers | On the Border | In Depth |
|---|---|---|---|
| Morocco-North (Port Say to Teniet-Sassi) | 150 | 5 battalions | 21 battalions |
| Morocco-Center (Teniet-Sassi to Figuig) | 300 | 1 battalion | 5 battalions |
| Morocco-South (Figuig to Tindouf) | 1,000 | 3 battalions | 3 battalions |
| Tunisia-North (La Calle to Negrine) | 400 | 8 battalions | 21 battalions |
| Tunisia-South (Negrine to Ghadames) | 500 | 2 companies | – |
| Libya | 1,000 | 1 company | – |

*Source*: 10e Région Militaire, État-Major Interarmées, 3e Bureau, *Le problem militaire en Algérie* (Alger, avril 1957), 8, in folder ''10e Région Militaire, État-Major Interarmées, 3e Bureau—Le problem militaire en Algérie, avril 1957,'' Dossier 1H1933 d. 2, Fonds Algérien, SHAT.

erable proportion of the scarce French manpower resources. In April 1957 some sixty-eight French battalions or their equivalent, out of a total of 278 battalion equivalents available in the 10th Military Region, were deployed on the frontiers, as shown in Table 7.2.

In April 1957 the northern border with Morocco was somewhat better equipped that the northern Tunisian frontier. As noted above, considerable work had already been accomplished on both the Moroccan and Tunisian border barriers (shown in finished form on Map 6). Several kilometers of the northern Moroccan barrage had been sketched out in 1956, and by April 1957 a continuous network of barbed wire and mines on the northern section impeded infiltration, channelized crossing attempts, and provided warning of intrusions.[21] There were an average of five attempted crossings each night: three would be stopped, and the remaining two groups of rebels would be pursued after daylight.[22] The northern Tunisian frontier presented a greater problem, both by its greater length and by the better organization of the FLN in Tunisia. The French authorities regarded the Tunisian frontier as insufficiently guarded and assigned a high priority to providing additional troops, constructing obstacles, and installing surveillance equipment. The installation of ground radar sets capable of detecting a single individual at fifteen kilometers and groups of men at forty kilometers was scheduled to be completed before 1 June 1957.[23] The southern sections of the Moroccan and Tunisian frontiers received a lower priority for men and equipment, due to the general absence of vegetation, which made observation much easier. In April 1957 the Libyan frontier was quiet, but plans were afoot to increase surveillance and improve infrastructure (trails, outposts, and airfields) to facilitate its defense.[24]

Even the rather primitive French border defenses that existed in April 1957 were fairly effective in impeding the free flow of ALN troops and matériel across the Algerian borders. On 15 June 1957 French forces killed over 205 rebels of an ALN battalion that attempted to cross the border from Tunisia into Algeria.[25] However, the existing border fortifications were still insufficient to prevent the infiltration of considerable amounts of arms and equipment to the ALN forces of the interior. During 1957 an average of two thousand men and one thousand weapons passed into Algeria each month.[26] In the last half of 1957 alone the rebels in Algeria received at least 2,617 weapons, including twenty mortars, thirteen machine guns, twenty-two antitank weapons, 162 automatic rifles, 151 submachine guns, and 2,246 rifles.[27]

## The Construction and Operation of the Eastern and Western Barrages

By the fall of 1957 the French forces in Algeria had begun to gain the upper hand. French manpower and equipment resources had been considerably augmented, and the ALN bands operating in Algeria were being pushed back into their mountain strongholds, where they no longer offered easy targets. The French thus began to contemplate the possibility of sealing off the borders with Tunisia and Morocco entirely and thereby suffocating the remaining ALN forces of the interior. French efforts to create ''*un champ clos algérien*'' (''a closed field in Algeria'') that would isolate the ALN forces of the interior from their sources of men and supplies in Tunisia and Morocco had already begun when the decision was made by the new French Minister of Defense, André Morice, in June 1957 to construct truly formidable barrages of electrified barbed wire and mines backed by air and ground surveillance, flexible artillery, and highly mobile ground troops capable of detecting and eliminating any rebel penetration of Algeria's borders.[28] Initial priority was given to the Tunisian frontier, and by 15 September 1957 some 320 kilometers of what would come to be called the ''Morice Line'' were completed from La Calle to El-Ma-el-Abiod, just northeast of Negrine, where the Sahara desert begins.[29] During 1959 General Maurice Challe added an entire second line (the so-called Challe Line) paralleling the Morice Line on the Tunisian frontier, and continuous improvements were made on all of the barrages right up to the end of the Algerian war in April 1962.[30] Colonel René Laure, who commanded the Bône sector on the northern end of the Morice Line, later recalled that new technical devices were constantly being added to the barrier, his superiors being graduates of technical schools and ''very keen to improve the technical efficiency of the fence. They liked those devices and were always inventing something new. The cost of the fence was, of course, commensurate with the cleverness of the generals!''[31]

The 320-kilometer Morice Line was sited on favorable defensive terrain along the general trace of the Medjerda River corridor between the Bône-Tebessa railroad line and the actual border with Tunisia, some fifteen to thirty-five miles to the east.[32] The barrier itself (sketched in Figure 7.1) consisted of a heavily mined strip a hundred meters wide, in the center of which was an eight-foot, five-thousand-volt electric fence. The fence was backed by two parallel barbed

Figure 7.1
Schematic of the Morice Line

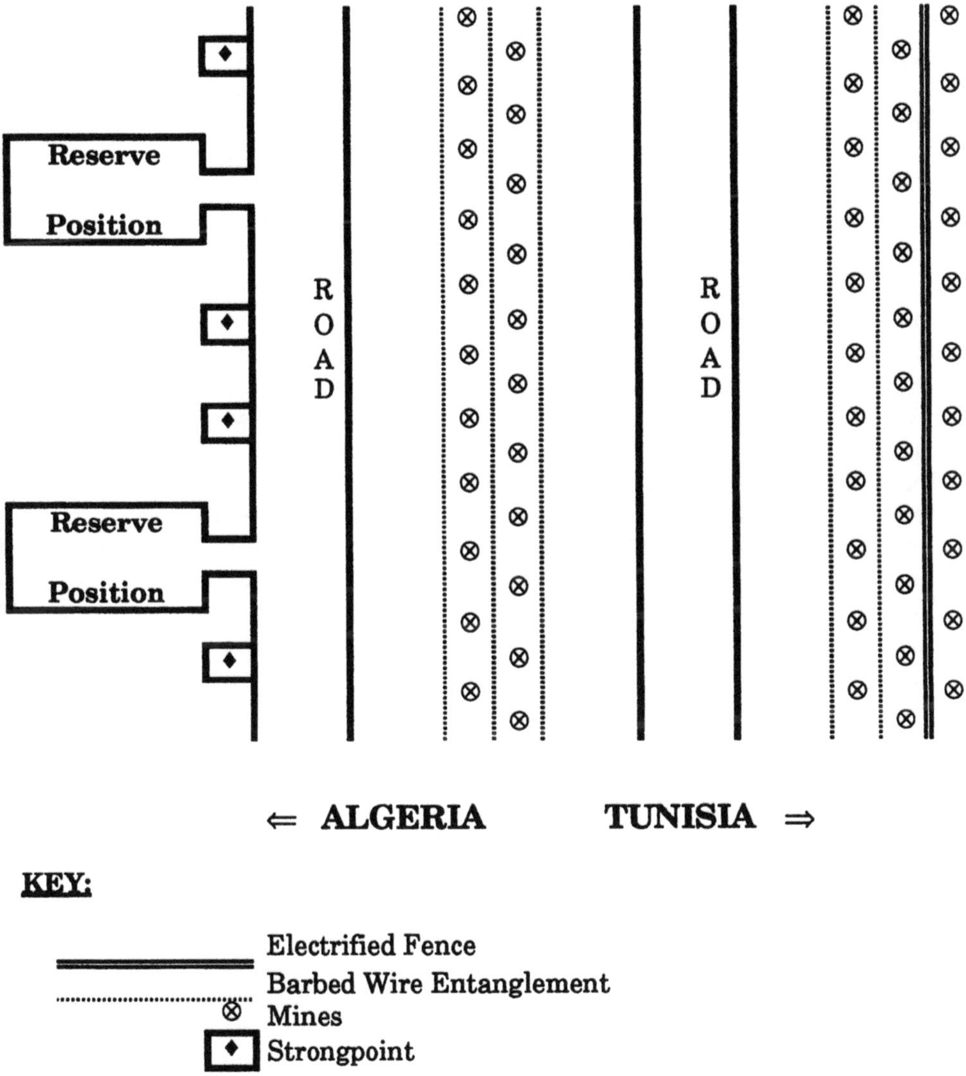

wire entanglements heavily seeded with mines, some 1.2 million in all.[33] Behind the barbed wire barriers was a road to permit the circulation of mechanized patrols, and west of the road was another triple row of barbed-wire barriers. West of the triple barbed-wire barrier was another road, along which at two-hundred to three-thousand-yard intervals were located strongpoints (pillboxes, bunkers, etc.) occupied by the troops manning the barrage. The entire barrage was illuminated by searchlights at thirty-meter intervals and covered by ground radar surveillance. Along about 180 kilometers of the Morice Line radar was linked directly to the artillery, permitting almost immediate fire on any group penetrating the barrier.[34] The ground radars were augmented by twenty-four-hour visual and radar aerial surveillance over the barrage, using B-26s, P-47s, and various types of light observation aircraft during the day and long-range Neptune patrol bombers with radar and flares during the night.[35] The French Air Force master air defense radar on the Tunisian border was linked to the artillery radar system to provide an integrated air and ground detection capability. In addition to the mechanized units manning the barrage itself, other mobile units and some two dozen isolated outposts were located in the "no man's land" between the barrage and the Tunisian border to detect and break up rebel units before they even reached the barrier. To the rear of the barrage were located sector troops, with the mission of tracking, and if possible attacking, any ALN units that managed to cross the barrage. Very mobile elite "para" and Foreign Legion units were held in reserve to finish off any rebel groups that managed to elude the sector forces.

Although the operation of the barrages was expensive in terms of the manpower required, the actual costs of construction and maintenance represented only a minuscule part of overall French defense expenditures between 1954 and 1962. In February 1959 some eleven thousand men were required to man the seven hundred kilometers of the Western Barrage between Oujda and Figuig, and another fifteen thousand men were needed to man the seven hundred and seventy kilometers of the Eastern Barrage between Souk-Ahras and Negrine.[36] In all, some eighty thousand French troops were committed to the defense of the barrages, about forty thousand of them on the Morice Line.[37] The maintenance of the barrages placed a particularly heavy burden on the French *Service du Matériel*; the upkeep of the Morice Line alone required some 2,500 men.[38] In 1958 the SM had to provide some 2,970,000 mines to be used on the barrages, and the following year the SM reclaimed over two thousand kilometers of barbed wire.[39] One SM maintenance team operating on the Western Barrage between Port-Say and Ben-Amar traveled over 40,000 kilometers in March and April 1961 repairing weapons for the troops manning the barrier.[40]

One former French commander in Algeria put the cost of constructing the Morice Line at about a million dollars per kilometer.[41] In fact, between 1956 and 1962, the French Government spent about 245 million FF on construction and maintenance of the barrages in Algeria: about 117 million for labor, 79 million for materials, and 49 million for mines.[42] However, 245 million FF represented only .55 percent of the Army budgets and .23 of the overall defense budgets during the period.

The barrages were not intended to exclude totally any rebel penetration but to slow rebel movement and force the rebels to take longer, more difficult, and more exposed routes. As warning devices, they provided the French defenders the opportunity to engage and annihilate any rebel group that attempted to cross. Even after the barrages were completed refugees and small groups of ALN infiltrators continued to find their way across in both directions, but on the whole by mid-1958 the sophisticated barrage system had proven very effective in preventing the movement into Algeria of any substantial number of men or any significant quantity of matériel.[43] However, not all French military leaders were capable of grasping the technical sophistication of the barrages or the real role they played. During an inspection of the Moroccan barrage, General de Gaulle grilled one sector commander, growling, "How do you know the number of passages if you don't see them?" Of course, the detection devices on the barrage were capable of fixing the location of a single goat passing through the barrier.[44]

The creation of nearly impermeable barrages separating the ALN forces of the interior from their training and supply bases in Morocco and Tunisia created two major problems for the FLN; both of them threatened to retard, if not end, the march of the rebels toward national independence.[45] First, the barrages interrupted the flow of supplies to the already hard-pressed rebel forces inside Algeria. Second, the barrages interrupted the flow of communications between the military and political leaders inside Algeria and the GPRA and other FLN governing bodies outside the country. In some respects this interruption of communications was an even greater threat than the reduced flow of supplies, inasmuch as it exacerbated the tendencies of the *wilaya* commanders to operate independently of any outside control and to quarrel among themselves. Given the threat to their vital interests, the FLN leaders had to find a way through or around the French barrages. Three different methods were attempted: infiltration, avoidance, and frontal assault. All failed.

## The "Battle of the Barrages," January–July 1958

The first response of the ALN to the completion of the Morice Line in September 1957 was to attempt to infiltrate directly through the heavily fortified barrier, and the rebels quickly developed a number of techniques for doing so.[46] Almost all attempts were made in darkness or periods of bad weather, and they emphasized speed and stealth. Typically, the ALN massed two forces on the Tunisian side of the border: an assault force, whose mission it was to breach the barrier, and a transport force of men carrying arms and other supplies for the interior. Under cover of diversionary attacks carried out by other forces, both elements approached the barrage on zigzag courses. Using wire-cutters, insulated blankets, and bangalore torpedoes, the assault element cut through the electric and barbed-wire fences and attempted to hold a "beachhead" on the

Algerian side of the barrier long enough for the transport element to move through and for its components to scatter toward their final destinations in Algeria. Once the transport element was safely through the barrage, the assault force withdrew into Tunisia. Of course, the detection devices imbedded in the barrier alerted French reaction forces the moment the first cut in the wire was made, and all available forces, including at need the zone commander's mistress and dog, were deployed to find, fix, and destroy the rebel forces.[47] Thus, each penetration quickly turned into a firefight, and some escalated into all-out battles lasting several days and involving several battalion-size units on both sides.

At first the rebels achieved some success in infiltrating the barrages, but the French soon perfected the coordination of their barrier forces, and rebel losses in men and matériel mounted as they increased the size of the units they threw against the Morice Line. The defense of the barrages was hard and dangerous duty for the French soldiers. French casualties were quite high, and one French historian has characterized the ''Battle of the Barrages'' (sometimes called the ''Battle of the Frontiers'') as one of the hardest endured by the French Army during the Algerian war.[48] The dangers were even more pronounced for the soldiers of ALN, some of whom sought to be interned by the Tunisians rather than return to battle on the Morice Line.[49] The wisdom of such a course of action is borne out by the heavy rebel losses in engagements on the Morice Line in late 1957 and early 1958.[50] On 4 November 1957 a rebel convoy intercepted near Tebessa lost forty-five men killed, and another forty-eight rebels were killed in a similar action near Souk-Ahras on 1 December. On 7 December 121 rebels were killed near El-Ma-el-Aboid, seventy-five more near Guelma on 10 December, and seventy more near Tebessa on 19 December. After the turn of the new year the toll rose more rapidly. On 4 January 1958 near Tebessa French forces intercepted a group of ALN recruits en route to Tunisia for training and killed eighty-one of them and captured forty-one. Another group of ALN troops returning from training in Tunisia were nearly annihilated on 10 January, when 116 were killed and thirty-one captured. In one of the hardest fights on the Morice Line, 285 rebels were killed in a series of engagements in the Guelma area between 27 January and 11 February. On 29 March a 450-man rebel battalion was caught breaking through the barrier, and 120 rebels were killed. The ALN suffered similarly heavy losses on the Western Barrage, where on 21 February one rebel unit was wiped out with the loss of 145 men killed and captured.[51] French authorities claimed at the end of March 1958 that none of the ALN forces crossing the Moroccan barrage since February 1958 had been able to penetrate more than ten miles into Algeria before being detected and subsequently wiped out or dispersed.[52]

### *Engagement in the Djebels Ergou, el-Aloui, and el-Azega, 31 March 1958*

Most of the rebel attempts to infiltrate through the barrages failed and resulted in heavy losses for the rebel infiltrators. French air power, artillery, and tactical mobility in the form of both mechanized and airmobile (helicopter-borne) forces proved able to find, fix, fight, and finish most rebel forces attempting to cross the barrages, either at the attempted crossing point or within twenty-four hours of the first indication that a crossing had taken place. Three factors explain this success: mobility, speed, and flexibility.[53] Superior French mobility permitted the movement of forces long distances on difficult terrain, the recall of elements pursuing unfruitful trails, and surprise. Speed was essential in making decisions, issuing orders, and reacting to changing situations. Speed was also an essential element in ensuring that most operations were concluded before nightfall. Flexibility in the concept of operations permitted quick reaction to changing tactical situations or receipt of additional intelligence. Flexibility was also required to change the mission of committed units, alter the composition of engaged units, and change the point of convergence of an effort.

The three essential factors of mobility, speed, and flexibility, as well as the value of good intelligence and the characteristic use of heliborne forces to overcome difficult terrain, are well illustrated in the destruction of a rebel company in the Djebels Ergou, el-Aloui, and el-Azega on 31 March 1958.[54] The area of operations, depicted in Map 7, lay in the very active Zone East Constantine some 1,500 meters south of the village of Kef el-Aks and approximately ten kilometers northeast of Sedrata. The region is characterized by steep hills (*djebels*), seasonal streams (*oueds*) with steep banks, and partially wooded areas extending from one to four hundred meters from the streambeds.

French forces planned an operation for 31 March 1958 in the vicinity of Ras el-Alia, an area well suited for harboring rebel bands that had made their way through the Tunisian barrage. Violent rainstorms on the night of 30–31 March caused a delay in the operation, and the units designated to take part were kept at their departure base. At 0700 hours, 31 March 1958, the lieutenant colonel commanding the 9th Parachute Chasseur Regiment (*9e Régiment de Chasseurs Parachutistes*; 9e RCP) arrived by helicopter at Sedrata, where he learned that a rebel had been captured who reported that he had been stationed in the Djebel Ergou. The weather conditions having improved, it was decided to proceed with the planned operation, and the 2nd Company of the 152nd Motorized Infantry Regiment (*152e Régiment d'Infanterie Motorisée*; 152e RIM), escorting the prisoner, moved toward Kremissa, where a battery of 155 mm artillery was already in position. The operational commander (Commander, 9e RPC) flew by helicopter to Kremissa, where he learned that a young shepherd sent by his father had reported that a band of approximately one hundred rebels had passed through the Ras el-Alia area during the night. Another shepherd arrived soon thereafter who reported forty rebels in the Djebel Ergou.

As a result of the shepherds' reports, the original operation was modified, and it was decided that the main effort would be undertaken by the 9e RPC on the axis Ras el-Ouen Kel–Djebel Ergou–Djebel el-Azega, with a secondary effort by the 3rd Battalion, 60th Infantry Regiment (*III/60e Régiment d'Infanterie*; III/60e RI) coming from the north and a company of the 152e RIM from the east.

**Map 7**
**Engagement in the Djebels Ergou, el-Aloui, and el-Azega, 31 March 1958**

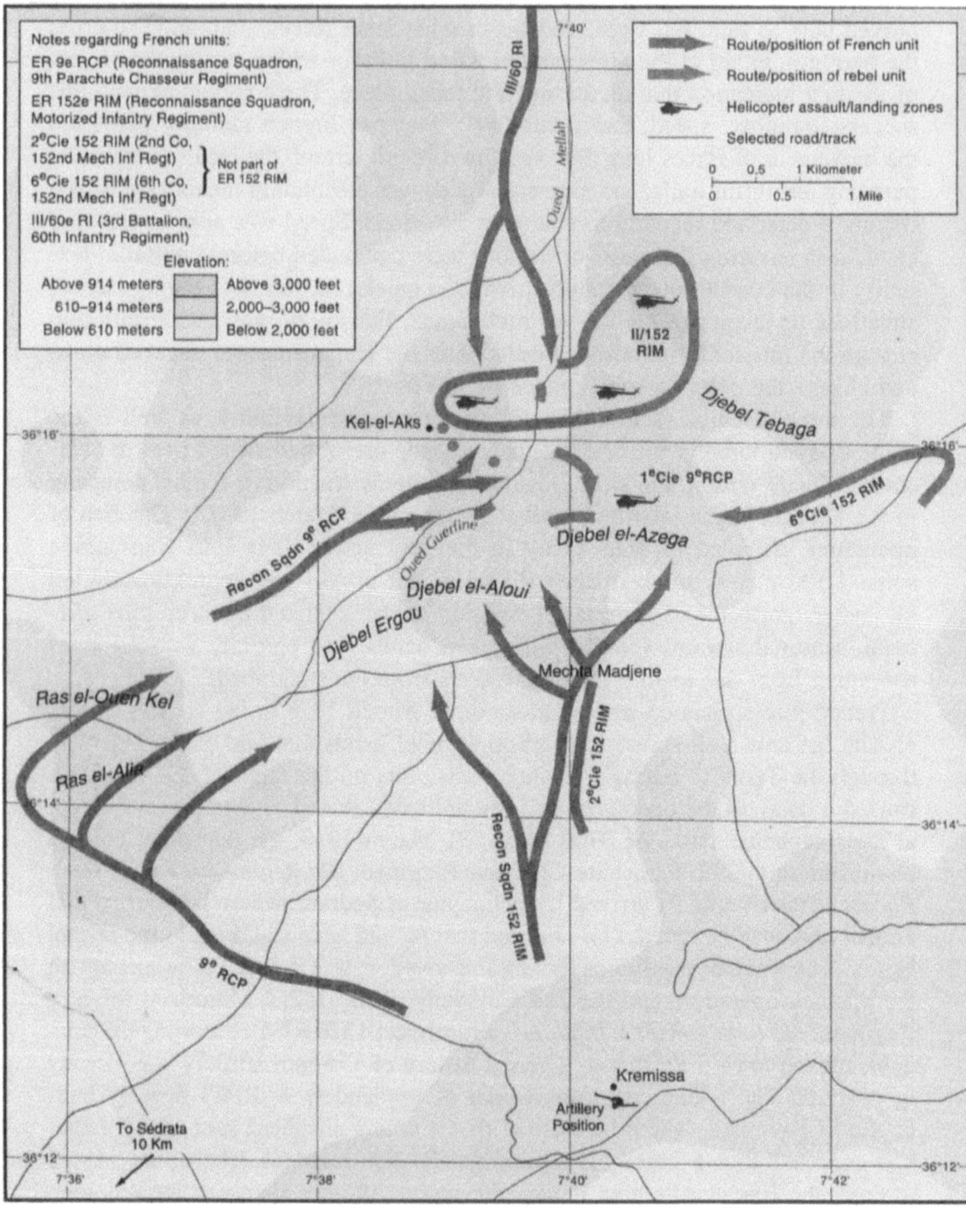

The mission of establishing a blocking position (*un bouclage*) to the north of the area of operations was assigned to the Reconnaissance Squadron of the 9e RPC. The Reconnaissance Squadron and 2nd Company of the 152e RIM, reinforced by a section of 106 mm recoilless rifles from the 9e RPC, was charged with establishing a blocking position to the south. The 2nd Battalion, 152nd Motorized Infantry Regiment (II/152e RIM) was ordered to conduct a helicopter-borne assault into the center of the operational zone. Artillery support for the operation was provided by a battery of 155 mm howitzers and another battery of 105 mm howitzers located at Kremissa.

At 1100 hours the 2nd Company of the 152e RIM reported the rebels to be fleeing north from the Mechta Madjene. At the same time the Reconnaissance Squadron of the 9e RPC arrived in the area of Kef el-Aks and engaged. By noon the 2nd Company of the 152e RIM was in contact with the rebels on the southern slopes of the Djebel el-Aloui, and the Reconnaissance Squadron of the 152e RIM had moved elements onto the Djebel el-Azega. The 1st Company, II/152e RIM, having been helicoptered into position, stopped rebel elements in flight in the bed of the Oued Mellah. Meanwhile, the 3rd Company, III/60e RI, completed the cordon to the north.

At 1230 hours, the companies of the 9e RPC were in contact in the Djebel Ergou and Kef el-Aks, and by 1300 hours the combat action extended throughout the area of operations. One hour later two companies of the III/60e RI arrived to reinforce the northern blocking position, and at 1500 hours the helicopter movement of the II/152e RIM was completed. By 1600 hours, the French forces were engaged in the Djebels Ergou, el-Aloui, and el-Azega and along the Oued Guerfine and Oued Mellah.

At 1700 hours part of the rebel force took refuge on the northern slopes of the Djebel Azega, and the 1st Company of the 9e RPC carried out a helicopter assault into that area. The fighting continued until nightfall; the rebels in the Djebel el-Azega were annihilated. French forces remained in position overnight, and at dawn on 1 April the operation was resumed, but without additional contact. However, eight hours later the remainder of the rebel company was intercepted by the II/60e RI attempting to pass back into Tunisia; the rebels lost twenty-seven men killed, four men captured, one machine gun, three automatic rifles, three submachine guns, seventeen rifles, and three automatic pistols.

Overall, rebel losses in the engagement totaled sixty-one men killed, eight men captured, two machine guns, three automatic rifles, seven submachine guns, forty-one rifles, and three automatic pistols. French losses included four men killed and two men wounded. The engagement in the Djebels Ergou, el-Aloui, and el-Azega demonstrated clearly the mobility, speed, and flexibility that gave the French forces such a tremendous advantage over the rebels. Unable to match the mobility and speed of French mechanized and heliborne forces in the often difficult terrain of northern Algeria, even rebel bands that managed to cross the barrage were soon intercepted and annihilated.

***Engagement in the Djebel Tenouchfi, 15–16 April 1958***

While air power, particularly light observation aircraft, helicopters, and propeller-driven fighters, was essential, perhaps the key factor in French success was superior ground mobility. French mechanized combat units played a most important role in preventing successful rebel crossings of both the Tunisian and Moroccan barrages. The destruction of a rebel band in the Djebel Tenouchfi on 15–16 April 1958 was typical of the French ability to respond rapidly to breaches in the barrage with mechanized forces.[55] The area of operations, depicted on Map 8, was some twelve kilometers southwest of Sebdou and some ten to twelve kilometers east of the Moroccan barrage. The very mountainous terrain of the area of operations is broken by numerous deep ravines, rocks, and cliffs. The mountain ridges in the area generally run in an east-west direction, and at the time of the battle the slopes were heavily wooded, particularly to the west and northwest.

At 0130 hours, 15 April 1958, a platoon of the 3rd Squadron, 3rd Cuirassier Regiment (*III/3e Régiment de Cuirassiers*; III/3e Cuirassiers), charged with defense of a sector of the Moroccan barrage, discovered a breach and signs of the passage of a rebel force estimated at fifteen men from Morocco into Algeria. The platoon commander immediately notified his company commander and set out to follow the trail of the infiltrators. The commander of the III/3e Cuirassiers immediately reported the incident to the commandant of the Sebdou Sub-Sector, and at 0145 hours a task force consisting of one squadron of the 3rd Cuirassiers and a company of the 22nd Colonial Infantry Regiment (22e *Régiment d'Infanterie Coloniale*; 22e RIC) was formed to hunt down the rebels. Two squadrons of the 3rd Cuirassiers and one squadron of the 2nd Algerian Chasseur Regiment (*2e Régiment de Chasseurs Algériens*; 2e RCA) were placed on alert to intervene at daybreak.

At 0745 hours, 15 April, the commander of the 4th Squadron, 3rd Cuirassiers (IV/3e Cuirassiers) reported that he had picked up the trail of the rebel force at Si Jilali and that the rebels were apparently headed toward the Djebel Tenouchfi. At 0800 hours, the operational commander (a major assigned to the Sebdou subsector headquarters) requested a light observation aircraft and ordered the 2nd Squadron, 3rd Cuirassiers (II/3e Cuirassiers), to the Djebel el-Rhar and the trail north of the Djebel Tenouchfi to intercept the rebels. At 0945 hours the commander of the IV/3e Cuirassiers reported his forces heavily engaged with a rebel force of approximately thirty men with automatic weapons between Hill 1784 and Hill 1608. Weather conditions prohibited aerial intervention at that time. Ten minutes later all the remaining elements of the 3e Cuirassiers, until then on stand-by for inspection by the visiting Inspector-General of Armored Forces, were put at the disposal of the operational commander.

At 1145 hours, the operational command post was located in the vicinity of Hill 1381. The 1st Squadron, 3rd Cuirassiers (I/3e Cuirassiers), was in blocking positions north and east of the Djebel Tenouchfi, and the 3rd Squadron, 2nd

**Map 8**
**Engagement in the Djebel Tenouchfi, 15–16 April 1958**

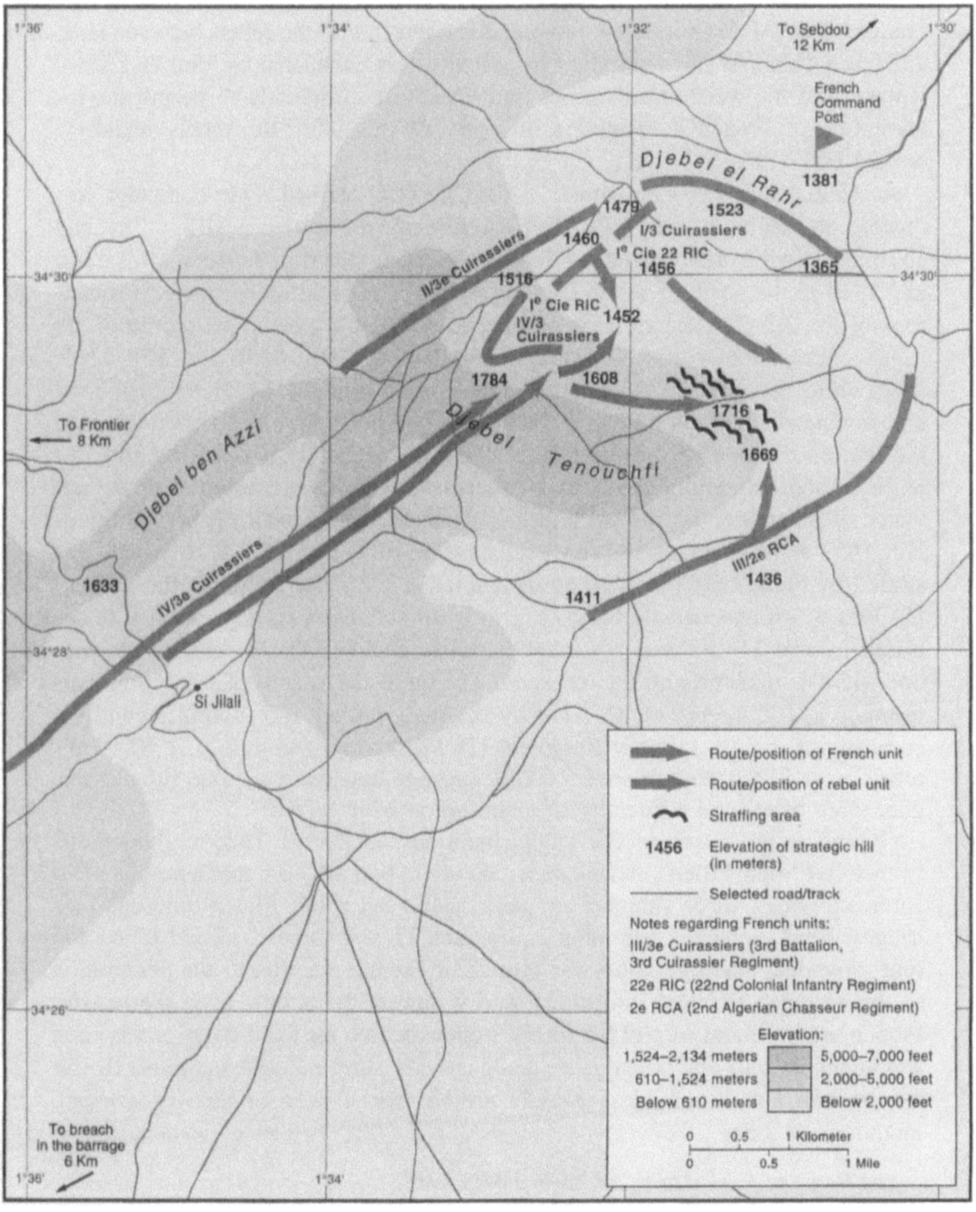

Algerian Chasseurs (III/2e RCA), blocked the southern end of the area of operations. The IV/3e Cuirassiers remained in contact with the rebels between Hills 1784 and 1608. At noon one French casualty was evacuated by Bell H-13 helicopter, and the weather conditions had improved sufficiently to permit the intervention of French T-6 fighters in force. At this point the rebels withdrew toward Hill 1716.

At 1230 hours two companies of the 22e RIC arrived. One company was ordered to reinforce the IV/3e Cuirassiers deployed between Hills 1784 and 1516, the other to reinforce the I/3e Cuirassiers, in position between Hills 1516 and 1479 to the north of the IV/3e Cuirassiers. Thus reinforced, the two squadrons of the 3rd Cuirassiers were ordered to conduct a meticulous search of the rugged terrain to the east as far as a line between Hill 1436 and Hill 1365. Meanwhile, the III/2e RCA maintained the cordon south of the Djebel Tenouchfi and advanced a platoon on foot to Hill 1669, while the II/3e Cuirassiers guarded the line on the north from the Djebel ben-Azzi to the Djebel el-Rhar and east to Hill 1365, extending the cordon progressively in coordination with the advance of other friendly forces. Following a short engagement in the vicinity of Hill 1452 at 1645 hours, the rebel forces took refuge on Hill 1716 which was strafed by French T-6s at 1700 hours and taken soon afterwards by the advancing French ground troops. Following mop-up activities, the operation was declared over at 2145 hours, but rebel prisoners declared that several rebels with one MG-34 machine gun had escaped. Therefore the search of the terrain was resumed at 1000 hours on 16 April by the same French forces, reinforced by a company of the 7th Infantry Regiment (*7e Régiment d'Infanterie*; 7e RI). Four rebels were killed, five prisoners taken, and one machine gun, one submachine gun, three rifles, and a quantity of ammunition were seized.

Overall rebel losses in the engagement in the Djebel Tenouchfi included twenty-five men killed, eleven men captured, one MG-34 machine-gun, two automatic rifles, three submachine guns, seventeen rifles, fifteen thousand cartridges, hand grenades, and other equipment. French forces lost one killed and four wounded. The operation was typical of French reaction to the penetration of the barrages by rebel infiltrators, and it demonstrates how even the heavier French armored and motorized forces, supplemented by fixed-wing observation and attack aircraft and helicopters, could quickly surround and annihilate the far less mobile and usually less heavily armed rebels, even in heavily wooded, mountainous areas.

### *The Battle of Souk-Ahras, 28 April–3 May 1958*

The largest battle of the Algerian war was fought between 28 April and 3 May 1958 near Souk-Ahras in the very active Zone East Constantine, where an ALN force of over eight hundred men attempted to break through the Morice Line.[56] The ALN threw into the battle at least seven companies (*katibas*) intended for the reinforcement of *wilayas* 2 and 3; over the course of the first three nights rebel penetrations were made north and south of Souk-Ahras. A number of rebels got through the wire before being pinned down by the French reaction forces and artillery and subsequently surrounded by a much superior force of French airborne troops moved in by helicopter. Overall, in the first two days of the battle of Souk-Ahras the rebels lost 436 men killed and 100 captured as well as a considerable amount of arms and equipment. The battle raged intensely for several more days, and in the end the ALN had lost 620 men killed or captured—including a battalion commander—out of 820 who had entered the barrage. The rebels also left behind on the battlefield forty-six crew-served and 412 individual weapons. French losses were also heavy: thirty-eight killed and thirty-five wounded. The battle of Souk-Ahras was a decisive defeat for the rebels, and for all intents and purposes it marked the end of the "Battle of the Barrages," although the ALN lost another 169 men killed in the Souk-Ahras area later in May 1958, and in the last major engagement of 1958 on the Morice Line, on 26 June, a three-hundred-man ALN battalion lost forty-six killed and sixty-four captured in a futile attempt to cross the Morice Line near Tebessa.[57]

The "Battle of the Barrages" was a terrible contest for both sides. Between the end of January and the end of May 1958 the ALN lost 4,800 men killed and 588 men captured, while an additional eighty-four rebels rallied to the French, whose own losses totaled 279 killed and 758 wounded.[58] In February 1958 alone the ALN lost in the frontier zones 3,410 men killed in action, fifty-two machine guns, twenty-eight automatic rifles, 286 submachine guns, and 885 rifles and shotguns.[59] In all, between September 1957 and May 1958 the ALN lost over 6,000 men and 4,300 weapons in a succession of futile attempt to cross the barrages.[60] By any standard the "Battle of the Barrages" was a major victory for the French, who thus effectively sealed off the rebel bands inside Algeria from their exterior bases.

### Attempts to Skirt the Barrages through the Sahara

Unable to infiltrate through the barrages, the ALN turned to the use of small convoys of men and camels to attempt to skirt the barrages to the south through the open desert of the Sahara (see Map 6). French defenses in the Sahara included a combination of fortified positions, roads, airfields, and temporary operational bases to support specialized, mobile forces, backed by air power and capable of reacting quickly and efficiently to find and eliminate any intruder.[61] Nevertheless, the great expanse of the desert offered some hope to the rebels for infiltrating at least small amounts of arms and ammunition into Algeria, and in fact some rebel convoys got through. However, most were detected by French aerial surveillance, attacked, and either destroyed or turned back with significant losses. Between 1 August 1958 and 1 January 1959 the French intercepted no fewer than seven ALN convoys attempting to resupply rebel forces in *Wilaya* 1 by skirting the Morice Line south of Negrine. The resulting loss to the rebels included four men killed, eight men captured, six camels killed and eleven camels captured, three Beretta submachine guns, eleven rifles, five 81 mm mortar rounds, twenty-four 51 mm mortar illumination rounds, twenty-four antitank rockets, 14,000 rifle cartridges, 3,200 blocks of explosives, six rolls of fuze, one box of detonators, over 470 kilograms of other munitions, and a large quantity of medical supplies, surgical instruments, clothing, and documents.[62]

The routes in the eastern Sahara south of Tebessa lacked water and were somewhat better covered by French aerial observation, making the western Sahara routes preferable. The Moroccan bases of the ALN thus took on importance; most of the camel convoys attempting to transit the Sahara into Algeria were launched from there, particularly from Bou-Denib.[63] Some camel caravans carrying arms for the ALN also originated in Libya, but the number was apparently quite small.[64] In April and May 1959 eight caravans crossed the Moroccan border headed for the ALN forces in Algeria. The composition and disposition of those eight caravans is outlined in Table 7.3. Later convoys were even less successful. Of five rebel caravans entering Algeria from Morocco in October 1959, only one got through to *Mintaka* 3, *Wilaya* 5; one was turned back; one laden with ten thousand cartridges and a hundred grenades for *Mintaka* 4, *Wilaya* 5, was intercepted on 30 October; one laden with ten thousand cartridges and a hundred grenades for *Mintaka* 7, *Wilaya* 5, was intercepted on 9 October fifteen kilometers northeast of Tarhit; and one laden with ten thousand cartridges and clothing was intercepted on 14 October 30 kilometers north of Abadla.[65]

On the whole, few rebel caravans through the Sahara were able to avoid detection and destruction or capture. In all, between 1 January 1959 and 1 August 1960 a total of twenty-two rebel caravans departed Bou-Denib in Morocco for Algeria; nineteen of those caravans were intercepted and their cargoes captured by the French.[66] Those that did get through carried only minuscule amounts of supplies for the ALN forces of the Interior. Skirting the barrages was thus no more effective a means of breaking the French blockade than direct infiltration of them.

## The Attempt to Force the Eastern Barrage

Unable to infiltrate through or go around the barrages, the ALN made one last attempt to overcome the eastern barrier, in late 1959 and early 1960, by committing the carefully husbanded battle corps in a series of frontal attacks on the Morice Line. The assaults produced only more rebel casualties and additional losses of matériel. Over 95 percent of the ALN companies (*katibas*) attempting to cross into Algeria in the first months of 1959 were destroyed.[67] The ALN leaders in Tunis thus decided to delay any further attempts to breach the Morice Line in force until their units were better prepared and could participate in a well planned and well coordinated major assault on the barrage.[68]

The first of the major frontal attacks against the Morice Line was launched on 11 September 1959, with the twin objectives of relieving ALN forces in

**Table 7.3**
**ALN Supply Caravans Originating in Morocco, April–May 1959**

| Designation | Dates | Composition | Disposition |
|---|---|---|---|
| Caravan DEBRA | 4 April 1959 | 6,060 cartridges and 60 grenades for W6; 6,010 cartridges and 60 grenades for W4; 4,000 cartridges and 50 grenades for M4, W5 | Munitions for W4 and M4, W5, arrived in M7, W5, on 30 May en route to final consignees |
| Caravan KADDA BEN LAKDAR combined with Caravan ZAOUI and elements of Caravan BRECK guided by Bahous M'Taouch | Departed Bouanane and crossed the Algerian border on 27 May 1959 | 17 men (1 AR, 2 SMG, 14 rifles); 26 camels; 59,260 cartridges; 300 Mills grenades; 50 kg of explosives; 3,500 flares; 400 kg of medicines; 54 SCR-300 radio batteries; rations | Intercepted on 7 June 25 km southwest of Tarhit; all cargo captured by the French |
| Caravan BEN MABROUK (with elements of Caravan BRECK) guided by Mohamed Ben Dida | Departed Bouanane on 25 May 1959 and crossed into Algeria on 30 May 1959 | 8 men; 18 camels; 17,377 cartridges; 391 Mills grenades; medicine; rations | Left Figuig in early April and entered Algeria but was detected and returned to Bouanane; intercepted again on 9 June and all cargo captured by the French |
| Caravan KAMITA | 28 April 1959 | Munitions, explosives, rations | |
| Caravan BEN AROUS | After 28 April 1959 | Munitions, explosives, rations | |
| Caravan GHRISS | May 1959 | 200 kg of explosives, mines, and detonators for M3, W5; medicine and batteries for W4 and W6 | En route vicinity. Bouanane on 29 June |

*Source*: Commandment en Chef des Forces en Algérie, État-Major Interarmées, 2e Bureau, *Logistique Rebelle au cours du mois de juin 1959*, No. 4036/EMI/2/OPE/PLIT, (Alger, 3 juillet 1959), Annexe 7, in folder "10e Région Militaire, État-Major Interarmées, 2e Bureau—Fiches sur la logistique rebelle, decembre 1958–decembre 1959," Dossier 1H1689 d. 2, Fonds Algérien, SHAT.

*Wilaya* 3 hard pressed by French Operation JUMELLES and of favorably affecting international opinion before the discussion of the Algerian question in the United Nations.[69] Designated Operation DIDOUCHE, the offensive involved four battalions in a series of three attempts (on 11–12 September, 18–19 September, and 15–16 October) to breach the Morice Line at night in the areas of Lamy, Lacroix, and Le Kouif. The offensive was designed to proceed in two phases: (1) the harassment of French outposts, an attempt to destroy the barrage in various places, and firing on French vehicles drawn to the action; and (2) an attempt to pass relatively weak forces through the barrier. The results of the operation were negligible; only twenty armed men succeeded in penetrating into Algerian territory, and they did so by skirting the barrage south of Negrine. However, Operation DIDOUCHE did succeed in polishing the ALN units and introducing about a hundred men into the area between the Morice Line and the newer Challe Line to the west.

The second major ALN operation against the Morice Line was more substantial. On 26 November 1959 the ALN launched Operation AMIROUCHE, with the primary objective of passing at least 1,300 men to *Wilaya*s 1, 2, 3, and 4 and the secondary objective, again, of creating a favorable impression of the strength of the FLN prior to the United Nations debates. The plan of the operation called for ten battalions (five to six thousand men) to breach the barrage to permit the subsequent passage of the 1,300 replacements for the *wilaya*s. The principal areas of attack were in the vicinity of Lamy, Lacroix, and El Meridj-Bekkaria, and the operation was proceeded in three phases: (1) harassment of French outposts and attempts to destroy the barrage; (2) introduction between the two barriers of the forces intended to breach the rear barrage (the Challe Line); and (3) a series of actions on both the Morice Line and the Challe Line to disorient the French forces and aid the crossings. Again, the results were dismal. Of 280 rebel soldiers who crossed the Morice Line, 140 were killed or captured, ninety to a hundred fled back into Tunisia, ten succeeded in crossing the Challe Line into Algeria, and thirty-forty were stranded between the two barriers. The failure of Operation AMIROUCHE enhanced the deterrent power of the barrage in the eyes of the rebels and lowered ALN morale. More importantly, Krim Belkacem and Mohammedi Said *Si Nacer* were relieved as the civilian and military leaders of the FLN and were replaced by Abdelhafid Boussouf and Houari Boumedienne.

Operations DIDOUCHE and AMIROUCHE caused heavy casualties and severe losses of matériel for the ALN. Such losses were not acceptable under the new strategy put in place by the new Chief of Staff of the ALN, Houari Boumedienne. Boumedienne's principal objective was to build up and preserve the ALN battle corps for the internal political struggle that was sure to come once independence was gained. Wasting men and matériel in futile frontal assaults on the heavily fortified frontiers simply for the sake of providing the already suffocating ALN forces of the interior a few recruits and cartridges was not part of Boumedienne's plan. However, he could not totally ignore the needs of the *wilaya*s, and thus devised a strategy that called for the reorganization of the *corps de bataille* in Tunisia, the destruction of the barrages from a distance, and the penetration into Algeria at several points simultaneously.[70] The idea was to harass the French forces on the barrages by repeated shelling from the safety of Tunisia and Morocco and to attempt the crossing of the barrage only with small, highly trained units under the most favorable conditions. Accordingly, the *wilaya* commanders were instructed to reduce the tempo of their operations to practically nil, to conserve men and supplies.

Under Boumedienne's supervision two "generalized actions" were undertaken with the apparent objective of crossing some small packets of men and matériel through the Morice and Challe lines.[71] The first operation took place between 13 March and 31 March 1960, and the second between 15 July and 31 July 1960. Both operations involved the major part of the ALN forces in Tunisia, the later operation engaging the units in rotation. Neither operation achieved any significant result. In the March operation the ALN lost some forty men, and only a few individuals succeeded in entering Algeria. Losses were heavier during the July operation (about three hundred men), and not a single rebel succeeded in crossing the rear barrage (Challe Line) into Algeria.

## CONCLUSION

The failure of the ALN to develop an effective counter to the French barrages contributed significantly to the slow strangulation of the rebel forces inside Algeria after the middle of 1958. The measure of that strangulation was the ever-decreasing numbers of arms entering Algeria from the ALN depots in Tunisia and Morocco. Between 1 December 1957 and 28 February 1958 the FLN in Tunisia received from its foreign suppliers some 18,388 weapons, of which about 2,700 passed over the Tunisian border into Algeria.[72] During the same period French forces in the Constantinois recovered some 1,300 weapons, the net gain for the ALN being thus only 1,400 weapons, or about 50–60 percent of the arms passed through the Tunisian barrage. Of 23,000 weapons delivered to the FLN in Tunisia in the first three quarters of 1958, only 4,300 passed into Algeria—3,000 in the first quarter, 1,300 in the second quarter, and none in the third quarter.[73] In all, during the first half of 1958 the French forces lost 1,280 weapons to the rebels and seized 6,000 weapons from them, for a net loss to the rebels of 4,720 weapons.[74]

During the third quarter of 1958 no rebel convoy succeeded in passing through the barrage from Tunisia into Algeria, and only six feeble camel caravans (a total of some twenty-five to thirty camels) passed around the barrage south of Negrine.[75] During the same period deliveries to rebel forces in *Wilaya* 5 from Morocco amounted to only one machine gun, two 81 mm mortars, 310 mines, 1,200 grenades, 500 kilograms of explosives, 78,000 cartridges, seventy-five 81 mm mortars rounds, some clothing and medicines, and perhaps another forty tons of other supplies. With no substantial deliveries from either Tunisia or Morocco, the rebel armament inside Algeria decreased by 850 weapons in the third quarter of 1958. The situation was nearly as bleak in the fourth quarter of 1958, when no arms or other supplies crossed the Morice Line, only forty camels (of which twenty-two were intercepted) assayed the dangerous route between Negrine and Chott-el-Rhassan, and a meager sixty mines, 650 grenades, 26,000 cartridges, and some clothing and medicine reached Algeria from Morocco. The balance of rebel armament in Algeria thus decreased by some seven hundred weapons in the fourth quarter of 1958.

During the first quarter of 1959, nonetheless, the ALN logistical situation improved marginally. Some matériel filtered through the Morice Line, a few convoys successfully turned the barrage south of Negrine, and the *wilaya*s in

**Table 7.4**
**ALN Receipt of Arms and Ammunition in Algeria, April–December 1959**

| Month | Arms and Ammunition Received in Algeria |
|---|---|
| April 1959 | 238 weapons (including 5 MG & 4 mortars); 500 grenades; ca. 70,000 rounds of small arms ammunition |
| May 1959 | 291 weapons (including 3 MG, 4 mortars, 5 RL, and 7 AR); 500 grenades; ca. 120,000 rounds of small arms ammunition |
| June 1959 | 8 weapons (including one 51 mm mortar and 1 AR); an unknown quantity of ammunition |
| July 1959 | 55 weapons; ca. 25,000 rounds of small arms ammunition |
| August 1959 | 30 weapons (including 1 AR, 3 SMG, and 17 rifles); ca. 21,000 rounds of small arms ammunition |
| September 1959 | 40 weapons; ca. 30,000 rounds of small arms ammunition |
| October 1959 | 34 weapons; 30 mortar rounds; 400 grenades; ca. 24,000 rounds of small arms ammunition (all from Morocco) |
| November 1959 | 59 weapons (including 3 rifles and 46 automatic pistols); ca. 20,000 rounds of small arms ammunition |
| December 1959 | 20 weapons; less than 5,000 rounds of small arms ammunition |

*Sources*: Commandment en Chef des Forces en Algérie, État-Major Interarmées, 2e Bureau, *Logistique Rebelle au cours du mois de juin 1959*, No. 4036/EMI/2/OPE/PLIT (Alger, 3 juillet 1959), Annexe 5; *Logistique Rebelle au cours du mois de septembre 1959*, No. 6025/EMI/2/OPE/PLIT (Alger, 5 octobre 1959), Annexe 2; and *Logistique Rebelle au cours du mois de decembre 1959*, No. 0101/EMI/2/OPE/PLIT (Alger, 8 janvier 1960), Annex 3; all in folder ''10e Région Militaire, État-Major Interarmées, 2e Bureau—Fiches sur la logistique rebelle, decembre 1958–decembre 1959,'' Dossier 1H1689 d. 2, Fonds Algérien, SHAT.

western Algeria were resupplied by a few small convoys crossing the western barrage in the vicinity of Ksours, and by two camel caravans through the Sahara.[76] The situation for the remainder of 1959 is shown in Table 7.4.

Following the disastrous Operation AMIROUCHE in November 1959, Boumedienne's policy of refusing to force the barrages in strength in order to preserve the ALN *corps de bataille* was adopted at the expense of the rebel forces inside Algeria. Only the assaults in March and July 1960 approached the status of a general engagement, and they produced no substantial results. Thus, in the last half of 1960 only forty men and forty rifles entered Algeria.[77] The flow of men and weapons to the rebels in the *wilayas* was not stopped, but after mid-1960 it was reduced to insignificance, and the fate of the rebel forces in Algeria was sealed.

## NOTES

1. Edgar O'Ballance, *The Algerian Insurrection, 1954–62* (Hamden, CT: Archon Books, 1967), 139.
2. Ibid.
3. The secret war against the arms dealers is described in some detail by Alistair Horne in *A Savage War of Peace* (New York: Viking Press, 1977), 261–263. I have followed his account closely here.
4. Ibid., 262.

5. O'Ballance, 139.
6. Horne, 263. Puchert's nemesis was probably "The Killer."
7. Ibid.
8. Ibid., 262.
9. Ibid., 261–262.
10. Ibid., 241–242.
11. 10e Région Militaire, État-Major Interarmées, 3e Bureau, *Le problem militaire en Algérie* (Alger, avril 1957), Annexe 3, 14, in folder "10e Région Militaire, État-Major Interarmées, 3e Bureau—Le problem militaire en Algérie, avril 1957," Dossier 1H1933 d. 2, Fonds Algérien, SHAT.
12. Lettre (No. 2/EM.3/OPS), Vice-Amiral Préfet Maritime Commandant la IVe Région Maritime d'Oran [Contre-Amiral Geli] au Général de Corps d'Armée Commandant la 10e Région Militaire [Général Lorillot], Mers-el-Kebir, 11 janvier 1956, sujet: Répression du trafic d'armes, 3, in folder "10e Région Militaire, État-Major Interarmées, 2e Bureau—Répression du trafic d'armes, 1955–62," Dossier 1H1536bis d. 1, Fonds Algérien, SHAT.
13. Arslan Humbaraci, *Algeria: A Revolution That Failed* (New York: Frederick A. Praeger, 1966), 35.
14. A. H. Peterson, G. C. Reinhardt, and E. E. Conger (eds.), *Symposium on the Role of Airpower in Counterinsurgency and Unconventional Warfare: The Algerian War*, RAND Memorandum RM-3653-PR (Santa Monica, CA: Rand Corporation, July 1963), 11.
15. Humbaraci, 35.
16. Lettre (No. 47/Ve R.A./3.S), Commandement de la 5e Région Aérienne [Général de Division Aérienne Frandon] au Général de Corps d'Armée Commandant la 10e Région Militaire [Géneral Lorillot], Alger, 13 janvier 1956, sujet: Répression du trafic d'armes, 1, in folder "10e Région Militaire, État-Major Interarmées, 2e Bureau—Répression du trafic d'armes, 1955–62," Dossier 1H1536bis d. 1, Fonds Algérien, SHAT.
17. Humbaraci, 35.
18. Horne (*A Savage War of Peace*, 158–159) describes the incident in detail.
19. *Le problem militaire en Algérie*, 6 and 11.
20. Lettre (No. 5156/RM.10/3.OPE), Général Commandant Supérieur Interarmées et de la 10e Région Militaire aux Généraux Commandants des Corps d'Armées de Constantine et d'Oran et le Général Commandant Interarmées Saharienne, Alger, 31 decembre 1956, sujet: Exercise du droit de poursuite, in folder "10e Région Militaire, État-Major Interarmées, 2e Bureau—Répression du trafic d'armes, 1955–62," Dossier 1H1536bis d. 1, Fonds Algérien, SHAT.
21. Luc Bourgeois, *Le Matériel pendant la guerre d'Algérie* (Paris: Inspection du Matériel de l'Armée de Terre, September 1987), vol. I, Part 1, 27.
22. *Le problem militaire en Algérie*, 9–10. The route from Figuig through Géryville to Aflou was deemed the most significant and received particular attention from French forces.
23. Ibid., 9.
24. Ibid., 10.
25. O'Ballance, 90.
26. Claude Carré, "Aspects opérationnels du conflit Algérien, 1954–1960," *Revue Historique des Armées* 166 (March 1987), 87.
27. Commandement Supérieur Interarmées, 10e Région Militaire, État-Major Interarmées, 2e Bureau, *Étude générale sur l'evolution du potentiel de l'A.L.N. (armement-*

*personnel) au cours du 2° semestre 1957*, No. 1655/EM.10/2/OPE/EXPL (Alger, 30 janvier 1958), 4, in folder "10e Région Militaire, État-Major Interarmées, 2e Bureau—Étude générale sur l'évolution du potentiel de l'A.L.N. (armements et effectifs), 1957–62," Dossier 1H1682 d. 2, Fonds Algérien, SHAT.

28. Horne, 240.

29. Humbaraci, 41–42; O'Ballance, 92.

30. Bourgeois, I, Part 1, 28.

31. Quoted in *Symposium on the Role of Airpower*, 59.

32. The physical characteristics of the Morice Line are described in Horne, 263–264; Humbaraci, 41–43; Bourgeois, I, Part 1, 27; Carré, 87; and Peter Braestrup, "Partisan Tactics—Algerian Style," *Army*, August 1960, 41. The Moroccan barrage, which was not continuous, was also somewhat less developed, with two rows of barbed wire about ten meters apart, with the electric fence in the center. The Moroccan barrage was also covered by mines, powerful lights, aerial surveillance, and ground radar stations located about twenty-five to forty kilometers apart. It was patrolled by mechanized forces every eleven minutes from dusk to thirty minutes before dawn (see *Symposium on the Role of Airpower*, 63).

33. Carré, 87.

34. Bourgeois, I, Part 1, 27.

35. Concepts Division, Aerospace Studies Institute, Air University, *Guerrilla Warfare and Airpower in Algeria, 1954–1960*, Project No. AU-411-62-ASI (Maxwell Air Force Rose, AL: Concepts Division, Aerospace Studies Institute, Air University, March 1965), 53.

36. 10e Région Militaire, État-Major Interarmées, 3e Bureau, *Exposé du Général Allard sur la situation militaire en Algérie (Forces Terrestres) à la date du 9 février 1959* (Alger, 9 février 1959), Charts 4 and 5, in folder "10e Région Militaire, État-Major Interarmées, 3e Bureau—Le problem militaire en Algérie, avril 1957," Dossier 1H1933 d. 2, Fonds Algérien, SHAT. The number of troops cited does not include sector troops and elite forces backing up the barrier forces.

37. Horne, 264; Melnik, 43; *Symposium on the Role of Airpower*, 13.

38. Constantin Melnik, *The French Campaign against the FLN*, RAND Memorandum RM-5449-ISA (Santa Monica, CA: Rand Corporation for the Office of the Assistant Secretary of Defense–International Security Affairs, September 1967), 43.

39. Bourgeois, I, Part 1, 27–28.

40. Ibid., 28.

41. *Symposium on the Role of Airpower*, 63.

42. Paule Arnaud-Ameller and Raymond Bourgerie, "Le cout des barrages terrestres en Algérie," *Revue Internationale d'Historie Militaire*, no. 76 (1997), 141.

43. Alf Andrew Heggoy, *Insurgency and Counterinsurgency in Algeria* (Bloomington: Indiana University Press, 1972), 227; Humbaraci, 42.

44. Norman C. Walpole et al., *U.S. Army Area Handbook for Algeria*, Department of the Army Pamphlet No. 550–44 (Washington, DC: Headquarters, Department of the Army, January 1965), 43.

45. Information Section, Secrétariat Général pour les Affaires Algériennes, *Algérie 1960—Documentation* (Paris: Segedo, octobre 1960), 2.

46. The rebel techniques are described in O'Ballance, 117–118; Braestrup, 42; and Vertol Aircraft Corporation (T. R. Pierpont, Director, European Operations), *French Army Helicopter Operations in Algeria, June 1956–September 1959*, Report No. SM-

406 (Morton, PA: Vertol Aircraft Corporation, 1 November 1959), 31–32. Although only the passage of the Morice Line is described here, the methods used by the rebels in crossing the Moroccan barrage were practically the same.

47. O'Ballance, 117–118; *Symposium on the Role of Airpower*, 58.

48. Erwan Bergot, *Algérie: Les Appelés au Combat* (Paris: Presses de la Cité, 1991), 12: "Cette 'Bataille des frontières' fut, sans conteste, l'une des plus dures qu'eut à subir l'armée française pendant la guerre d'Algérie."

49. O'Ballance, 133.

50. The toll is recapitulated by O'Ballance, 92 and 118.

51. Ibid., 164.

52. Ibid.

53. Noted in a précis of the engagement in the Djebels Ergou, el-Aloui, and el-Azega on 31 March 1958 prepared for instructional purposes, probably at the École Supérieur de Guerre, sometime after the battle. The précis is included in a folder of similar material found in the basement of the U.S. Army Military History Institute at Carlisle Barracks, Pennsylvania, and identified hereafter as "Miscellaneous File-Algeria-Operations-1958," USAMHI.

54. "Miscellaneous File-Algeria-Operations-1958."

55. Details of the engagement in the Djebel Tenouchfi on 15–16 April 1958 are based on a précis filed as "Miscellaneous File-Algeria-Operations-1958," USAMHI.

56. The battle of Souk-Ahras is described in O'Ballance, 119–120; Horne, 265–266; and "Miscellaneous File-Algeria-Operations-1958."

57. O'Ballance, 119–120. O'Ballance also notes that on 26 June 1958 thirty ALN recruits trying to cross into Algeria from Tunisia were killed by French forces near Bir-el-Ater.

58. Bergot, 282. During the same period the rebels lost 350 crew-served weapons and about three thousand individual weapons.

59. Melnik, 15.

60. Horne, 266.

61. 10e Région Militaire, Commandement des Troupes et Services des Territories du Sud Algérie, *Directive sur l'organisation défensive des frontières Sahariennes*, No. 5197/CTS/OPE.3 (Alger, 17 decembre 1956), 1–2, in folder "10e Région Militaire, État-Major Interarmées, 2e Bureau—Répression du trafic d'armes, 1955–62," Dossier 1H1536bis d. 1, Fonds Algérien, SHAT.

62. Commandement des Forces Terrestres en Algérie et de la 10e Région Militaire, État-Major, 2e Bureau, *Logistique rebelle du 1er octobre au 31 decembre 1958*, No. 320/RM/10/2/OPE/PLIT (Alger, 16 janvier 1959), Annexe I, in folder "Fiches sur la logistique rebelle, decembre 1957–decembre 1958," Dossier 1H1689 d. 1, Fonds Algérien, SHAT.

63. Commandant en Chef des Forces en Algérie, État-Major Interarmées, 2e Bureau, Section "Opérations," *Evolution des structures et des méthodes de la rébellion de 1957 à 1960*, No. 4515/EMI/2/OPE (Alger, 18 août 1960), 15, in folder "10e Région Militaire, État-Major Interarmées, 3e Bureau—Enseignements de la lutte contre les rebelles d'AFN-ESG, 10 mars 1958," Dossier IH1942 d. 3, Fonds Algérien, SHAT.

64. O'Ballance, 88.

65. Commandement en Chef des Forces en Algérie, État-Major Interarmées, 2e Bureau, *Logistique Rebelle au cours du mois d'octobre 1959*, No. 6550/EMI/2/OPE/PLIT (Alger, 5 novembre 1959), 10–11, in folder "10e Région Militaire, État-Major Interar-

mées, 2e Bureau—Fiches sur la logistique rebelle, decembre 1958–decembre 1959,'' Dossier 1H1689 d. 2, Fonds Algérien, SHAT.

66. *Evolution des structures et des méthodes,* 15.

67. Melnik, 42.

68. Humbaraci, 42.

69. Operations DIDOUCHE and AMIROUCHE are described in *Evolution des structures et des méthodes,* Annexe 17. The two ALN operations were named for FLN heroes.

70. *Evolution des structures et des méthodes,* 16 and Annexe 17; Horne, 412–413.

71. The two actions are also described in *Evolution des structures et des méthodes,* Annexe 17.

72. Commandement Supérieur Interarmées pour la 10e Région Militaire, État-Major Interarmées, 2e Bureau, *Fiche: Logistique de decembre 1957 à février 1958* (Alger, 24 février 1958), in folder ''Fiches sur la logistique rebelle, decembre 1957–decembre 1958,'' Dossier 1H1689 d. 1, Fonds Algérien, SHAT.

73. *Logistique rebelle du 1er octobre au 31 decembre 1958,* Annexe III.

74. Commandement Supérieur Interarmées pour la 10e Région Militaire, État-Major Interarmées, 2e Bureau, *Logistique rebelle du 1er janvier au 30 juin 1958,* No. 5082/RM/10/2/OPE/PLIT (Alger, 10 juillet 1958), 7, in folder ''Fiches sur la logistique rebelle, decembre 1957–decembre 1958,'' Dossier 1H1689 d. 1, Fonds Algérien, SHAT.

75. The rebel logistical situation in the second half of 1958 is summarized in Commandement Supérieur Interarmées pour la 10e Région Militaire, État-Major Interarmées, 2e Bureau, *Logistique rebelle du 1er juillet au 30 septembre 1958,* No. 6969/RM.10/2/PLIT (Alger, 10 octobre 1958), 3, in folder ''Fiches sur la logistique rebelle, decembre 1957–decembre 1958,'' Dossier 1H1689 d. 1, Fonds Algérien, SHAT.

76. 10e Région Militaire, État-Major Interarmées, 2e Bureau, *Fiche sur la Rébellion Algérienne* (Paris, 16 mars 1959), 9, in folder ''10e Région Militaire, État-Major Interarmées, 2e Bureau—Étude générale sur l'évolution du potentiel de l'A.L.N. (armements et effectifs), 1957–62,'' Dossier 1H1682 d. 2, Fonds Algérien, SHAT.

77. Melnik, 42.

# 8

# Logistics and Mobility in Algeria, 1954–1962

Although surprised by the outbreak of the Algerian rebellion on 1 November 1954, French civilian and military leaders soon regained their composure, rushed massive reinforcements to Algeria, and instituted effective programs to deal with the rebellion. Applying the defensive technique known as the *quadrillage*, the French divided Algeria into small sectors and blanketed the country with French forces to control key areas and root out the rebel bands. In the fall of 1957 the increasingly effective *quadrillage* was supplemented by the completion on Algeria's Tunisian and Moroccan frontiers of strong barriers composed of electric fences, barbed wire, mines, and sophisticated detection devices backed by mobile reaction forces and covered by constant aerial surveillance. These "barrages" isolated the rebel forces in Algeria almost completely from their bases of support in Tunisia and Morocco and threatened to strangle the rebel bands in the *wilayas*. To prevent the gradual asphyxiation of the rebellion, the leaders of the Algerian National Liberation Front (FLN) were obliged to find a way through or around the barrages. The "Battle of the Barrages" in the first half of 1958 ensued, as the Algerian Army of National Liberation (ALN) made a succession of bloody futile attempts to infiltrate the eastern barrage, known as the Morice Line. Unable to go through the barrages, the rebels next sought to go around them. However, the attempt to resupply the ALN forces in Algeria by small camel caravans skirting the barrages to the south through the Sahara desert proved equally ineffective. Exposed to French aerial surveillance and attack, most of the caravans were detected and destroyed or dispersed. Increasingly desperate to provide men and matériel to their forces inside Algeria in late 1959 and early 1960, the ALN leaders ordered a series of fruitless and costly efforts to force the barrages by frontal assault, and again they failed.

The successful French defense of the barrages cut off the rebel forces inside Algeria from their sources of trained manpower, arms, ammunition, and other supplies in Tunisia and Morocco, and the ALN in the *wilayas* regressed to a more primitive stage of revolutionary warfare. Units were broken up, automatic weapons were hidden and rebel operations were postponed, canceled, or transformed into small-scale terrorist attacks, sabotage, or occasional raids on isolated French outposts. In many areas the continued existence of an armed rebel presence hung by a very thin and frayed logistical thread. Most of those threads were finally broken during the series of French offensives conducted between February 1959 and April 1960 under the direction of French air force general Maurice Challe, who replaced General Raoul Salan as French commander in chief in Algeria in December 1959. Noting the successes achieved by the French forces through the *quadrillage* program of point and area defense as well as by the effective sealing of the Algerian borders, General Challe devised a plan to destroy the remaining military capability of the ALN inside Algeria. Collecting the necessary mobile reserves for such offensive actions by the greater use of draftees and loyal Algerian Moslems, thereby leaving in place the major part of the forces required for maintaining the *quadrillage* and the barrages, Challe began his campaign in western Algeria in February 1959. Operations under Plan Challe continued until the spring of 1961, each successive offensive moving the principal area of operations farther to the east and then repeating the "steamroller" over again from the west. By the time General Challe left Algeria on 23 April 1960, some 26,600 rebels had been killed, 11,000 captured, and nearly 21,000 weapons taken. The ALN of the interior, reduced to as few as nine thousand men scattered in small groups throughout Algeria, had been destroyed as a significant military force, and the hope of establishing an independent Algeria by military means had to be abandoned.

The relative logistical superiority of the French forces in Algeria, particularly in the critical area of mobility, and their ability to interdict the flow of men, arms, and equipment to the nationalist rebels ensured that the humiliating defeat of the French Army at the hands of the Viet Minh was not repeated. Despite an apparent technological superiority, the logistical system of the French in Indochina had not been appreciably more effective than that of the Viet Minh. The French were far from their bases of supply, while the Viet Minh had lived among theirs and had a safe base of support nearby in Communist China. Massive U.S. aid to the French had been offset by Communist Chinese aid to the Viet Minh. Moreover, the climate and terrain of Indochina had favored the manpower-rich Viet Minh over the technology-rich French forces, French air power notwithstanding. Given the dense jungles, rugged terrain, and limited road network of Indochina, porters had proved more efficient than trucks. The plentiful cover and concealment provided by the jungles of southeast Asia had done much to offset the French advantage of observation and attack from the air, and poor flying weather and the lack of adequate airfields had reduced the effectiveness of French air transport. In any event the French forces in Indochina had never had a sufficient number of the right kind of aircraft to make a critical difference. Thus, with the effectiveness of their respective logistical systems closely matched, the French and Viet Minh had had to decide the issue on other grounds, and the superior will, leadership, and operational finesse of the Viet Minh had ultimately produced the disastrous French defeat at Dien Bien Phu and subsequent victory and independence for the Viet Minh.

The situation in Algeria was quite different. In the first instance, the two sides were unequal in leadership, a condition that had not been obtained in Indochina, where Ho Chi Minh and Vo Nguyen Giap had been more than a match for their French opponents. The civilian and military leadership of the FLN/ALN did not approach that of the Viet Minh in strategic vision, operational acumen, or tactical skill. On the other hand, France was blessed by the presence in Algeria of her most skilled, experienced, and effective soldiers: Massu, Bigeard, Ducornau, Trinquier, Challe, Salan, and the gallant Jeanpierre, among many others. The French leaders had learned many lessons about counter-insurgent warfare in Indochina, and having experienced the shame of defeat, they were determined that the mistakes of Indochina not be repeated in Algeria. This commitment to victory extended to lower levels as well. Algeria was a part of France, inhabited in large part by Algerians of European descent, and thus stakes were readily understood by French soldiers and civilians at all levels. Such had not been the case in Indochina. Given an important French advantage in leadership and a high level of commitment to the task, French leaders were able to capitalize on three factors that made French military success in the Algerian war possible: a physical environment that in most respects favored the French rather than the rebels; an effective French logistical system capable of providing superior logistical and tactical mobility; and an ineffective rebel logistical system dependent on bases of supply outside Algeria and extremely vulnerable to French sea, air, and ground interdiction.

The physical environment of Algeria conferred several clear advantages on the French forces, particularly with respect to the crucial factor of logistical and tactical mobility. At the operational level and given superior French technology, the Algerian geography, topography, and climate favored the French more than the rebels. French air transport, and the helicopter in particular, mastered the great distances and rugged terrain in a way men on foot or pack animals could not. Although the theater of operations in Algeria covered an area of over 2.2 million square kilometers (compared to only some 740,000 square kilometers in Indochina), flying conditions in Algeria were far superior to those encountered in Indochina. In Algeria the French forces also enjoyed proximity to their bases of support in metropolitan France, whence reinforcements and critical matériel could be obtained with ease once the necessary political support had been obtained. For the nationalist rebels, however, the sources of supply were distant and required extensive cultivation, and the routes over which men and supplies for the ALN forces of the interior were channeled were fraught with the dangers of diplomatic and military interdiction, covert and overt. Nor could the rebels claim a special knowledge of the hidden by-ways and shortcuts, which was the usual advantage granted an insurgent force; far too many loyal Moslem *harki*s and French *pieds noirs* knew the mountains, plateaus, and deserts of Algeria quite as well as any ALN *djound*.

Algerian demographic and economic factors also favored the French. Despite being overwhelmingly outnumbered by the Moslem population of Algeria, the French were able to retain the support of a large part of the indigenous population. More important was the fact that most of the skilled manpower of Algeria was of European descent and thus not inclined to the rebel cause. French control of Algerian industry and commerce also conveyed a special advantage by denying, or at least restricting, rebel access to goods and services available on the commercial market. Moreover, from the beginning of the rebellion the French firmly controlled the relatively modern transportation facilities in Algeria. Thus, the rebels were limited in their ability to produce their own military equipment and supplies or to obtain them locally and so were forced to rely heavily on outside support.

The debilitation of World War II had had a profound negative impact on the French logistical system in Indochina, making it entirely different from but rather evenly matched with that of the Viet Minh. In Algeria the logistical systems of the opposing sides were nowhere near comparable. While the coordination of political and diplomatic policies and initiatives with the conduct of military operations in Algeria left something to be desired, the high command structure and territorial organization of the French forces in Algeria were adequate to the task of conducting a counter-guerrilla campaign in which the objective was to seal Algeria's international borders, clear populated areas of rebel forces, and protect the population against renewed rebel attack and exploitation. The French were also reasonably successful in adapting the somewhat heavy, Europe-oriented tactical organization of their combat forces to the needs of the operational situation in Algeria. Most importantly, the French military command in Algeria benefited from a full range of modern, well equipped logistical services, long accustomed to service in Algeria. Backed by the resources of nearby metropolitan France, the French supply services in Algeria were organized on a territorial basis corresponding to that of the combat forces, and they operated in accordance with proven principles of logistics. Although plagued by rapidly escalating requirements, the wide dispersion of forces, a chronic insufficiency of trained personnel and funds, an aging logistics infrastructure, and antiquated regulations ill suited to the tasks at hand, the French logistical system in Algeria overcame their deficiencies by adopting effective policies of supply economy and decentralized control.

The official records of the French logistical services in Algeria and most of the personal accounts of the participants in the Algerian war constitute a litany of shortages, problems, and inefficiencies. Such sources almost always emphasize the negatives rather than the positives. In reality, the French combat forces in Algeria seldom lacked the matériel and services necessary to prosecute the war actively. Shortages of ammunition, fuel, transport, or maintenance personnel did occur from time to time in certain areas, but such problems were usually overcome quickly and without any substantial decrease in the tempo of operations. When one compares the logistical resources available to the French forces in Algeria to those available to their Algerian nationalist opponents, the "problems" and "shortages" are of different orders entirely. In short, the French combat forces in Algeria had everything they needed to defeat the Algerian rebels militarily. What the French lacked was not the logistical wherewithal but an effective political program and the will to carry it out.

As is generally the case with rebels, the Algerian nationalists faced seemingly impossible odds. Their French opponent had the distinct advantage of holding the contested ground with long-established and fully functioning political, administrative, military, and logistical systems in place. Even that most sacred of guerrilla logistical birthrights—a sea of friendly local supporters in which to swim—was for the Algerian rebels a pond polluted by *pieds noirs* and Moslems loyal to *Algérie Française*. Like the Viet Minh, the Algerian rebels were confronted with all of the problems inherent in welding together a heterogeneous collection of factions with often conflicting goals, and of creating completely new organs of civil government and administration capable of winning the support of the Algerian people and directing their affairs. At the same time they had to establish both a military force capable of defeating a significant portion of one of the world's most modern, experienced armed forces and the means and methods for sustaining that military force logistically. Thus, the path to independence was strewn with many obstacles, not all of which the leaders of the FLN were able to overcome. Nevertheless, they succeeded in creating *ex nihilo* an effective political structure able to suppress internal dissent and gain the support of outside forces. That same newly created political entity also proved itself able to raise, organize, equip, train, and direct military forces capable of posing a significant threat to French control inside Algeria. At the same time it could create a conventional army of imposing presence which, although retained in Tunisia and never fully committed in battle, played an important role in achieving the overall victory.

From a logistical point of view, the rebel leaders faced three main tasks: to obtain funds, arms, ammunition, and other supplies from whatever source, principally through the good offices of friendly Arab and Eastern bloc governments; to move the matériel obtained abroad into the network of safe bases established in Egypt, Libya, Tunisia, and Morocco and to stage that matériel on the Algerian border for subsequent movement to the rebel forces fighting in Algeria; and to move the accumulated supplies, along with personnel reinforcements, from the bases in Morocco and Tunisia into Algeria and distribute them to the ALN forces there. On the whole the rebels were remarkably successful in accomplishing the first two of their logistical tasks. The necessary outside financial support and significant quantities of weapons, ammunition, and other war matériel were obtained quickly in the early months of the rebellion and continued to increase as time went on. Nor did the establishment of bases in Tunisia and Morocco and the staging of supplies for the interior at those bases prove very difficult. However, the crucial third and final logistical task of the ALN proved very difficult. With the completion of the eastern and western barrages in September 1957 the French effectively cut off the ALN fighting forces in Algeria from their bases of support in Tunisia and Morocco. Only the barest trickle of men, weapons, ammunition, and other supplies could be forced through or around the barrages, and after May 1958 the ALN forces inside Algeria were living on their capital. Without the trained men, arms, ammunition, and other supplies forwarded from FLN bases in Tunisia and Morocco the rebel commanders were forced to reduce the size of their units, stockpile many of their automatic weapons, and reduce their operations to sabotage, terrorist attacks, and the occasional guerrilla raid. The French not only cut the rebels off from their outside bases but seriously disrupted what little internal movement of supplies and equipment was taking place between the rebel districts (*wilayas*) and within each *wilaya*. The failure of the ALN commanders and logisticians to master the third and most important of their tasks led to the near-total collapse of the armed resistance to the French inside Algeria and certainly precluded any hope of an ALN military victory over the French. That victory had to be achieved by other means.

In no other area were the French forces in Algeria and their rebel opponents more mismatched than in the area of logistical and tactical mobility. The terrain and relative lack of French resources in Indochina allowed the Viet Minh to restrict the French use of the rather primitive transportation infrastructure. In Algeria, however, the French forces enjoyed full possession of a relatively modern system of ports, railroads, highways, and airfields. They also had the use of large numbers of modern, well maintained ships and boats, wheeled and tracked vehicles, railroad equipment, and transport aircraft, including significant numbers of the transport vehicle that would soon revolutionize logistical and tactical mobility on the battlefield—the helicopter. In Indochina the French had never had more than a hundred helicopters, few of them capable of carrying any significant number of troops or quantity of cargo. By the end of the war in Algeria in 1962, the French Army, Navy, and Air Force had over six hundred helicopters, most of which were far more useful for tactical and logistical movements than had been the primitive models available in Indochina. In Indochina, Viet Minh mobility had been enhanced by large numbers of porters, well suited to the jungle terrain and supplemented by considerable numbers of motor vehicles. By comparison with either the Viet Minh or their French opponents, the Algerian rebels were poor indeed. Only in the safety of their exterior bases did the rebels possess a meager number of often poorly maintained cars, trucks, and buses. Inside Algeria they were largely restricted to means of transport common two thousand years earlier: the feet and back of a man, a donkey, a horse, or a camel. Even these limited means were subject to nearly continuous French air and ground surveillance and attack.

Nonetheless, an efficient logistical system and superior logistical and tactical mobility, even coupled with an effective program for the sea, land, and air interdiction of the rebel logistical system, did not guarantee the French a decisive overall victory in the war against the Algerian insurgents. Although by mid-1958 the military outcome of the war in Algeria had indeed been determined by the relative advantage in strategic, operational, and tactical mobility possessed by the French and by the effectiveness of French counter-mobility operations, the final decision was determined by psychological and political factors. Inspired by the hope of national independence and determined not to abandon their dream under any circumstances, the leaders of the FLN/ALN vowed to continue the struggle at all costs and by any means. Given the political disarray and war weariness of their French opponent and the determination of the French leader, General Charles de Gaulle, to end the drain of resources in Algeria, the FLN was able eventually to achieve on the political and diplomatic front what it clearly had not been able to achieve on the battlefield—an end to the war on its terms and the total independence of Algeria from French rule. Nevertheless, the final triumph of the Algerian nationalists at the negotiating table in no way diminishes the accomplishments of the French military forces on the battlefields of Algeria, where superior mobility and effective interdiction had been the dominant factors and had determined the military outcome of the conflict if not its ultimate result.

# Appendices

# Appendix A
# Major French Combat Forces in Algeria in 1960

| Unit Designation | French Acronym | Assigned Zone—Location |
|---|---|---|
| **10th Military Region** | 10e RM | HQ at Algiers |
| **General Reserve** | RG | HQ at Algiers |
| 11th Infantry Division | 11e DI | Bône |
| 10th Parachute Division | 10e DP | Algiers |
| 25th Parachute Division | 25e DP | Philippeville |
| **Oran Corps Area** | RT/CAO | HQ at Oran |
| 5th Armored Division | 5e DB | Zone North Oran—Mostaganem |
| 4th Motorized Infantry Division | 4e DIM | Zone East Oran—Tiaret |
| 12th Infantry Division | 12e DI | Zone West Oran—Tlemcen |
| 13th Infantry Division | 13e DI | Zone South Oran—Méchéria |
| 29th Infantry Division | 29e DI | Zone Central Oran—Sidi-bel-Abbès |
| 102nd Army Light Aviation Group | GALAT 102 | Sidi-bel-Abbès |
| **Algiers Corps Area** | RT/CAA | HQ at Algiers |
| 27th Alpine Infantry Division | 27e DIA | Zone East Algiers—Tizi-Ouzou |
| 9th Infantry Division | 9e DI | Zone West Algiers—Orléansville |
| 20th Infantry Division | 20e DI | Zone South Algiers—Médéa |
| 105th Army Light Aviation Group | GALAT 105 | HQ at Algiers |
| 114th Army Light Aviation Group | GALAT 114 | HQ at Algiers |
| **Constantine Corps Area** | RT/CAC | HQ at Constantine |
| 2nd Motorized Infantry Division | 2e DIM | Zone East Constantine—Bône |
| 7th Light Mechanized Division | 7e DMR | Zone East Constantine—Tebessa |
| 14th Infantry Division | 14e DI | Zone North Constantine—Constantine |
| 19th Infantry Division | 19e DI | Zone West Constantine—Sétif |
| 21st Algerian Infantry Division | 21e DIA | Zone South Constantine—Batna |
| 101st Army Light Aviation Group | GALAT 101 | HQ at Ain Arnat (Sétif) |

| Unit Designation | French Acronym | Assigned Zone—Location |
| --- | --- | --- |
| **5th Air Region** | 5e RA | HQ at Algiers |
| 1st Tactical Air Group | GATAC 1 | Constantine |
| 2nd Tactical Air Group | GATAC 2 | Oran |
| 3rd Tactical Air Group | GATAC 3 | Algiers |
| 4th Tactical Air Group | GATAC 4 | Colomb-Béchar |
| 5th Tactical Air Group | GATAC 5 | Laghouat |
| Military Air Transport Sub-Group–Algeria | SGMMTA | Algiers |
| **4th Maritime Prefecture** | 4e PRÉMAR | HQ at Algiers |
| Maritime Command—Algiers | COMAR—Alger | Algiers |
| Maritime Command—Oran | COMAR—Oran | Mers-el-Kébir |
| Maritime Command—Bône | COMAR—Bône | Bône |
| Demi-Brigade of Marines | DBFM | Under 10e RM operational control |
| Naval Commando Group—Algeria | GCNA | Under 10e RM operational control |

# Appendix B
## Characteristics of French Transport Aircraft and Helicopters

| Aircraft | Type | Speed | Range | Combat Load/Remarks |
|---|---|---|---|---|
| Junkers 52 *Toucan* | 3–engine fixed wing transport | 190 KPH | 250 KM loaded 1,200 KM max | 18 troops; 12 litters; 2.4 tons @ 250 KM; 3–man crew; requires 700–meter landing field |
| Douglas C–47 *Dakota* (DC–3) | 2–engine fixed wing transport | 250 KPH | 500 KM loaded 1,800 KM max | 28 troops; 18 litters; 7 containers; 4 tons @ 500 KM; 5,850 lbs cargo max; one jeep or one howitzer; 3–man crew; requires 500–1,000–meter landing field |
| Fairchild C–119 *Packet*; *Flying Boxcar* | 2–engine fixed wing transport | 320 KPH | 800 KM loaded 6,400 KM max | 42 troops; 33 litters; 8 tons @ 2,400 KM (landed); 10 tons @ 2,400 KM or 13 tons @ 800 KM (para drop); 35 tons max; 27 tons landing max; requires 1,150–meter landing field |
| Nord 2501 *Noratlas* | 2–engine fixed wing transport | 335 KPH | 1,000 KM loaded 4,000 KM max | 42 troops; 32 paras; 20 tons max takeoff wt; requires 1,300–meter take off field |
| Bell H–13 *Sioux* | single rotor helicopter | 128 KPH | 360 KM | one passenger; two external litters |
| Sikorski H–19 *Chickasaw* | single rotor helicopter | 137 KPH | 650 KM | 10 troops; 8 litters; ca. 2,400 lbs. of cargo |
| Piasecki H–21 *Flying Banana* | twin rotor helicopter | 158 KPH | 800 KM | 14 troops; 12 litters; ca. 4,700 lbs. of cargo |
| Vertol H–34 *Choctaw* | single rotor helicopter | 167 KPH | 435 KM | 12 troops; 8 litters; ca. 4,200 lbs. of cargo |

*Source*: Compiled by the author from various sources. Performance characteristics are only approximate due to variance in models and dates of manufacture. Speed given is approximate cruising speed.

# Appendix C
# Common French Map Symbols

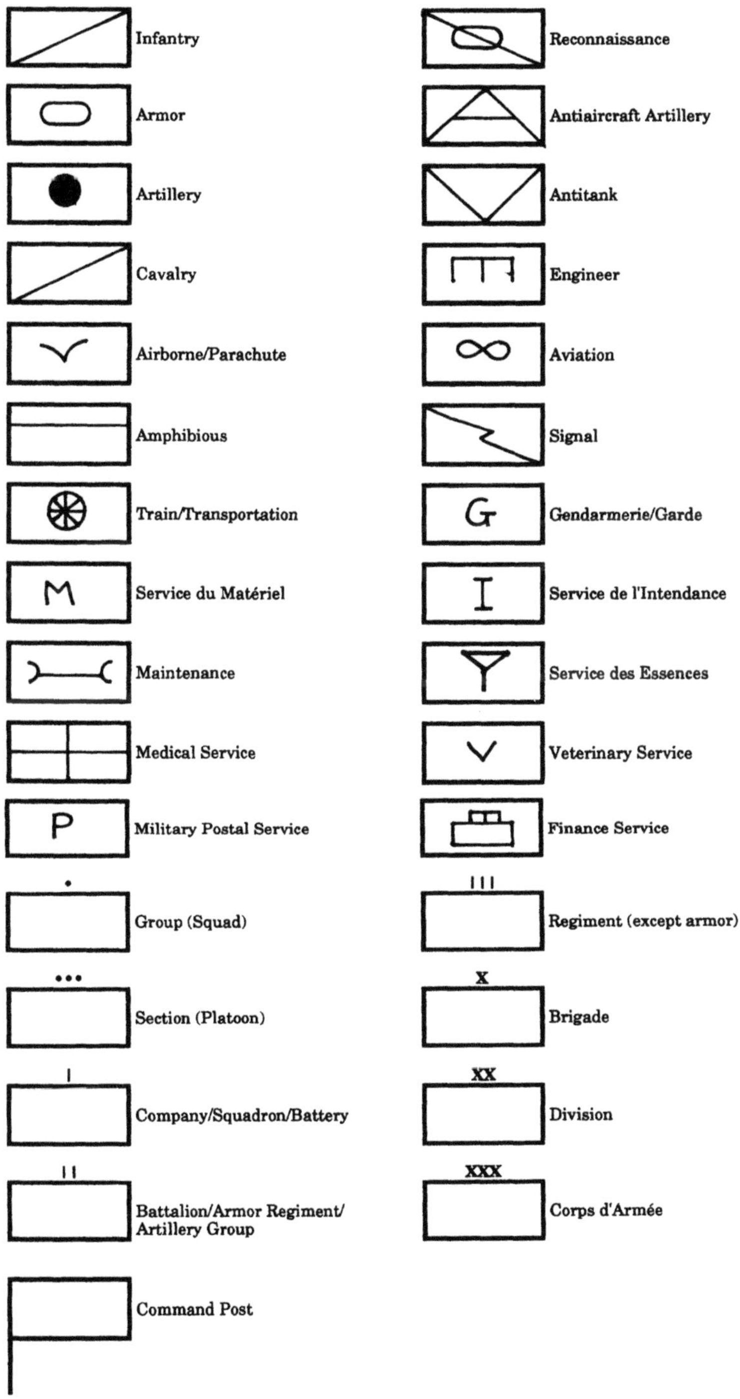

# Glossary

| | |
|---|---|
| AA | Antiaircraft |
| AAA | Antiaircraft artillery |
| AAC | Advanced Air Command |
| AAMG | Antiaircraft machine-gun |
| abn | Airborne |
| ADP | Air Directing Post |
| AEROTEC | Société Civil Aérotechnique (French commercial air line) |
| AFN | Afrique Française du Nord (French North Africa) |
| AG-PR | General Administration and Procurement |
| ALAT | Aviation Legère de l'Armée de Terre (Army light aviation) |
| Algerois | The region of Algiers |
| ALMED | L'Amiral Commandant les Forces Maritimes en Méditerranée (Admiral Commanding French Naval Forces in the Mediterranean) |
| ALN | Armée de Libération Nationale (Army of National Liberation) |
| AML | Amis du Manifeste et de la Liberté (Friends of the Manifesto and Liberty) |
| AMX | Designation of French tank model |
| AR | Automatic rifle |
| Armd | Armored |
| *arrondissement* | A subprefecture (French administrative unit) |
| arty | Artillery |
| ASP | Ammunition supply point |
| *aspirant* | Officer candidate |

| | |
|---|---|
| AT | Antitank |
| BAN | Base Aéronatique Navale (naval air station) |
| BAP | Bureau des Activités Psychologique (Psychological Activities Bureau); Base Aéroportée (airborne base) |
| barrage | Barrier of electrified wire, barbed wire, and mines |
| bbl | Barrel (42 U.S. gallons) |
| BCP | Bataillon de Chasseurs Parachutistes (Parachute Chasseurs Battalion) |
| BEP | Bataillon Étrangère Parachutiste (Foreign Legion Parachute Battalion) |
| BG | Bataillon du Genie (Engineer Battalion) |
| BI | Bataillon d'Infanterie (Infantry Battalion) |
| *bidon* | Tin can |
| *Bidonville* | Moslem village (suburb) around a major Algerian city |
| BIFM | Bataillon d'Intervention de Fusiliers-Marins (Marine Intervention Battalion) |
| *bir* | Deep well |
| *bled* | Back country; "the bush" |
| BMRG | Bataillon du Matériel de la Réserve Générale (General Reserve Ordnance Battalion) |
| BMT | Bureau de Mouvements et Transports (Movements and Transport Office) |
| bn | Battalion |
| *bordj* | Isolated post |
| BP | Bataillon de Protection (Defense Battalion) |
| *brouettage* | Housekeeping transport |
| BS | Bataillon du Secteur (Sector Defense Battalion) |
| BT | Bataillon de Marche du Train (Train Infantry Battalion) |
| *Bureau* 24 | French covert action agency |
| C. | Centigrade (Celsius) |
| CA | Corps d'Armée (Army Corps); Compagnie Auto (Car Company) |
| CAA | Corps d'Armée d'Alger |
| CAC | Corps d'Armée de Constantine |
| CADR | Compagnie Auto-Défense Rurale (Rural Self-Defense Company) |
| CAO | Corps d'Armée d'Oran |
| CAST | Compagnie Automobile Saharienne de Transport (Saharan Motor Transport Company) |
| cbt | Combat |
| CCAS | Compagnie de Commandement, d'Appui et de Services (Headquarters and Service Company) |

| | |
|---|---|
| CCBS | Compagnie de Commandement de Bataillon du Secteur (Sector Defense Battalion Headquarters Company) |
| CCBSP | Compagnie de Commandement de Bataillon du Secteur de Pacification (Pacification Sector Defense Battalion Headquarters Company) |
| CCE | Comité de Coordination et d'Exécution (Committee of Coordination and Execution) |
| CCFA | Commandement en Chef des Forces en Algérie (Commander in Chief Algeria) |
| CCR | Compagnie de Circulation Routière (Traffic Control Company) |
| CCS | Compagnie de Commandement et de Services (Headquarters and Service Company) |
| CEMN | Comité d'Entente des Mouvements Nationaux (Agreement Committee of National Movements; also known as the Comité d'Entente) |
| CENA | Congres des Étudiants Nord Africains (Congress of North African Students; Algerian nationalist organization) |
| CGPA | Compagnie de Gestion du Parc Auto (Automotive Park Management Company) |
| *chef* | Chief |
| *chott* | Salt lake/desert |
| CIA | Central Intelligence Agency (U.S.) |
| CIDA | Centre Interarmées des Opérations Amphibies (Joint Amphibious Operations Center) |
| Cie | Compagnie (Company) |
| CIE | Centre d'Importation des Essences (POL Importation Center) |
| Cie Mag | Compagnie de Magasin (Depot Company) |
| Cie Mu | Compagnie des Munitions (Ammunition Company) |
| Cie Mule | Compagnie Muletière (Pack Mule Company) |
| Cie Parc | *See* CGPA |
| CIR | Centre d'Instruction de Reléves (Reserve Training Center) |
| CIS | Commandement Interarmées au Sahara (Joint Sahara Command) |
| CISM | Centre d'Instruction de Service du Matériel (Ordnance Training Center) |
| CLA | Compagnie de Livraison par Air (Aerial Delivery Company) |
| CLRA | Compagnie Lourde de Réparation Auto (Heavy Automotive Maintenance Company) |
| cm | Centimeter |
| CM | Compagnie Mixte (Composite Transport Company) |
| Cmd | Command |

| | |
|---|---|
| CME | Compagnie Mixte des Essences (Composite POL Company) |
| CMEC | *See* CCE |
| Cmndo | Commando |
| CMRA | Compagnie Moyenne de Réparation Auto (Medium Automotive Maintenance Company) |
| CMRM | Compagnie Moyenne de Réparation du Matériel (Medium Maintenance Company) |
| CMRP | Compagnie Moyenne de Réparation de Parachutes (Medium Parachute Maintenance Company) |
| CNRA | Conseil National de la Révolution Algérienne (National Council of the Algerian Revolution) |
| co | Company |
| CO | Commanding Officer |
| *colon* | Algerian farmer of European descent |
| COMA | Compagnie de Commis et Ouvriers Militaires d'Administration (Intendance Personnel Company) |
| COMAR | Commandement du Marine (Naval Command) |
| *commando de chasse* | Group of lightly armed loyal Moslems loyal to the French |
| *commissar* | Political deputy |
| Constantinois | The region of Constantine |
| *corps de bataille* | Battle corps (the ALN conventional army in Tunisia and Morocco) |
| *correspondant du matériel* | Maintenance/matériel expediter |
| COSM | Compagnie d'Ouvriers du Service du Matériel (Ordnance Personnel Company) |
| *couscous* | North African wheat staple |
| CP | Command Post |
| CPA | Compagnie de Protection d'Algérie (Algerian Defense Company) |
| CQG | Compagnie de Quartier-Générale (Headquarters Support Company) |
| CR | Corps répartiteur (POL distribution center) |
| CRA | Compagnie de Ravitaillement par Air (Aerial Resupply Company) |
| CRALAT | Compagnie de Réparation d'Aviation Légère de l'Armée de Terre (Army Light Aviation Maintenance Company) |
| CRD | Compagnie de Réparation Divisionnaire (Divisional Maintenance Company) |
| CRDE | Compagnie de Ravitaillement et de Distribution en Essences (POL Resupply and Distribution Company) |

| | |
|---|---|
| CRE | Compagnie de Ravitaillement en Essences (POL Resupply Company) |
| CREMP | Compagnie de Ravitaillement en Essences "Moyenne Porteur" (POL Medium Carrier Company) |
| CRI | Compagnie de Ravitaillement de l'Intendance (Intendance Resupply Company) |
| CRO | Centre de renseignements opérationnels (operational intelligence center) |
| CRT | Compagnie Régionale de Transport (Regional Transport Company) |
| CRUA | Comité Révolutionnaire pour l'Unité et l'Action (Revolutionary Committee for Unity and Action) |
| CS | Compagnie du Secteur (Sector Defense Company) |
| CSA | Commandement Supérieur en Algérie (Commander in Chief Algeria) |
| CSI | *See* CIS |
| CSIA | Commandement Supérieur Interarmées en Algérie (Joint Senior Command-Algeria) |
| CSM | Compagnie Saharienne du Matériel (Saharan Ordnance Company) |
| CSQP | Compagnie Support de Quartier de Pacification (Pacification Quarter Support Company) |
| CST | Compagnie Saharienne de Transport (Saharan Motor Transport Company) |
| CT | Compagnie de Transport (Motor Transport Company) |
| CTAC | Corps de Troupes Administration Centrale (Troop Administration Center) |
| CTB | Compagnie de Transbordement (Transfer Company) |
| CTDT | Commandant du Train et Directeur des Transports (Train Commander and Director of Transport) |
| CTGP | Compagnie de Transport "Gros Porteur" (POL Bulk Carrier Company) |
| ctr | Center |
| CTSTS | *See* CIS |
| CTSTSA | *See* CIS |
| DA | Department of the Army (U.S.) |
| DB | Division Blindée (Armored Division) |
| DBFM | Demi-Brigade de Fusiliers Marins (Demi-Brigade of Naval Infantry) |
| DBLE | Demi-Brigade de la Légion Étrangère (Demi-Brigade of the Foreign Legion) |

| | |
|---|---|
| def | Defense |
| *denrées d'ordinance* | French issue fruits and vegetables |
| DI | Division d'Infanterie (Infantry Division); Direction/Directeur de l'Intendance |
| DIA | Division d'Infanterie Alpine (Mountain Infantry Division) |
| DIH | Détachement d'Intervention Hélicoptère (Helicopter Intervention Detachment) |
| DIM | Division d'Infanterie Motorisée (Motorized Infantry Division) |
| DIOG | Direction/Directeur de l'Intendance Officier Général (General Direction/Director of Intendance) |
| DIRMAT | Direction/Directeur du Matériel (Ordnance Direction/Director) |
| div | Division |
| div'l | Divisional |
| *djebel* | Hill; mountain |
| *djebellah* | North African outer garment |
| *djemaa* | Algerian village popular assembly |
| *djound* | ALN private soldier |
| DLB | Division Legère Blindée (Light Armored Division) |
| DMA | Division Militaire d'Alger |
| DMC | Division Militaire de Constantine; Dépôt de Matériels Complets (Ordnance End Item Depot) |
| DMO | Division Militaire d'Oran |
| DMR | Division Mécanique Rapide (Mechanized Infantry Division) |
| DO | Délégue Opérationnel (Operational Delegate of the French Petroleum Service) |
| DOP | Détachements Opérationnels de Protection (French Army psychological warfare detachments) |
| DOS | Days of supply |
| *dotation* | Authorization (of equipment or supplies) |
| *dotation practique* | Special authorization of equipment or supplies |
| douar | Mountain village |
| DP | Division Parachutiste (Parachute Division) |
| DRA | Dépôt de Réchanges Automobiles (Automotive Parts Depot) |
| *draisine* | Railroad sabotage detection device |
| DRM | Direction Régionale du Matériel (Regional Ordnance Direction) |
| DS | Direct support |

| | |
|---|---|
| DSEA | Direction/Directeur/Délégué du Service des Essences en Algérie (Direction/Director/Delegate of the Petroleum Service in Algeria) |
| DSM | Direction/Directeur du Service du Matériel |
| DST | Direction de la Surveillance du Territoire (French Security Police) |
| DT | Division Territoriale (Territorial Division) |
| DTA | Division Territoriale d'Alger |
| DTC | Division Territoriale de Constantine |
| DTO | Division Territoriale d'Oran |
| DZ | Drop Zone |
| EBR | Engin Blindé de Reconnaissance (armored reconnaissance vehicle) |
| Economat | French Army commissary/post exchange |
| EEAC | Escadron d'Éclairage et Antichar (Surveillance and Antitank Squadron) |
| EHO | Escadron d'Hélicoptères (Helicopter Squadron) |
| elm | Element |
| EM | État-Major (staff; General Staff) |
| EMI | État-Major Interarmées (Joint General Staff) |
| ENA | Étoile Norde Africaine (North African Star; Algerian nationalist organization) |
| *engin de balayage* | Railroad minesweeping machinery |
| engr | Engineer |
| EOD | Explosive ordnance disposal |
| *erg* | Sand desert |
| ERGM | Étabilissement de Réserve Générale du Matériel (General Reserve Ordnance Establishment) |
| ERM | Étabilissement Régionale du Matériel (Regional Ordnance Establishment) |
| ERMu | Étabilissement Régionale des Munitions (Regional Ammunition Establishment) |
| est | East; estimated |
| ET | Escadron du Train (Train Squadron) |
| *établissement* | Fixed logistical facility |
| F. | Fahrenheit |
| FA | Field artillery |
| FAF | French Air Force |
| *faïlek* | ALN battalion |
| *faoudj (fauj)* | ALN group (squad) of eleven men |

| | |
|---|---|
| *fellagha* | "Bandit" (derogatory term for ALN soldier) |
| FEMA | Fédération des Elus Musulmans d'Algérie (Federation of Elected Muslims in Algeria; Algerian nationalist organization) |
| *ferka* | An ALN section (platoon) of about thirty-five men |
| FF | French franc (17 "old" francs = $1 U.S.; 5 "new" francs [introduced in 1957] = $1 U.S.) |
| *fidayine* | Terrorist |
| FLN | Front de Libération Nationale (National Liberation Front) |
| *foggara* | Oasis water system/underground water channel |
| *force de frappe* | Striking force (General de Gaulle's term for a French nuclear force) |
| 4 × 4 (etc.) | four (etc.) wheels, four-wheel (etc.) drive |
| Fr | French; France |
| GALAT | Groupement d'Aviation Legére de l'Armée de Terre (Army Light Aviation Group) |
| GATAC | Groupement Aérien Tactique (Tactical Air Group) |
| GCNA | Groupement des Commandos Navales en Algérie (Naval Commando Group-Algeria) |
| Gendarmerie | French paramilitary police; police forces of the sub-prefects |
| GGR | Groupement de Gestion-Réparation (Ordnance Management and Repair Group) |
| GH | Groupement d'Hélicoptère (Helicopter Group) |
| GHAN | Groupement d'Hélicoptères Aéro-Navales (or d'Aéronatique Navale) (Naval Helicopter Group) |
| GM | Garde Mobile |
| GMC | General Motors Corporation |
| GMMTA | Groupement des Moyens Militaire de Transports Aériens (French Military Air Transport Command) |
| Gpe | Groupe (group) |
| GPRA | Gouvernement Provisoire de la République Algérienne (Provisional Government of the Algerian Republic) |
| *gros colon* | Major Algerian landowner of European descent |
| GROUCO | Groupement de Commandos Marine (Naval Command Group) |
| GRST | Groupement Reconnaissance Saharienne et Transport (Saharan Reconnaissance and Transport Group) |
| GS | General support |
| GSM | Groupement Mobile de Sécurité (Mobile Security Group) |
| GST | Groupement Saharienne de Transport (Saharan Motor Transport Group) |
| GT | Groupement de Transport |

| | |
|---|---|
| GTM | Groupement Territoriale du Matériel |
| H | Habillement (clothing) |
| *harka* | An armed Moslem village |
| *harki* | Member of a self-defense force organized by the French in each Moslem village; Moslem soldier in the French forces |
| HLL | Hors la loi (outlaws) |
| HMG | Heavy machine gun |
| howz | Howitzer |
| HQ | Headquarters |
| HQDA | Headquarters Department of the Army (U.S.) |
| hvy | Heavy |
| IGAME | Inspecteur Général de l'Administration en Mission Extraordinaire (Inspector General of the Administration on Extraordinary Mission; Super-Prefect) |
| inf | Infantry |
| Intel | Intelligence |
| ITR | Interfederation des Transports Routières (Algerian commercial trucking organization) |
| ITST | *See* CIS |
| ITSTSA | *See* CIS |
| IWC | Inter-Ministerial War Committe (FLN political body) |
| JOC | Joint Operations Center |
| jt | Joint |
| *kasma* | Sector |
| *katiba* | An ALN company of 90–130 men |
| kg | Kilogram |
| km | Kilometer |
| KPH | Kilometers per hour |
| lbs | Pounds |
| LCM | Landing Craft, Mechanized |
| ldd | Loaded |
| *lentisque* | A type of high-growing brush found in Algeria |
| LMG | Light machine gun |
| lt | Light |
| LVT | Landing Vehicle, Tracked |
| LZ | Landing Zone |
| M1, M2, etc. | Mintaka 1, Mintaka 2, etc. (*see mintaka*) |
| $m^2$ | Square meter |

| | |
|---|---|
| $m^3$ | Cubic meter |
| MA | Magasin-Atelier (depot-workshop) |
| *maghzen* | Lightly armed forces controlled by officers of the Bureau des Affaires Algériennes |
| maint | Maintenance |
| MALG | Ministère d'Armements et Liaisons Générales (GPRA Ministry of Armaments and General Liaison; see MARG) |
| *maquis* | Scrub forest; ''the bush'' |
| MARG | Ministère d'Armements et Ravitaillement Générale (GPRA Ministry of Armaments and Supply; later MALG) |
| M.A.S.H. | Mobile Army Surgical Hospital (U.S.; Korean War era) |
| *matériel du secteur* | Special garrison equipment |
| MDAP | Mutual Defense Assistance Program (U.S.) |
| mdm | Medium |
| mech | Mechanized |
| *mechta* | Village |
| metric ton | 1,000 kg; about 2,200 lbs. |
| MG | Machine gun |
| *mintaka* | FLN territorial subdivision (zone) |
| mm | Millimeter |
| MMH | Magasin Militaire d'Habillement (Military Clothing Depot) |
| MNA | Mouvement National Algérien (Algerian National Movement) |
| mo | Month |
| mort | Mortar |
| *moudjahidine* | ALN regular soldier |
| *moussebiline* | ALN part-time soldier |
| MRP | Mouvement Républicain Populaire (Popular Republican Movement) |
| mt | Maintenance |
| MTLD | Mouvement pour le Triomphe des Libertés Démocratiques (Movement for the Triumph of Democratic Liberties) |
| Mtn | Mountain (Alpine) |
| MTOE | Modified Table of Organization and Equipment (*see dotation practique*) |
| Mtrz | Motorized |
| Mtz | Motorized |
| *nahia* | Region |
| NATO | North Atlantic Treaty Organization |

| | |
|---|---|
| NCO | Noncommissioned officer |
| *nord* | North |
| OACSI | Office of the Assistant Chief of Staff for Intelligence (U.S. Army) |
| OAS | Organisation Armée Secrète (Secret Army Organization) |
| Obsv | Observation |
| OEM | Officers and enlisted men |
| off | Offensive; officer |
| OPA | Organisation Politico-Administrative (FLN underground political structure) |
| Oranais | The region of Oran |
| ORU | Organisation Rurale et Urbaine (Rural and Urban Organization of the FLN) |
| OS | Organisation Secrète (Secret Organization); Organisation Spéciale (Special Organization of the MTLD) |
| *oued* | Stream; usually torrential |
| *ouest* | West |
| Oulema | Association of the Reformist Oulemas (Algerian Moslem political party) |
| oz | Ounce(s) |
| para | Parachute; parachutist |
| *pataugas* | Rubber and canvas field shoes |
| PC | Poste de Commandement (Command Post) |
| PCA | Parti Communiste Algérien (Algerian Communist Party); Post de Control Aérien (Air Control Post) |
| *penetrant* | A main north-south highway |
| *pied noir* | Algerian of European descent |
| PIN | Parc d'Interest Nationale (Algerian reserve motor transport organization) |
| *piste* | Trail; track |
| POA | Poste d'observation aérien (aerial observation post) |
| POL | Petroleum, oil, and lubricants |
| PPA | Parti du Peuple Algérien (Party of the Algerian People) |
| PRA | Principal Régulateur Aérien (Principal Air Coordinator) |
| PRC | People's Republic of China |
| *Préfecture* | Prefecture (French administrative sub-division) |
| *Préfet* | Prefect |
| PRÉMAR IV | 4e Préfecture Maritime (4th Maritime Prefecture) |
| prep | Preparation |

| | |
|---|---|
| PRLE | Peleton de Réparation de la Légion Étrangère (Foreign Legion Maintenance Platoon) |
| Ptn | Peloton (Platoon) |
| qtr | Quarter |
| *quadrillage* | French defensive technique for clearing and controlling territory; involved dividing geographical areas into squares to facilitate search and control operations |
| RA | Région Aérienne (Air Region) |
| RAR | Régulateur Aérien Régionale (Regional Air Coordinator) |
| ration | Supplies for one man for one day |
| *ratissage* | French counter-guerrilla technique; involved ''raking over'' (searching) the terrain meticulously |
| RCP | Régiment de Chasseurs Parachutistes (Parachute Chasseurs Regiment) |
| RD | Régiment de Dragons (Dragoon Regiment) |
| Recon | Reconnaissance |
| regt | Regiment |
| REI | Régiment Étrangère d'Infanterie (Foreign Legion Infantry Regiment) |
| *responsable* | Responsible party; leader; chief |
| RFMA | Rassemblement Franco-Musulman Algérien (Rally of Algerian French Muslims) |
| RI | Régiment d'Infanterie (Infantry Regiment) |
| RIA | Régiment d'Infanterie Algérienne (Algerian Infantry Regiment) |
| RICM | Régiment d'Infanterie et de Chars de Marine (Marine Infantry-Tank Regiment) |
| RIM | Régiment d'Infanterie Motorisée (Motorized Infantry Regiment) |
| RL | Rocket launcher; bazooka |
| RM | Région Militaire |
| *rocade* | A main east-west highway |
| RPF | Rassemblement du Peuple Français (Rally of French People) |
| RR | Railroad |
| RT | Région Territoriale |
| RT/CA | Région Territoriale et de Corps d'Armée (Territorial Region/Corps Area) |
| RT/CAA | Région Territoriale et de Corps d'Armée d'Alger |
| RT/CAC | Région Territoriale et de Corps d'Armée de Constantine |
| RT/CAO | Région Territoriale et de Corps d'Armée d'Oran |
| S | Subsistence |

| | |
|---|---|
| SAA | Small Arms Ammunition |
| SAM | Secteur Autonomie de Méchéria (Autonomous Sector of Méchéria) |
| SAS | Sections Administratives Specialisées (Special Administrative Sections) |
| SAU | Sections Administratives Urbaines (Urban Administrative Sections) |
| SCA | Section de Commandement et d'Appui (Command Section) |
| SDECE | Service d'Espionage et Countre-Espionage; French secret intelligence service |
| SEA | Service des Essences en Algérie (Petroleum Service in Algeria) |
| *sebkha* | Marsh |
| *section* | Platoon |
| SEI | Section d'Exploitation d'Intendance (Operational Intendance Section) |
| SFP | Section de Fabrication de Pain (Bakery Section) |
| S/GMMTA | Sub-Groupement des Moyens Militaire de Transports Aériens (French military air transport command in Algeria) |
| *shadoof* | Shallow well |
| SHAT | Service Historique de l'Armée de Terre (French Army Historical Service) |
| SI | Service de l'Intendance |
| *sirocco* | Hot, dust-laden wind from the Sahara |
| SM | Service du Matériel |
| SMG | Submachine gun |
| SNCFA | Société Nationale des Chemins de Fer Algériens (Algerian National Railways) |
| *Société en Najah* | GPRA purchasing agency |
| sqd | Squad |
| sqdn | Squadron |
| SRI | Section de Ravitaillement de l'Intendance (Intendance Resupply Section) |
| *sud* | South |
| svc | Service |
| T | Ton |
| TA | Table of Allowances |
| TEDG | Table d'Équipements et Dotations Générales (TOE) |
| tk | Tank |

| | |
|---|---|
| tkr | Tanker |
| Tn | Train |
| TOE | Table of Organization and Equipment |
| *tonne* | *See* metric ton |
| *tranche* | "Slice"; concept used in ammunition management |
| Trans | Transportation |
| trk | Truck |
| UDMA | Union Démocratique du Manifesto Algérien (Democratic Union of the Algerian Manifesto) |
| UE | Unité d'Essence (arbitrary quantity of POL used in planning) |
| UF | Unité de Feu (unit of fire; arbitrary quantity of ammunition using in planning) |
| UGCA | Union Générale des Commerçants et des Artisans (General Union of Algerian Tradesmen) |
| UGEMA | Union Générale des Etudiants Musulmans Algériens (General Union of Algerian Muslim Students) |
| UGTA | Union Générale des Travailleurs Algériens (General Union of Algerian Workers) |
| *ulema* | Moslem religious teachers |
| UPA | Union Populaire Algérienne (Popular Algerian Union) |
| U.S. | United States (of America) |
| USTA | Union Syndicale des Travailleurs Algériens (Syndicated Union of Algerian Workers) |
| UT | Unités Territoriales |
| W1, W2, etc. | Wilaya 1, Wilaya 2, etc. (*see wilaya*) |
| *wadi* | Watercourse; draw (*see oued*) |
| *wilaya* | FLN military-political-administrative province |
| wk | Week |
| wpn | Weapon |
| wt | Weight |
| yr | Year |
| ZCO | Zone Centrale Oranais (Zone Central Oran) |
| ZEA | Zone Est Algérois (Zone East Algiers) |
| ZEC | Zone Est Constantine (Zone East Constantine) |
| ZEO | Zone Est Oranais (Zone East Oran) |
| ZES | Zone Est Saharien (Zone East Sahara) |
| ZNA | Zone Nord Algérois (Zone North Algiers) |
| ZNC | Zone Nord Constantine (Zone North Constantine) |
| ZNO | Zone Nord Oranais (Zone North Oran) |

| | |
|---|---|
| ZO | Zone opérationnel (operational zone) |
| ZOA | Zone Ouest Algérois (Zone West Algiers) |
| ZOC | Zone Ouest Constantine (Zone West Constantine) |
| ZOO | Zone Ouest Oranais (Zone West Oran) |
| ZOS | Zone Ouest Saharien (Zone West Sahara) |
| ZSA | Zone Sud Algérois (Zone South Algiers) |
| ZSC | Zone Sud Constantine (Zone South Constantine) |
| ZSO | Zone Sud Oranais (Zone South Oran) |

# Selected Bibliography

## PRIMARY MATERIALS

This work is based primarily on official French military documents which form part of the Fonds Algérien preserved by the Ministère de la Défense, État-Major de l'Armée de Terre, Service Historique, at the Château de Vincennes outside Paris. The principal cartons/dossiers used are indicated below. Individual items are cited in full in the text.


1 H 1536 bis d.1 (10e Région Militaire, État-Major Interarmées, 2e Bureau. Répression du trafic d'armes, 1955–62).

1 H 1682 d.2 (10e Région Militaire, État-Major Interarmées, 2e Bureau. Étude générale sur l'évolution du potentiel de l'A.L.N. (armements et effectifs), 1957–62).

1 H 1684 d.3 (Commandement Supérieur des Forces Françaises en Algérie, Région Territoriale et Corps d'Armée d'Oran, État-Major, 2e Bureau. Statistiques; Ordre de Bataille du ALN, 10 février 1962).

1 H 1689 d.1 (10e Région Militaire, État-Major Interarmées, 2e Bureau. Fiches sur la logistique rebelle, decembre 1957–decembre 1958).

1 H 1689 d.2 (10e Région Militaire, État-Major Interarmées, 2e Bureau. Fiches sur la logistique rebelle, decembre 1958–decembre 1959).

1 H 1692 d.1 (10e Région Militaire, État-Major Interarmées, 2e Bureau. Renseignements reçus concernant l'organisation et le potentiel des bandes rebelles, janvier–octobre 1958).

1 H 1692 d.2 (10e Région Militaire, État-Major Interarmées, 2e Bureau. Renseignements reçus concernant l'organisation et le potentiel des bandes rebelles, janvier–octobre 1959).

1 H 1692 d.3 (10e Région Militaire, État-Major Interarmées, 2e Bureau. Renseignements reçus concernant l'organisation et le potentiel des bandes rebelles, janvier–octobre 1960).

1 H 1874 d.7 (10e Région Militaire, État-Major Interarmées, 3e Bureau. Étude sur l'aide exterieure susceptible d'être apportée à la rébellion algérienne, 1956).

1 H 1881 d.1 (10e Région Militaire, État-Major Interarmées, 3e Bureau. Ordre de Bataille [Française], 1955–63).

1 H 1881 d.3 (10e Région Militaire, État-Major Interarmées, 3e Bureau. Organigrammes du Commandement Supérieur Interarmées [CSIA] et Commandant en Chef des Forces en Algérie [CCFA], février 1959–février 1960).
1 H 1905 d.1 (10e Région Militaire, État-Major Interarmées, 3e Bureau. Rapports et synthèses des armes, 1956).
1 H 1905 d.2 (10e Région Militaire, État-Major Interarmées, 3e Bureau. Rapports et synthèses des services, 1956).
1 H 1917 d.1 (10e Région Militaire, État-Major Interarmées, 3e Bureau, Commandement du Train et Direction des Transports. Notice technique sur le Commandement du Train, etc., 1958).
1 H 1917 d.6 (10e Région Militaire, État-Major Interarmées, 3e Bureau, Commandement du Train et Direction des Transports. Le Train en Algérie et au Sahara, 1960–61).
1 H 1933 d.2 (10e Région Militaire, État-Major Interarmées, 3e Bureau. Le problem militaire en Algérie, avril 1957).
1 H 1942 d.3 (10e Région Militaire, État-Major Interarmées, 3e Bureau. Enseignements de la lutte contre les rebelles d'AFN-ESG, 10 mars 1958).
1 H 2100 d.1 (10e Région Militaire, État-Major Interarmées, 3e Bureau. Documentation sur l'Algérie, 1954–62).
1 H 2665 d.1 (Aide-Major Général en AFN. Organisation du commandement et coordination des services et moyens logistiques, 1946–57).
1 H 2692 d.1 (Organisation du Service des Essences, 1945–61).
1 H 2762 d.3 (Corps d'Armée d'Alger, État-Major, 4e Bureau. Instructions sur le soutien logistique de CAA, 1957–62).
1 H 2989 d.1 (Corps d'Armée de Constantine, État-Major, 4e Bureau. Organisation et directives logistiques, 1957–62).
1 H 3215 d.1 (Corps d'Armée d'Oran, État-Major, 4e Bureau. Organisation logistique des forces en Algérie et au Sahara, 1959–60).
France, Ministère de l'Armée de Terre, École d'État-Major, École Supérieure de Guerre et École d'État-Major. *Aide-Mémoire pour les Travaux d'État-Major*. Provisional ed. 2 volumes in 1. Paris: École Supérieure de Guerre et École d'État-Major, February 1952. (French Army staff officer's manual. Includes doctrine, organizational diagrams, and Tables of Equipment.)
France, Ministère des Armées. *Employment of Helicopter-Borne Units in Maintaining Order in French North Africa [Manual T.T.A. 152: Emploi d'unités hélicoportées en maintien de l'ordre en A.F.N.*]. Translated and edited by Office of the Assistant Chief of Staff for Intelligence, Headquarters, Department of the Army. Washington, DC: Headquarters, Department of the Army, Office of the Assistant Chief of Staff for Intelligence, 4 May 1956.
France, Ministère des Armées. *Notice on Air Support in Operations for the Maintenance of Order in Algeria [Manual T.T.A. 172: Notice sur l'appui aérien dans les opérations de maintien de l'ordre en Algérie*]. Translated and edited by Office of the Assistant Chief of Staff for Intelligence, Headquarters, Department of the Army. Washington, DC: Headquarters, Department of the Army, Office of the Assistant Chief of Staff for Intelligence, 9 December 1960.
France, Ministère des Armées, École Supérieure de Guerre. *Opérations en Algérie—1958*. Paris: École Supérieure de Guerre, 1958. (A series of detailed loose-leaf sketches of sixteen operations in Algeria, 1957–1958, with maps. Includes a list of the

forces engaged, a résumé of the action, a map, and lessons learned for each engagement.)

France, Ministère de la Défense, État-Major de l'Armée, Service Historique. Lettre (No. 1135/DEF/EMAT/SH/DEP), Chef du Service Historique [Le Général Delmas] to Chef d'État-Major des Armées, Division Organisation-Logistique, Château de Vincennes, 11 février 1983, sujet: Statistiques des effectifs et pertes en A.F.N.

France, Ministère de la Défense Nationale et des Forces Armées. *Counter Guerrilla Operations for Maintaining Order in French North Africa [Manual T.T.A. 123: Opérations de contre guerilla dans le cadre du maintien de l'ordre en A.F.N.]*. Translated and edited by Office of the Assistant Chief of Staff for Intelligence, Headquarters, Department of the Army. Washington, DC: Headquarters, Department of the Army, Office of the Assistant Chief of Staff for Intelligence, 24 August 1956.

France, National Defense Committee for Scientific Action, Operations Research Group. *Report of the Operations Research Mission on H-21 Helicopter Operations in Algeria*. Morton, PA: Vertol Aircraft Corporation, April 1957. 102 pages. (Report presented to the President of the French National Defense Committee for Scientific Action by the Operations Research Group.)

France, Secrétariat Général pour les Affaires Algériennes, Information Section. *Algérie 1960—Documentation*. Paris: Segedo, octobre 1960. (Packet of maps and fact sheets on political, economic, social, and military affairs in Algeria during 1960.)

Gouvernement Général de l'Algérie. *Notions essentielles sur l'Algérie*. Paris: Georges Lang [ca. 1956]. (Pamphlet containing official statement of basic data concerning Algeria.)

United States Army Medical Service, Medical Information and Intelligence Agency. *Health and Sanitary Data for Algeria (A Special Study)*. Intelligence Research Project No. PO 2306. Washington, DC: Office of the Assistant Chief of Staff for Intelligence, Headquarters, Department of the Army, 11 January 1960.

United States Department of the Army, Office of the Assistant Chief of Staff, Intelligence. *The French Motor Vehicle and Ground Munitions Industry*. Intelligence Research Project No. 10,027. Washington, DC: Office of the Assistant Chief of Staff, Intelligence, Headquarters, Department of the Army, 1 August 1957.

Vertol Aircraft Corporation (T. R. Pierpont, Director, European Operations). *French Army Helicopter Operations in Algeria, June 1956–September 1959*. Report No. SM-406. Morton, PA: Vertol Aircraft Corporation, 1 November 1959.

## SECONDARY WORKS

Bergot, Erwan. *Algérie: Les Appelés au Combat*. Collection "Troupes de Choc," ed. Jeannine Ballard. Paris: Presses de la Cité, 1991.

Bourgeois, Luc. *Le Matériel pendant la guerre d'Algérie*. 2 volumes in 3. [Paris]: Inspection du Matériel de l'Armée de Terre, September 1987.

Braestrup, Peter. "Partisan Tactics-Algerian Style." *Army* (August 1960), 33–44. (Based on Braestrup's time with ALN rebels in Algeria and Tunisia in September 1958.)

Brocart, Rémi. "L'Organisation politico-administrative et militaire du F.L.N.: vue à travers les archives du 5e bureau de l'E.M.I." *Revue Historique des Armées* 187, no. 2 (June 1992), 44–53.

Carré, Claude. "Aspects opérationnels du conflit Algérien, 1954–1960." *Revue Historique des Armées* 166 (March 1987), 82–91.

Commission Internationale d'Histoire Militaire. *La guerre d'Algérie: La défense des frontières—Les barrages algéro-marocain et algéro-tunisien, 1956–1962. Revue Internationale d'Histoire Militaire*, no. 76. Vincennes: Commission Française d'Histoire Militaire 1997.

Croizat, Victor J. "The Algerian War." *Marine Corps Gazette* 41, no. 12 (December 1957), 44–51. (Based on observations made by Croizat during a trip to Algeria in 1957.)

Dubois, Jean-Pierre. "L'Aéronautique navale et les opérations d'Algérie, 1954–1962." *Revue Historique des Armées* 187, no. 2 (June 1992), 109–117.

Etcheverry, [Major]. "Refléxions sur la guerre subversive d'Algérie." *Revue des Forces Terrestres* 17 (July 1959), 41–48. (Brief discussion of Algeria as a war of counter-insurgency.)

France, Service Historique de l'Armée de Terre. *Introduction à l'étude des archives de l'Algérie*. Château de Vincennes: Service Historique de l'Armée de Terre, 1992.

Furniss, Edgar Stephenson, Jr. *De Gaulle and the French Army: A Crisis in Civil-Military Relations*. New York: Twentieth Century Fund, 1964. (Scholarly analysis of the de Gaulle–Army clash during and immediately after the Algerian War.)

Gouineau, Gaston. "Trente-cinq ans d'inspection du Matériel de l'armée de Terre." *Revue Historique des Armées*, Special Issue (no. 3 for 1980) (1980), 126–138.

Gourmen, Pierre. "Les bataillons du Train en Algérie." *Revue Historique des Armées*, Special Issue (1978), 172–184.

Heggoy, Alf Andrew. *Insurgency and Counterinsurgency in Algeria*. Bloomington: Indiana University Press, 1972.

Horne, Alistair. *A Savage War of Peace*. New York: Viking Press, 1977. (Classic study of all aspects of the Algerian war with emphasis on political aspects.)

Humbaraci, Arslan. *Algeria: A Revolution That Failed*. New York: Frederick A. Praeger, 1966. (By a pro-FLN Turkish journalist. Focuses on period after independence.)

Kelly, George Armstrong. *Lost Soldiers: The French Army and Empire in Crisis, 1947–1962*. Cambridge, MA: MIT Press, 1965. (Sympathetic account of French Army's experiences in Indochina and Algeria, based largely on French military journals.)

Kessler, Jean. "Témoignage: La surveillance des frontières maritimes de l'Algérie, 1954–1962." *Revue Historique des Armées* 187, no. 2 (June 1992), 94–101.

Leulliette, Pierre. *St. Michael and the Dragon: Memoirs of a Paratrooper*. Boston: Houghton Mifflin, 1964. (Personal account of service in a French parachute regiment in Algeria, 1954–1957. Original edition: *Saint Michel et le dragon: souvenirs d'un parachutiste*. Paris: Editions du Minuit, 1961.)

Mans, Anne-Marie. "Évolution du Matériel, 1945–1980." *Revue Historique des Armées*, Special Issue (no. 3 for 1980) (1980), 111–125.

Melnik, Constantin. *The French Campaign against the FLN*. RAND Memorandum RM-5449-ISA. Santa Monica, CA: Rand Corporation for the Office of the Assistant Secretary of Defense—International Security Affairs, September 1967. (Analysis of French strategy in Algeria, May 1958 to November 1960 by a former member of the staff of French Prime Minister Michel Debré.)

O'Ballance, Edgar. *The Algerian Insurrection, 1954–62*. Hamden, CT: Archon Books, 1967. (Brief military and political history of the Algerian War by a well known military writer.)

Peterson, A. H., G. C. Reinhardt, and E. E. Conger (eds.). *Symposium on the Role of Airpower in Counterinsurgency and Unconventional Warfare: The Algerian War.* Santa Monica, CA: Rand Corporation, July 1963.

United States Air Force, Air University, Aerospace Studies Institute, Concepts Division. *Guerrilla Warfare and Airpower in Algeria, 1954–1960.* Project No. AU-411-62-ASI. Maxwell Air Force Base, AL: Concepts Division, Aerospace Studies Institute, Air University, March 1965. (USAF study of employment of airpower [including helicopters] in Algeria.)

Vial, Philippe, and Pascal Tanchoux. "Les Archives 'Algérie' de l'Air." *Revue Historique des Armées* 187, no. 2 (June 1992), 66–76.

Walpole, Norman C., et al. *U.S. Army Area Handbook for Algeria.* DA Pamphlet No. 550–44. Ed. Special Operations Research Office, Foreign Areas Studies Division, American University. Washington, DC: Headquarters, Department of the Army, January 1965. (Reference work on pre- and post-independence Algeria.)

Waraschitz, Bruno. "Le Cadre territoriale et Opérationnelle." In *Introduction à l'étude des archives de l'Algérie.* Ed. France, Service Historique de l'Armée de Terre. Château de Vincennes: Service Historique de l'Armée de Terre, 1992, 71–108. (Article covers territorial organization of military and civil government in Algeria.)

# Index

## About the Author

CHARLES R. SHRADER, an independent historian and consultant, currently serves as the executive director of the Society for Military History. His other books include *U.S. Military Logistics, 1607–1991: A Research Guide* (Greenwood, 1992), and *Communist Logistics in the Korean War* (Greenwood, 1995).

www.ingramcontent.com/pod-product-compliance
Ingram Content Group UK Ltd.
Pitfield, Milton Keynes, MK11 3LW, UK
UKHW041130071026
15587UKWH00009B/398

9 780275 963880